PROCESS TECHNOLOGY EQUIPMENT

Center for the Advancement of Process Technology

Prentice Hall

Boston   Columbus   Indianapolis   New York   San Francisco   Upper Saddle River   Amsterdam
Cape Town   Dubai   London   Madrid   Milan   Munich   Paris   Montreal   Toronto   Delhi
Mexico City   Sao Paulo   Sydney   Hong Kong   Seoul   Singapore   Taipei   Tokyo

Editor in Chief: Vernon Anthony
Acquisitions Editor: David Ploskonka
Editorial Assistant: Nancy Kesterson
Director of Marketing: David Gesell
Senior Marketing Coordinator: Alicia Wozniak
Marketing Assistant: Les Roberts
Associate Managing Editor: Alexandrina Benedicto Wolf
Project Manager: Alicia Ritchey
Senior Operations Supervisor: Pat Tonneman
Operations Specialist: Laura Weaver
Art Director: Diane Ernsberger

Cover Designer: Jayne Conte
Cover Art: Center for the Advancement of Process Technology, Bayport Technical and Training Center, Design Assistance Corporation, and Eyewire
Lead Media Project Manager: Karen Bretz
Full-Service Project Management: Lisa Garboski, bookworks
Composition: TexTech
Printer/Binder: LSC Communications
Cover Printer: LSC Communications
Text Font: Times Ten Roman

Credits and acknowledgments borrowed from other sources and reproduced, with permission, in this textbook appear on appropriate page within text.

Library of Congress Cataloging-in-Publication Data
Process technology equipment / CAPT.
    p. cm.
  Includes index.
  ISBN 0-13-700412-5
1. Chemical processes—Equipment and supplies.   I. CAPT (Organization)
  TP155.7.P763 2010
  660'.283—dc22   2009003587

Prentice Hall
is an imprint of

www.pearsonhighered.com

ISBN-10: 0-13-700412-5
ISBN-13: 978-0-13-700412-6

Contents

Preface   *xi*

Chapter 1   **Introduction to Process Equipment**   **1**
Objectives   1
Key Terms   2
Introduction   3
How Process Industries Operate   3
Process Technician's Role in Operation and Maintenance   3
Equipment   5
Safety and Environmental Hazards   10
Summary   11
Checking Your Knowledge   11
Student Activities   12

Chapter 2   **Process Drawings and Equipment Standards**   **13**
Objectives   13
Key Terms   14
Introduction   15
Common Process Drawings and Their Uses   15
Common Information Contained on Process Drawings   22
Symbols   22
Equipment Standards   31
Summary   33
Checking Your Knowledge   34
Student Activities   35

Chapter 3   **Piping, Tubing, Hoses, and Fittings**   **36**
Objectives   36
Key Terms   37
Introduction   37
Piping, Tubing, Hoses, and Blinds   37
Materials of Construction   42
Pipe Selection and Sizing Criteria   45
Connecting Methods   46
Fitting Types   48
Leak Testing   51
Piping Protection   51
Potential Hazards   53

Process Technician's Role in Operation and Maintenance   55

Piping Symbols   55

Summary   55

Checking Your Knowledge   56

Student Activities   57

Chapter 4   Valves   58

Objectives   58

Key Terms   59

Introduction   60

Valve Components   60

Types of Valves   62

Potential Problems   74

Hazards Associated with Improper Valve Operation   74

Process Technician's Role in Operation and Maintenance   77

Summary   77

Checking Your Knowledge   78

Student Activities   79

Chapter 5   Tanks and Vessels   80

Objectives   80

Key Terms   81

Introduction   82

Types of Tanks and Vessels   82

Common Components of Vessels   86

Auxiliary Equipment   90

Potential Problems with Tanks and Vessels   94

Typical Procedures   95

Hazards Associated with Improper Operation   96

Process Technician's Role in Operation and Maintenance   97

Vessel and Reactor Symbols   97

Summary   98

Checking Your Knowledge   98

Student Activities   99

Chapter 6   Pumps   100

Objectives   100

Key Terms   101

Introduction   102

Selection of Pumps   103

Types of Pumps   103

Dynamic Pumps   104

Positive Displacement Pumps   110

Operating Principles   117

Associated Utilities/Auxiliary Equipment   119

Purpose of a Pump Curve   121

Potential Problems   124

Safety and Environmental Hazards   126

Process Technician's Role in Operation and Maintenance   128
Typical Procedures   128
Summary   129
Checking Your Knowledge   130
Student Activities   131

Chapter 7   Compressors   133
Objectives   133
Key Terms   134
Introduction   134
Selection of Compressors   135
Types of Compressors   135
Positive Displacement Compressors   138
Operating Principles   143
Associated Utilities and Auxiliary Equipment   144
Potential Problems   148
Safety and Environmental Hazards   151
Process Technician's Role in Operations and Maintenance   151
Typical Procedures   152
Summary   154
Checking Your Knowledge   155
Student Activities   156

Chapter 8   Turbines   157
Objectives   157
Key Terms   158
Introduction   159
Common Types and Applications of Turbines   159
Operating Principles   166
Auxiliary Equipment Associated with Turbines   168
Potential Problems   169
Safety and Environmental Hazards   171
Typical Procedures   171
Process Technician's Role in Operation and Maintenance   173
Summary   174
Checking Your Knowledge   175
Student Activities   176

Chapter 9   Electrical Distribution and Motors   177
Objectives   177
Key Terms   178
Introduction   179
What is Electricity?   179
Types of Current   183
Electrical Transmission   185
Components of an Electrical Distribution System   185
Purpose of Motors   187
Types of Motors   187

Operating Principles of Electric Motors   189
Potential Problems   190
Safety and Environmental Hazards   191
Typical Procedures   193
Process Technician's Role in Operation and Maintenance   195
Summary   195
Checking Your Knowledge   196
Student Activities   197

Chapter 10   Engines   198
Objectives   198
Key Terms   199
Introduction   199
Common Types of Engines   200
Uses of Engines in the Process Industry   201
Potential Problems   208
Safety and Environmental Hazards   208
Typical Procedures   209
Process Technician's Role in Operation and Maintenance   210
Summary   211
Checking Your Knowledge   212
Student Activities   212

Chapter 11   Heat Exchangers   214
Objectives   214
Key Terms   215
Introduction   215
Heat Transfer Overview   216
Theory of Heat Exchanger Operation   216
Types of Heat Exchangers   219
Exchanger Applications and Services   224
Potential Problems   226
Safety and Environmental Hazards   226
Typical Procedures   228
Process Technician's Role in Operation and Maintenance   230
Summary   230
Checking Your Knowledge   231
Student Activities   231

Chapter 12   Cooling Towers   233
Objectives   233
Key Terms   234
Introduction   235
Types of Cooling Towers   235
Components and Their Purposes   238
Applications of Cooling Towers   240
Theory of Operation   240

Factors that Affect Cooling Tower Operations   240

Safety and Environmental Hazards   244

Process Technician's Role in Operation and Maintenance   245

Summary   248

Checking Your Knowledge   248

Student Activities   249

Chapter 13   Furnaces   250

Objectives   250

Key Terms   251

Introduction   251

Applications   252

Common Furnace Designs   252

Furnace Draft Types   253

Furnace Sections and Components   254

Furnace Operating Principles   258

Safety and Environmental Hazards   261

Typical Procedures   263

Potential Problems   263

Summary   265

Checking Your Knowledge   266

Student Activities   266

Chapter 14   Boilers   267

Objectives   267

Key Terms   268

Introduction   269

General Components of Boilers   269

Water Tube, Waste Heat, and Fire Tube Boilers   272

Theory of Operation   275

Potential Problems   278

Safety and Environmental Hazards   280

Typical Procedures   281

Process Technician's Role in Operation and Maintenance   281

Summary   282

Checking Your Knowledge   283

Student Activities   284

Chapter 15   Auxiliary Equipment   285

Objectives   285

Key Terms   286

Introduction   286

Types of Auxiliary Equipment   286

Potential Problems   295

Typical Procedures   295

Process Technician's Role in Operation and Maintenance   295

Summary   296

Checking Your Knowledge   296

Student Activities   297

Chapter 16   Tools   298

Objectives   298

Key Terms   299

Introduction   300

Hand Tools   300

Power Tools   305

Lifting Equipment   305

Basic Hand and Power Tool Safety   309

Tool Care and Maintenance   313

Summary   313

Checking Your Knowledge   314

Student Activities   315

Chapter 17   Separation Equipment   316

Objectives   316

Key Terms   317

Introduction   317

Simple Separators   318

Distillation   318

Extraction   326

Absorption and Stripping   327

Adsorption   328

Evaporation   329

Crystallization   330

Potential Problems   330

Typical Procedures   331

Process Technician's Role in Operation and Maintenance   331

Summary   331

Checking Your Knowledge   332

Student Activities   332

Chapter 18   Reactors   333

Objectives   333

Key Terms   334

Introduction   334

Chemical Reactions   335

Types of Reactions and Reactors   335

Components of Reactors   341

Theory of Operation   342

Auxiliary Equipment Associated with Reactors   343

Potential Problems   344

Typical Procedures   344

Process Technician's Role in Operation and Maintenance   345

Summary 346

Checking Your Knowledge 346

Student Activities 347

Chapter 19 Filters and Dryers 348

Objectives 348

Key Terms 349

Introduction 349

Types of Filters 349

Filter Ratings 353

Types of Dryers 353

Consequences of Improper Operation 358

Potential Problems 359

Safety and Environmental Hazards 360

Typical Procedures 360

Process Technician's Role in Operation and Maintenance 362

Summary 363

Checking Your Knowledge 363

Student Activities 364

Chapter 20 Solids Handling Equipment 365

Objectives 365

Key Terms 366

Introduction 366

Types of Solids Handling Equipment 366

Potential Problems 386

Safety and Environmental Hazards 387

Typical Procedures 387

Process Technician's Role in Operation and Maintenance 387

Summary 388

Checking Your Knowledge 389

Student Activities 389

Chapter 21 Environmental Control Equipment 390

Objectives 390

Key Terms 391

Introduction 391

Types of Environmental Control Equipment 391

Federal Regulations 401

Potential Problems 402

Environmental Rules and Regulations 402

Typical Procedures 402

Process Technician's Role in Operation, Maintenance, and Compliance 402

Summary 403

Checking Your Knowledge 403

Student Activities 404

Chapter 22 **Mechanical Power Transmission and Lubrication** **405**

Objectives 405

Key Terms 406

Introduction 406

Operating Principles of Mechanical Transmission 406

Bearings 409

Gears 411

Principles of Lubrication 412

Potential Problems 417

Safety and Environmental Hazards 417

Process Technician's Role in Operation and Maintenance 418

Summary 419

Checking Your Knowledge 420

Student Activities 421

Glossary *423*

Index *435*

Preface

The Process Industries Challenge

In the early 1990s, the process industries recognized that they would face a major labor shortage due to the large number of employees retiring. Industry partnered with community colleges, technical colleges, and universities to provide training for their process technicians because it recognized that substantial savings on training and traditional hiring costs could be realized. In addition, the consistency of curriculum content and the exit competencies of process technology graduates could be ensured if industry collaborated with education.

To achieve this consistency of graduates' exit competencies, the Gulf Coast Process Technology Alliance and the Center for the Advancement of Process Technology identified a core technical curriculum for the Associate Degree in Process Technology. This core, consisting of eight technical courses, is taught in alliance member institutions throughout the United States. This textbook is intended to provide a common standard reference for the Process Technology I—Equipment course that serves as part of the core technical courses in the degree program.

Purpose of the Textbook

Instructors who teach the process technology core curriculum, and who are recognized in the industry for their years of experience and their depth of subject matter expertise, requested that a textbook be developed to match the standardized curriculum. Reviewers from a broad array of process industries and education institutions participated in the production of these materials so that the widest audience possible would be represented in the presentation of the content.

The textbook is intended for use in community colleges, technical colleges, universities, and corporate settings in which process technology is taught. However, educators in many disciplines will find these materials to be a complete reference for both theory and practical application. Students will find this textbook to be a valuable resource throughout their process technology career.

Organization of the Textbook

This textbook has been divided into 22 chapters. Chapter 1 provides an overview of the equipment used in the process industries. Chapter 2 provides an overview of process drawings and equipment standards. Chapters 3 to 14 provide an overview of the main pieces of equipment used in process industries. Chapter 15 provides an overview of auxiliary equipment, and Chapter 16 provides an overview of tools used in the process industries. Chapters 17 to 22 describe the various types of process equipment used in the process industries.

Each chapter is organized in the following way:

- Objectives
- Key Terms
- Introduction
- Key Topics
- Summary
- Checking Your Knowledge
- Student Activities

The **Objectives** for a chapter may cover one or more sessions in a course. For example, Chapter 2 may take two weeks (or two sessions) to complete in the classroom setting.

The **Key Terms** section is a list of important terms and their respective definitions that students should know and understand before proceeding to the next chapter.

The **Introduction** may be a simple introductory paragraph, or it may introduce concepts necessary to the development of the content of the chapter itself.

Any of the **Key Topics** can have several subtopics. Although these topics and subtopics do not always follow the flow of the objectives as stated at the beginning, all objectives are addressed in the chapter.

The **Summary** is a restatement of the learning outcomes of the chapter.

The **Checking Your Knowledge** questions are designed to help students test themselves on learning points from the chapter.

The **Student Activities** section contains activities that can be performed by students on their own or with other students in small groups, and activities that should be performed with instructor involvement.

Chapter Summaries

CHAPTER 1: INTRODUCTION TO PROCESS EQUIPMENT

This chapter describes the various types of equipment used in process industries and what they do. In addition, process technicians' duties, responsibilities, expectations, and working conditions are discussed.

CHAPTER 2: PROCESS DRAWINGS AND EQUIPMENT STANDARDS

This chapter explains the purpose of process system drawings and equipment standards in the process industries. Common drawing types and their components are identified.

CHAPTER 3: PIPING, TUBING, HOSES, AND FITTINGS

This chapter explains various types of piping, tubing, hoses, and fittings, and how they are used in the process industries. Materials of construction, hazards associated with improper operation, and monitoring and maintenance activities are also discussed.

CHAPTER 4: VALVES

This chapter explains various types of valves and how they are used in the process industry. Components of valves, purpose of valves, safety and environmental concerns, and the role of the process technician in maintenance are also discussed.

CHAPTER 5: TANKS AND VESSELS

This chapter explains various types of tanks and vessels and how they are used in the process industries. Components, associated safety and environmental hazards, typical procedures, the role of the process technician, and potential problems are also discussed.

CHAPTER 6: PUMPS

This chapter explains various types of pumps and how they are used in the process industries. Components, operating principles, associated safety and environmental hazards, typical procedures, the role of the process technician, and potential problems are also discussed.

CHAPTER 7: COMPRESSORS

This chapter explains various types of compressors and how they are used in the process industries. Components, operating principles, associated safety and environmental hazards, typical procedures, the role of the process technician, and potential problems are also discussed.

CHAPTER 8: TURBINES

This chapter explains various types of turbines and how they are used in the process industries. Components, operating principles, associated safety and environmental hazards, typical procedures, the role of the process technician, and potential problems are also discussed.

CHAPTER 9: ELECTRICAL DISTRIBUTION AND MOTORS

This chapter explains the components of electrical distribution systems in the process industries, including three-phase versus single-phase, one-line diagrams, transformers, motor control centers, circuit breakers, fuses, and switches. Components, operating principles, associated safety and environmental hazards, typical procedures, the role of the process technician, and potential problems are also discussed.

CHAPTER 10: ENGINES

This chapter explains the use and types of engines in the process industries. Components, operating principles, associated safety and environmental hazards, typical procedures, the role of the process technician, and potential problems are also discussed.

CHAPTER 11: HEAT EXCHANGERS

This chapter explains the common types, applications, and components of heat exchangers in the process industries. Methods of heat transfer, operating principles, associated safety and environmental hazards, typical procedures, the role of the process technician, and potential problems are also discussed.

CHAPTER 12: COOLING TOWERS

This chapter explains the principles of operation for cooling towers and how they are used in the process industries. Components, operating principles, associated safety and environmental hazards, typical procedures, the role of the process technician, and potential problems are also discussed.

CHAPTER 13: FURNACES

This chapter explains the principles of operation for furnaces and how they are used in the process industries. Components, operating principles, associated safety and environmental hazards, typical procedures, the role of the process technician, and potential problems are also discussed.

CHAPTER 14: BOILERS

This chapter explains the principles of operation for boilers and how they are used in the process industries. Components, operating principles, associated safety and environmental hazards, typical procedures, the role of the process technician, and potential problems are also discussed.

CHAPTER 15: AUXILIARY EQUIPMENT

This chapter explains the common types and applications of auxiliary equipment used in the process industries. Components, operating principles, associated safety and environmental hazards, typical procedures, the role of the process technician, and potential problems are also discussed.

CHAPTER 16: TOOLS

This chapter explains the types of tools used by process technicians in the process industries. Appropriate use, tool safety, and appropriate care of basic hand tools and power tools are covered.

CHAPTER 17: SEPARATION EQUIPMENT

This chapter explains the common types and applications of separation equipment used in the process industries. Components, operating principles, associated safety and environmental hazards, typical procedures, the role of the process technician, and potential problems are also discussed.

CHAPTER 18: REACTORS

This chapter explains the common types and applications of reactors used in the process industries. Components, operating principles, associated safety and environmental hazards, typical procedures, the role of the process technician, and potential problems are also discussed.

CHAPTER 19: FILTERS AND DRYERS

This chapter explains the common types and applications of filters and dryers used in the process industries. Components, operating principles, associated safety and environmental hazards, typical procedures, the role of the process technician, and potential problems are also discussed.

CHAPTER 20: SOLIDS HANDLING EQUIPMENT

This chapter explains the common types and applications of solids handling equipment used in the process industries. Components, operating principles, associated safety and environmental hazards, typical procedures, the role of the process technician, and potential problems are also discussed.

CHAPTER 21: ENVIRONMENTAL CONTROL EQUIPMENT

This chapter explains the common types and applications of environmental control equipment used in the process industries. Components, operating principles, associated safety and environmental hazards, the role of the process technician, and potential problems are also discussed.

CHAPTER 22: MECHANICAL POWER TRANSMISSION AND LUBRICATION

This chapter explains the common types and applications of mechanical power transmission and lubrication used in the process industries. Components, operating principles, associated safety and environmental hazards, typical procedures, the role of the process technician, and potential problems are also discussed.

Acknowledgments

The following organizations and their dedicated personnel voluntarily participated in the production of this textbook. Their contributions to the success of this project are greatly appreciated. Perhaps our gratitude for their involvement can best be expressed by this sentiment:

> The credit belongs to those people who are actually in the arena . . . who know the great enthusiasms, the great devotions to a worthy cause; who at best, know the triumph of high achievement; and who, at worst, fail while

daring greatly . . . so that their place shall never be with those cold and timid souls who know neither victory nor defeat. — Theodore Roosevelt

Process technicians, both current and future, will utilize the information within this textbook as a resource to understand the process industries more fully. This knowledge will strengthen these paraprofessionals by helping to make them better prepared to meet the ever challenging roles and responsibilities within their specific process industry.

Industry Content Developers and Reviewers

Chuck Baukal, John Zink Company LLC, Oklahoma
Ted Borel, Equistar, Texas
Linda Brown, Pasadena Refining System, Inc., Texas
Gayle Cannon, ConocoPhillips, New Jersey
Candy Carrigan, ConocoPhillips, New Jersey
Karl Diederich, Sterling Solutions, Inc., Texas
Larry Ely, ConocoPhillips, California
Steve Erickson, Gulf Coast Process Technology Alliance, Texas
Jimmy Greene, Eastman Chemical Company, Texas Operations, Texas
Debera Hanrahan, British Petroleum, Washington
Richard Honea, The Dow Chemical Company, Texas
Leslie Hunt, TailorMade Training, Tennessee
Glenn E. Johnson, The Sun Products Corporation, Texas
Alex Kharazi, The Dow Chemical Company, New Jersey
Susanne Kolodzy, Troubleshooting Resources, Texas
Steve Lagger, Citgo Petroleum Corporation, Illinois
John Leedy, The Dow Chemical Company, Texas
Diane McGinn, INEOS, Texas
Walter Eric Newby, BASF, Texas
Don Parsley, Valero Refining Corporation, Texas
Ray Player, Eastman Chemical Company, Texas Operations, Texas
Lyndon Pousson, Independent Reviewer, Louisiana
Brian Smith, CheveronTexaco, Louisiana
Barbara Tracy, ConocoPhillips, New Jersey
Mike Tucker, Eastman Chemical Company, Texas Operations, Texas
Roy Viator, CheveronTexaco, Louisiana
Norris Watt, British Petroleum, Louisiana
John E. Wilson, Training & Development Systems, Inc. (TDS), Texas

Education Content Developers and Reviewers

Chuck Beck, Red Rocks Community College, Colorado (formerly of Coors Brewing Company)
Tommie Ann Broome, Mississippi Gulf Coast Community College, Mississippi
Tom Carleson, Bellingham Technical College, Washington
Mike Cobb, College of the Mainland, Texas
Mike Connella, McNeese State University, Louisiana
David Corona, College of the Mainland, Texas
Mary Darden, Independent Reviewer, Texas
John Dees, d3 Consulting, Texas
Lisa Arnold Diederich, Independent Reviewer, Texas
Mark Demark, Alvin Community College, Texas
Eric Douglas, University of the Virgin Islands
Jerry Duncan, College of the Mainland, Texas
Jim Forthman, Calhoun Community College, Alabama (formerly with Monsanto/Solutia)

Gary Hicks, Brazosport College, Texas (formerly of The Dow Chemical Company)
Lauren Hightower, Baylor University, Texas
Jerry Layne, Baton Rouge Community College, Louisiana
Linton Lecompte, Independent Reviewer, Louisiana
Mike Link, Delaware Technical and Community College, Delaware
Jim Lockett, Lee College, Texas
Derrill Mallett, College of the Mainland, Texas
Martha McKinley, Texas State Technical College, Marshall, Texas (formerly of Eastman Chemical Company)
Larry Perswell, College of the Mainland, Texas
Kelsey Rexroat, Baylor University, Texas
Paul Rodriguez, Lamar Institute of Technology, Texas
Vicki Rowlett, Lamar Institute of Technology, Beaumont, Texas
Pete Rygaard, College of the Mainland, Texas
Dan Schmidt, Bismarck State College, North Dakota
Dale Smith, Alabama Southern Community College, Alabama
Robert (Bobby) Smith, Texas State Technical College, Marshall, Texas (formerly of Eastman Chemical Company)
Walter Tucker, Lamar Institute of Technology, Texas
Steve Wethington, College of the Mainland, Texas

Center for the Advancement of Process Technology Staff

Bill Raley, Principal Investigator
Jerry Duncan, Director
Melissa Collins, Associate Director
Angelica Toupard, Instructional Designer
Scott Turnbough, Graphic Artist
Cindy Cobb, Program Assistant
Joanna Perkins, Outreach Coordinator
Chris Carpenter, Web Application Developer

This material is based upon work supported, in part, by the National Science Foundation under Grant No. DUE 0202400. Any opinions, findings, and conclusions or recommendations expressed in this material are those of the author(s) and do not necessarily reflect the views of the National Science Foundation.

1

Introduction to Process Equipment

Objectives

After completing this chapter, you will be able to:

- Describe the process industries and what they produce.
- Describe the process technician's role in operation and maintenance.
- Describe the types of common equipment used in the process industries.
- Describe safety and environmental hazards associated with equipment usage in the process industries.

Key Terms

Boiler—a device in which water is boiled and converted into steam under controlled conditions.

Compressor—a mechanical device used to increase the pressure of a gas or vapor.

Cooling tower—a structure designed to lower the temperature of water using latent heat of evaporation.

Dryer—a device used to remove moisture from a process stream.

Engine—a machine that converts chemical (fuel) energy into mechanical force.

Filter—device that removes particles from a process, allowing the clean product to pass through the filter.

Fitting—system components used to connect together two or more pieces of piping, tubing, or other equipment.

Furnace—a piece of equipment that burns fuel in order to generate heat that can be transferred to process fluids flowing through tubes; also referred to as a process heater or reaction furnace.

Heat exchanger—a device used to transfer heat from one substance to another without the two physically contacting each other.

Hose—flexible tube that carries fluids; can be made of plastic, rubber, fiber, metal, or a combination of materials.

Lubrication—the application of a lubricant between moving surfaces in order to reduce friction and minimize heating.

Motor—a mechanical driver that converts electrical energy into useful mechanical work and provides power for rotating equipment.

Pipe—long, hollow cylinder through which fluids are transmitted; primarily made of metal, but can also be made of glass, plastic, or plastic-lined material.

Process—the conversion of raw materials into a finished or intermediate product.

Process drawing—illustration that provides a visual description and explanation of the processes, equipment, flows, and other important items in a facility.

Process industries—a broad term for industries that convert raw materials, using a series of actions or operations, into products for consumers.

Process technician—a worker in a process facility who monitors and controls mechanical, physical, and/or chemical changes throughout a process in order to create a product from raw materials.

Pump—a mechanical device that transfers energy to move materials through piping systems.

Reactor—vessel in which a controlled chemical reaction is initiated and takes place either continuously or as a batch operation.

Solids handling equipment—equipment that is used to process and transfer solid materials from one location to another in a process facility, and may also provide storage for those materials.

Tank—a container in which atmospheric pressure is maintained (i.e., they are neither pressurized nor placed under a vacuum; they are at the same pressure as the air around them).

Troubleshooting—the systematic search for the source of a problem so that it can be solved.

Tubing—small diameter (typically less than one inch) hose or pipe used to transport fluids.

Turbine—a machine that is used to produce power and rotate shaft-driven equipment such as pumps, compressors, and generators.

Unit—an integrated group of process equipment used to produce a specific product; may be referred to by the processes they perform, or named after their end products.

Valve—piping system components used to control, throttle, or stop the flow of fluids through a pipe.

Vessel—an enclosed container in which the pressure is maintained at a level that is higher than atmospheric pressure.

Introduction

This chapter provides an overview of the process industries, the role of the process technician, the types of equipment a process technician may encounter, as well as safety and environmental hazards.

Throughout this textbook, the term "**process industries**" is used as a broad term for industries that use processes to create products for consumers. A **process** is the conversion of raw materials into a finished or intermediate product. Process industries are some of the largest industries in the world, employing hundreds of thousands of people in almost every country. These industries, both directly or indirectly, create and distribute thousands of products that affect the daily lives of almost everyone on the planet.

Generally speaking, process industries involve process technologies that are processes that take specific quantities of raw materials and transform them safely into other products. The result might be an end product for a consumer or an intermediate product that is later converted to a different end product.

Each company in the process industries uses a system of people, methods, equipment, procedures and structures to create products. A **process technician** is a worker in a process facility who monitors and safely controls mechanical, physical, and/or chemical changes throughout a process in order to create a product from raw materials, while maintaining a healthy and safe environment.

While the processes and products may vary, process industries do share some basic equipment. For example, pumps are used to move liquids through a piping system in a production setting, motors are used to drive equipment, and furnaces are used to generate heat.

How Process Industries Operate

A variety of industries are classified as process industries; these include oil and gas exploration and production, chemical manufacturing, mining, power generation, water and waste water treatment, food and beverage production, pharmaceutical manufacturing, and paper and pulp processing While there are a wide range of processes and products associated with each industry, they all have some common operations:

1. Raw materials, sometimes called input or feedstock, are made available to a process facility or plant (Note: Most facilities have different units that perform a specific process. A **unit** is an integrated group of process equipment used to produce a specific product. Units can be referred to by the processes they perform [e.g., reforming unit], or named after their end products [e.g., olefin unit]).

2. The raw materials are sorted by process requirements.

3. The raw materials are processed using people, equipment, and methods.
Process technicians monitor and control the mechanical, physical, and/or chemical changes that occur during the process. Safety, health, environment, quality, and efficiency are key process elements.

4. A product (output), or desired component, is the result of a particular process. The product can either be an end product for consumers, or an intermediate product used as part of another process to make a different end product.
5. The product is then distributed to consumers.

Process Technician's Role in Operation and Maintenance

A process technician is a key member of a team responsible for planning, analyzing, and controlling the production of products. The production process may range from the acquisition of raw materials through the production and distribution of products to customers.

The job duties of a process technician may include a wide variety of tasks including: monitoring and controlling processes; assisting with equipment maintenance; communicating and working with others; performing administrative duties; performing with safety, health, and the environment in mind; troubleshooting; maintaining quality standards; and working in teams.

While monitoring and controlling a process, the process technician may be required to sample processes; inspect equipment; start, stop, and regulate equipment; view instrumentation readouts; analyze data; evaluate processes for optimization; make process adjustments; respond to changes, emergencies, and abnormal operations; and document activities, issues, and changes.

While the process technician is assisting with equipment maintenance, he or she may also be required to change or clean filters or strainers, lubricate equipment, monitor and analyze equipment performance, prepare equipment for repair, and return equipment to service.

One of the most valuable skills a process technician can master is the ability to communicate and work with others. While communicating and working with others is important for interpersonal reasons, it can also mean life or death in the process industries. This is especially true when the issue being communicated pertains to health, environmental impact, and/or safety of both plant personnel and the surrounding community.

Communication can be performed through written reports, performing analysis, writing and reviewing procedures, documenting incidents, listening to and training others, learning new skills and information, and working as part of a team. Strong computer, oral, and written communication skills are essential for process technicians operating within the organizational structure of a company. These skills are used on a daily basis when describing activities for relief personnel, maintaining data logs, preparing reports, and performing other tasks as needed.

Process technicians may also be required to perform additional duties which include housekeeping, and performing safety and environmental checks. Safety, health, environment, and security are typically the most spoken words in the process industry. Because of this, process technicians must keep safety, health, environmental, and security regulations and procedures in mind at all times. They must also look for unsafe or abnormal conditions and watch for signs of potentially hazardous situations.

Equipment troubleshooting is another skill that is both taught and learned through on-the-job training. **Troubleshooting** is the systematic search for the source of a problem so that it can be solved. A process technician working in a process unit may be required to apply troubleshooting techniques and principles to discover what is wrong with a piece of equipment, procedure, or process. Procedures for equipment are available and a list of potential problems and solutions may be listed. However, if an item is not listed, a process technician may be required to conduct a basic investigation and apply troubleshooting skills to try and figure out the problem.

In many facilities, the maintenance department repairs most problems associated with equipment. However, it is helpful when a process technician is able to inform the maintenance department of what they witnessed or observed prior to the equipment stoppage (for example, if a compressor goes down, a technician may report that a rattling sound was heard prior to shutdown). Observations and feedback such as this one may help the maintenance technician identify the cause of the problem more quickly and return the system to normal operations in a timelier manner.

Process technicians must have a basic understanding of quality and how it is important to the company's reputation and production. Quality has two major definitions: (1) a product or service free of deficiencies; and (2) the characteristics of a product or service that bear on its ability to satisfy stated or implied needs. Without quality measures, products and services could be deficient or unsatisfactory. Unsatisfactory products lead to the loss of customers, increased waste, inefficiencies, increased costs, reduced profits, and an inability to maintain a competitive edge.

In addition to maintaining a quality atmosphere, it is also essential for a process technician to have the ability to work effectively in a team-based environment. People are picked for teams because they have skills that complement the skills of other team members. Thus, everyone on the team is committed to a common purpose. All team members hold the other members of the team mutually accountable for the success of the project and the team.

When working as part of a team, process technicians must understand diversity and practice its principles. Process technicians must recognize and appreciate others for their contributions, while not discounting them because of their differences.

The life of a process technician must be flexible since it involves shift work in all types of weather. This career provides a variety of experiences for an individual looking for a challenging occupation. Employees in the process industries are generally rewarded for job excellence through salary increases, promotions, and bonuses. Job benefits usually include health and dental insurance, profit sharing, and retirement plans.

Equipment

Process technicians routinely work with many different types of equipment in a process unit as part of the daily tasks. It is critical for a process technician to have a clear understanding of the various types of equipment, their components, and how they work.

In addition to the types of equipment used in the various industries, process technicians must also know how to read process diagrams and have a basic understanding of equipment standards. The chapters in this textbook provide a detailed understanding of various types of equipment and how they are used in the process industries. The following is a brief explanation of the various pieces of equipment described in this textbook.

PROCESS DRAWINGS AND EQUIPMENT STANDARDS

Process drawings identify various parts of a process and a process facility. Each drawing includes a legend, title block, and application block. On each drawing are various symbols that represent the various components of a process. Each of these symbols is standardized to allow people from different facilities to read the process diagrams. Standardization reduces confusion and mistakes.

Some institutions that are responsible for creating these standardized symbols include:

ISA
67 Alexander Drive
Research Triangle Park, NC 27709
www.isa.org

Courtesy of ISA

- International Society of Automation (ISA)
- American National Standards Institute (ANSI)
- American Petroleum Institute (API)
- American Society of Mechanical Engineers (ASME)
- National Electric Code (NEC)

PIPES, TUBING, HOSES, AND FITTINGS

Pipes, tubing, hoses, and fittings are critical to the process industries and account for 30 to 40 percent of the initial investment when creating a new facility. **Pipes** are long, hollow cylinders through which fluids are transmitted. They are primarily made of metal, but they can also be made of glass, plastic, or plastic-lined material.

Tubing is small-diameter hose or pipe that is used to transport liquids or solids. Tubing is made from a variety of materials and is used in many applications, including sample systems and instrumentation.

Hoses are larger than tubing but less permanent than piping. **Hoses** are flexible tubes that carry fluids; they can be made of plastic, rubber, fiber, metal, or a combination of materials. They are used to make temporary connection, as in the case where facility air is used to drive equipment or when a rail car or tank truck is hooked into the piping system.

Fittings are system components used to connect two or more pieces of pipe, tubing, or other equipment. Fittings are selected based on the demands of the process.

VALVES

Valves are piping system components used to control, throttle, or stop the flow of fluids through a pipe. Valve types include:

- Gate
- Globe
- Check
- Ball
- Diaphragm
- Control valve
- Plug
- Relief and safety
- Butterfly
- Multi-port

Valves can be either manually or automatically actuated. The most commonly used type of valve in the process industries is the gate valve. Main valve components include the hand wheel, packing, bonnet, valve body, valve disc, and valve seat. Figure 1-1 shows a gate valve.

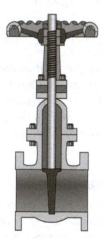

FIGURE 1-1 Gate Valve

TANKS AND VESSELS

Tanks are containers in which atmospheric pressure is maintained. They operate at pressures generally close to atmospheric, and are governed by API standards.

Vessels are enclosed containers in which the pressure is maintained at a level that is higher than atmospheric pressure. They typically operate at higher pressures and are governed by ASME standards. Tanks and vessels allow companies to more effectively and efficiently store products in large amounts, provide intermediate storage between processing steps, and provide residence time for reactions to complete or settle.

Companies use tanks and vessels to store and transport products throughout the United States and to foreign countries.

PUMPS

Pumps are mechanical devices that provide the energy required to move liquids through a piping system. Pumps are used in many applications, including filling or emptying tanks, providing water to boilers, supplying fire control water, and lubricating equipment. There are two main pump categories: dynamic and positive displacement. Figure 1-2 shows a centrifugal pump.

Dynamic pumps convert centrifugal force to dynamic pressure to move liquids. Positive displacement pumps use pistons, diaphragms, gears, or screws to deliver a constant volume. Positive displacement pumps deliver the same amount of liquid regardless of the discharge pressure. Dynamic pumps are classified as either centrifugal or axial. Positive displacement pumps are classified as either rotary or reciprocating.

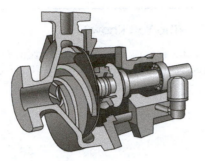

FIGURE 1-2 Centrifugal Pump

COMPRESSORS

A **compressor** is a mechanical device used to increase the pressure of a gas or vapor. There are two types of compressors: dynamic and positive displacement. Positive displacement compressors use screws, sliding vanes, gears, or pistons to deliver a set volume of gas. Dynamic compressors use a centrifugal or rotational force to move gasses. Positive displacement compressors can be classified as either reciprocating or rotary. Dynamic compressors are classified as either centrifugal or axial.

TURBINES

Turbines are machines that are used to produce power and rotate shaft-driven equipment such as pumps, compressors, and generators. Turbines can also be used to provide auxiliary power back to the process units they service. Figure 1-3 shows a steam turbine.

FIGURE 1-3 Steam Turbine

Turbines convert the kinetic and potential energy of a motive fluid into the mechanical energy required to drive a piece of equipment. Common types of turbines include: steam, gas, hydraulic, and wind.

MOTORS

Motors are mechanical drivers that convert electrical energy into useful mechanical work and provide power for rotating equipment. Power distribution is important in the process industries. Electricity is the most common source of energy for driving motors, and motors are typically the largest electrical loads in an industrial manufacturing facility. Thus the efficiency of the motors and their operations directly affects the cost structure of the process plant.

Motors and equipment use two forms of electricity: alternating current (AC) and direct current (DC). AC uses a changing flow of electrons in a conductor, while DC flows in a single direction.

ENGINES

Engines are machines that convert chemical (fuel) energy into mechanical force. Types of engines include internal combustion, external combustion, gasoline, and diesel.

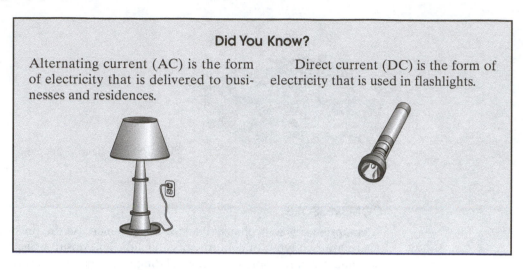

Did You Know?

Alternating current (AC) is the form of electricity that is delivered to businesses and residences.

Direct current (DC) is the form of electricity that is used in flashlights.

HEAT EXCHANGERS

Heat exchangers are devices used to transfer heat from one substance to another without the two physically contacting each other. Types of heat exchangers include double-pipe, shell and tube, and plate and frame. Each heat exchanger includes an inlet and outlet flow. Flow may be characterized as laminar or turbulent, and flow paths may be defined as countercurrent, co-current, parallel flow, or cross flow. Figure 1-4 shows a heat exchanger.

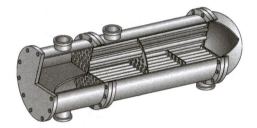

FIGURE 1-4 Heat Exchanger

COOLING TOWERS

Cooling towers are structures designed to lower the temperature of water using latent heat of evaporation. Cooling towers are classified based on how air flow moves through them and whether the air movement is natural, induced, or forced. Air flow inside a cooling tower can either be crossflow or counterflow. A very small amount of heat is directly transferred by heating the air (a process called convection). Most of the heat transferred is through evaporation. As the water evaporates (changes from liquid into a vapor) it absorbs a substantial amount of heat from the remaining water. This heat is called latent heat of vaporization, and is similar to the energy necessary to boil water on your stove. A small portion of the warm water evaporates into the air, while the remaining water gives up heat energy and is thereby cooled.

FURNACES

Furnaces (also referred to as process heaters) are pieces of equipment that burn fuel to generate heat that can be transferred to process fluids flowing through tubes; they are also referred to as process heaters or reaction furnaces. Furnace designs vary according to the function, heating duty, type of fuel, and the method of introducing combustion air. Types of furnaces include box, vertical, and cabin. Furnace components include burners, a purge system, radiant tubes, refractory lining, bridge wall, shock bank, convection tubes, soot blower, stack, damper, and plenum.

BOILERS

Boilers are devices in which water is boiled and converted into steam under controlled conditions. In process industries, steam is used to heat and cool process fluids, fight fires, purge equipment, and promote reactions. Types of boilers include water tube, waste heat, and fire tube. The main components of boilers include the fire box, burners, drums, tubes, economizer, steam distribution system, fuel system, and boiler feedwater system.

AUXILIARY EQUIPMENT

Auxiliary equipment is used to assist processes in mixing, transferring, and creating vacuum in mixers, eductors, centrifuges, and hydroclones.

TOOLS

The selection, knowledge, and use of hand and power tools are essential to the safety and well-being of the process technician. Process technicians use basic hand tools to perform various work functions, and it is imperative that the proper tool be selected to prevent injuries to the employee or the equipment.

At times, lifting equipment is used in the process industries to raise heavy items to elevated locations. Types of lifting equipment include hoists, cranes, forklifts, and personal lifts. Electric, pneumatic, hydraulic, and powder-activated power tools may also be used. Process technicians must know which tools can be used on each piece of equipment. If there is any doubt about which tool should be used, the process technician should review the standard operating procedures, consult with a supervisor, or consult with the safety department. As a general rule, a process technician should not operate any power tools or equipment without proper training or authorization.

SEPARATION EQUIPMENT

Often materials are not suitable in their original form and must be separated prior to processing. Separation equipment (e.g., a distillation system) is used to separate these raw materials. A few types of distillation include dual component, multicomponent, azeotropic, and extractive. Separation also occurs through extraction, evaporation, crystallization, adsorption, absorption, and stripping.

REACTORS

Reactors are vessels in which a controlled chemical reaction is initiated and takes place either continuously or as a batch operation. Within a reactor, raw materials are combined at various flow rates, pressures, and temperatures and are then reacted to form a product. Reactor processes can be batch or continuous. Batch processes involve a single charge to a reactor from the product stream (as opposed to a continuous feed) to make a final product. A continuous reaction is when products are continuously being formed, removed as raw materials, and fed into the reactor. Types of reactors include stirred tank, tubular, fixed bed, fluidized bed, hot wall, cold wall, and nuclear.

FILTERS AND DRYERS

Filters and dryers are used in the process industries to remove undesirable components from the process streams. **Filters** remove particles from a process, allowing the clean product to pass through; **dryers** are typically used to remove moisture in a process stream. Types of filters include cartridge, bag, leaf, rotary, pleated, and plate and frame. Types of dryers include fixed bed, dual bed, fluid bed, flash, rotary, and cyclone. Figure 1-5 shows a fluid bed dryer.

SOLIDS HANDLING EQUIPMENT

Solids handling equipment is used to process and transfer solid materials from one location to another in a process facility. Types of solids handling equipment include

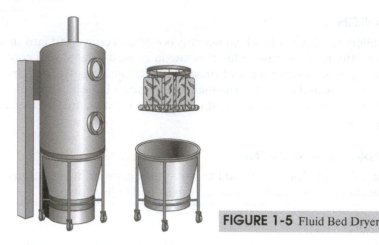

FIGURE 1-5 Fluid Bed Dryer

inducers, conveyors, feeders, extruders, bucket elevators, cyclones, screening systems, bins, silos, hoppers, dryers, trickle valves, bulk bag stations, and blowers.

ENVIRONMENTAL CONTROL EQUIPMENT

Environmental control equipment is equipment used to maintain the operation of process equipment and piping within the lowest level of the standards allowable by law. Types of air pollution control equipment include baghouses and precipitators, coal gasification, vapor and gas emissions, scrubbers, incinerators, and flare systems. Types of water and soil pollution controls include activated sludge, clarifiers, dikes, settling ponds, and landfills.

MECHANICAL POWER TRANSMISSION AND LUBRICATION

Mechanical power transmission transfers rotational energy from a driver to driven equipment with minimum loss of energy due to friction. **Lubrication** is the application of a lubricant between moving surfaces to reduce friction and minimize heating. Mechanical transmission includes the following components: couplings, belts, gearboxes, chains, magnetic or hydraulic drives, gears, and bearings.

Safety and Environmental Hazards

When working with equipment, process technicians must be aware of the following factors that can potentially affect safety:

- Abnormal sounds coming from equipment
- Excessive vibration
- Leaks around equipment
- Faulty equipment gauges
- Use of incorrect tools
- Not wearing proper "Personal Protective Equipment (PPE)" such as hearing protection, safety glasses, goggles, hard hats or flame retardant clothing
- Not following standard operating procedures (normal operating conditions of pressure and temperatures)

In addition, process technicians are also expected to:

- Avoid incidents and errors (i.e., work safely, in compliance with government, industry, and company regulation)
- Help with environmental compliance
- Work smarter and focus on business goals
- Look for ways to reduce waste and improve efficiency
- Keep up with industry trends and constantly improve skills

Summary

A process technician is a worker in a process facility that safely monitors and controls mechanical, physical, and/or chemical changes throughout a process to produce either a final product or an intermediate product made from raw materials. A variety of industries are classified as process industries, including oil and gas exploration and production, chemical manufacturing, mining, power generation, waste and water treatment, food and beverage production, pharmaceutical manufacturing, and paper and pulp processing.

The process technician's role in operation and maintenance is important because they are responsible for planning, analyzing, and controlling the production and distribution of products. The duties of a process technician include maintaining a safe work environment, as well as controlling, monitoring, and troubleshooting equipment.

Equipment is used throughout the process industries, so it is important for process technicians to have a clear understanding of the equipment, its components, and its operation. Common equipment used in the process industries includes valves, pumps, compressors, turbines, motors, engines, heat exchangers, cooling towers, furnaces, boilers, reactors, and filters. With each piece of equipment, process technicians must also have an understanding of potential problems, environmental and safety concerns, typical procedures, and their role according to standard operating procedures and company requirements.

Safety is a major expectation of each employee in the workforce today. It is important for all safety rules to be followed and for each employee to have a proactive attitude regarding their own safety and the safety of fellow employees.

Checking Your Knowledge

1. Define the following terms:
   a. Process drawing
   b. Process industries
   c. Process technician
   d. Process technology
   e. Troubleshooting
   f. Unit
2. *(True or False)* A process technician controls mechanical, physical, and/or chemical changes throughout many processes to produce a final product made from raw materials.
3. *(True or False)* Process technicians are required to analyze data and communicate those data to the appropriate employees.
4. *(True or False)* Equipment troubleshooting is a skill that can be learned only in the classroom environment.
5. Process drawings include: (*Select all that apply*)
   a. plot plans
   b. electrical diagrams
   c. schematics
   d. system flow diagrams
6. A process technician may be required to perform the following maintenance activities (select all that apply):
   a. Lubricate equipment
   b. Change or clean filters
   c. Monitor and analyze equipment performance
   d. Lift the equipment to install a new piece
7. *(True or False)* Selection of the proper tools for a job is the sole responsibility of the tool room clerk.
8. A mechanical piece of equipment used to move materials through a piping system within various processes is a:
   a. heat exchanger
   b. compressor
   c. burner
   d. pump
9. Drivers that convert electrical energy into useful mechanical work and provide power for rotating equipment like pumps, compressors, and conveyor drives are:
   a. reactors
   b. solids handling equipment
   c. motors
   d. furnaces

10. Equipment used to move material within a process facility includes:
    a. environmental control equipment
    b. solids handling equipment
    c. separation equipment
    d. auxiliary equipment

Student Activities

1. Write a two-page paper on one or more of the standards (e.g., ANSI, API, ASME, and ISA).
2. Work with a classmate to draw a flow sequence of a simple process and describe the basic steps associated with the process.

2

Process Drawings and Equipment Standards

Objectives

After completing this chapter, you will be able to:

- Explain the purpose of diagrams including why, when, and where they are used.
- Identify the major unit sections in flow sequence (block diagram).
- Identify symbols used for process equipment and instrumentation.
- Identify components on a typical Process Flow Diagram (PFD).
- Identify components on a typical Piping and Instrument Diagram (P&ID).
- Explain the purpose of equipment standards.
- Identify a plot plan and explain the purpose of equipment layout drawings (plot plan).

Key Terms

ANSI—American National Standards Institute; oversees and coordinates the voluntary standards in the U.S. ANSI develops and approves norms and guidelines that affect many business sectors. The coordination of U.S. standards with international standards allows American products to be used worldwide.

API—American Petroleum Institute; a trade association that represents the oil and gas industry in the areas of advocacy, research, standards, certification, and education for the petroleum industries, petrochemical industries, and municipalities.

Application block—the main part of a drawing that contains symbols and defines elements such as relative position, types of materials, equipment descriptions, flows, and functions.

ASME—American Society of Mechanical Engineers; specifies requirements and standards for pressure vessels, piping, and their fabrication.

Block Flow Diagram (BFD)—a simple illustration that shows a general overview of a process, indicating the parts of a process and their relationships.

Electrical diagram—illustration showing power transmission and how it relates to the process.

ISA—The Instrumentation, Systems, and Automation Society; a global, nonprofit technical society that develops standards for automation, instrumentation, control, and measurement.

Isometric drawing—an illustration showing objects as they would appear to the viewer (similar to a 3D drawing that appears to come off the page).

Legend—a section of a drawing that explains or defines the information or symbols contained within the drawing. Legends include information such as abbreviations, numbers, symbols, and tolerances.

NEC—National Electric Code; specifies electrical cable sizing requirements and installation practices.

OSHA—Occupational Safety and Health Administration; a U.S. government agency created to establish and enforce workplace safety and health standards, conduct workplace inspections and propose penalties for noncompliance, and investigate serious workplace incidents.

Piping and instrument diagram (P&ID)—detailed illustration that graphically represents the relationship of equipment, piping, instrumentation, and flows contained within a process in the facility.

Plot plan—illustration showing the layout and dimensions of equipment, units, and buildings. They are drawn to scale so that everything is of the correct relative size.

Process flow diagram (PFD)—basic illustration that uses symbols and direction arrows to show the primary flow of a product through a process. It includes information such as operating conditions, the location of main instruments, and major pieces of equipment.

Schematics—show the direction of electrical current flow in a circuit, typically beginning at the power source.

Symbol—simple illustration used to represent the equipment, instruments, and other devices on a PFD or P&ID. Some symbols are standard throughout the industry, while others may be specific to the individual manufacturing or engineering company.

Title block—a section of a drawing (typically located in the bottom right corner) that contains the drawing title, drawing number, revision number, sheet number, company information, process unit, and approval signatures. Identifies the drawing as a legend, system, or layout.

Utility flow diagram (UFD)—illustration that provides process technicians a PFD-type view of the utilities used for a process.

Introduction

Diagrams or process drawings are used to provide process technicians with a visual description and explanation of the processes, its associated equipment, and other important items in a facility. Process drawings are as critical to a process technician as a topographical map (a map that shows hills, streams, and trails) is to a hiker in the deep woods. Just as the hiker must be able to read a map to navigate travel, a process technician must be able to read process system drawings to understand what is happening at a process facility.

There are many different types of drawings. Each drawing type represents different aspects of the process and different levels of detail. Looking at combinations of these drawings provides a more complete picture of the processes and the facility. Without process drawings, it would be difficult for process technicians to understand the process and how it operates.

When examining process drawings, it is important to remember that all drawings have three common functions: which include simplifying (using common symbols to make processes easy to understand), explaining (describing how all of the parts or components of a system work together), and standardizing (using a common set of lines and symbols to represent components).

Diagrams are also used extensively for process technicians learning to troubleshoot at startup, at shutdown, and after initial commissioning.

Process drawings must meet several requirements to be considered a proper industrial drawing. These requirements include specific, universal rules on how lines are drawn, how proportions are used, what measurements are used, what components are included, and what industrial unit is targeted.

Common Process Drawings and Their Uses

Process technicians must recognize a wide variety of drawings and understand how to use them. The most commonly encountered drawings include:

- Block Flow Diagrams (BFDs)
- Process Flow Diagrams (PFDs)
- Piping and Instrument Diagrams (P&IDs)
- Utility Flow Diagrams (UFDs)
- Electrical Diagrams
- Schematics
- Isometrics

The following sections describe each of these drawings and their uses.

BLOCK FLOW DIAGRAMS (BFD)

Block Flow Diagrams (BFDs) are simple drawings that show a general overview of a process and contain few specifics. BFDs use blocks to represent sections of a process and use flow arrows to show the order and relationship of each component.

The BFD in Figure 2-1 is a raw water processing example. The numbers on the diagram indicate the order of the steps in the process. (Note: these numbers are used for illustrative purposes and are typically not found on a BFD.) The list below describes what is occurring at each step.

1. Raw Water—The raw water is imported from whichever source is available at the facility (e.g., river water). However, the water is not clean enough to be used as boiler feed water, so it is sent to step two so it can be clarified.

2. Clarification—The raw water is sent to a clarifier. This phase removes suspended solids from the water.

FIGURE 2-1 Block Flow Diagram (BFD)

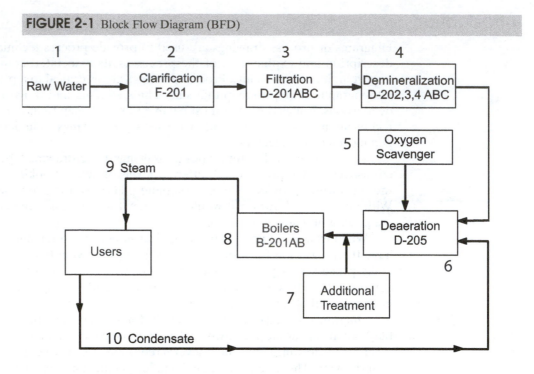

3. Filtration—In this phase clarified water is sent across a set of filters to remove any remaining solid matter that could accumulate in the boiler tubes and cause plugging or possible tube failure.

4. Demineralization—In this phase, most of the mineral content is removed.

5. Oxygen scavenging—The water that goes into the boiler cannot contain excess oxygen (O_2), so an oxygen scavenger is used to remove the oxygen from the water. This is necessary to prevent problems like bacterial growth and rust.

6. Deaeration—This section of the boiler feedwater process removes excess air or gas that may be contained in the water. The process is accomplished by heating (flashing) the water above steam temperature and removing any entrained gases or free air.

7. Additional Treatment—If any additional water treatment is required, it will be entered at this point.

8. Boilers—At this point, the water is ready to be sent to the boiler so it can be converted into steam. The amount of steam produced depends on the size and makeup of the particular boiler.

9. Steam—Steam feeds the users throughout the facility. The output from the users is cooled, collected, and removed as condensate.

10. Condensate—Condensate is collected and reused (the more condensate that can be retrieved, the less raw water that has to be introduced and treated. This translates to an economic savings for the company).

PROCESS FLOW DIAGRAMS (PFDs)

Process technicians are exposed to different types of industrial drawings on the job. The two most common types of drawings are Process Flow Diagrams (PFDs) and Piping and Instrument Diagrams (P&IDs).

Process Flow Diagrams (PFDs) are basic drawings that use symbols and direction arrows to show the primary flow of a product through a process. PFDs allow process technicians to trace the step-by-step flow of a process. PFDs contain symbols that represent the major pieces of equipment and piping used in the process and directional arrows to show the path of the process. Figure 2-2 shows an example of a PFD.

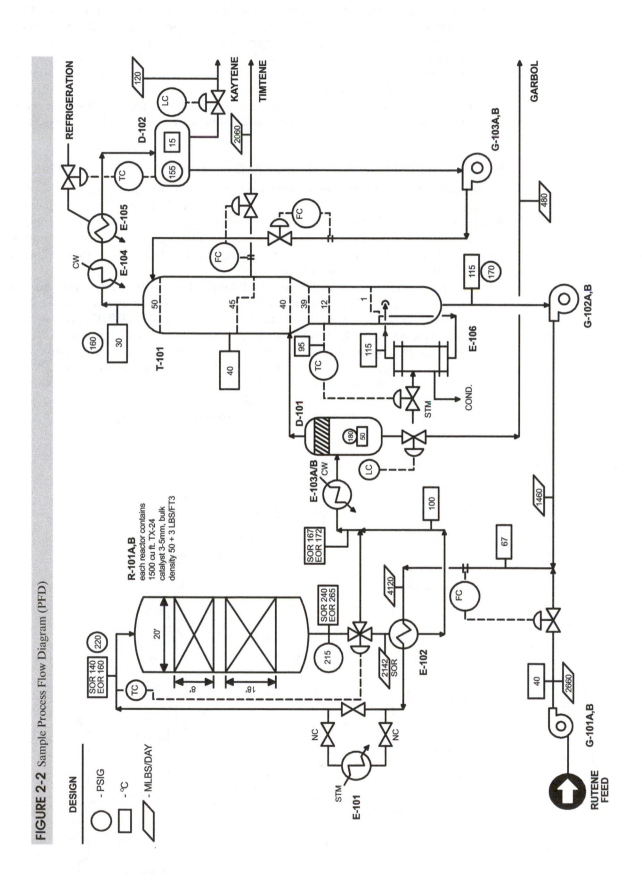

FIGURE 2-2 Sample Process Flow Diagram (PFD)

The process flow is typically drawn from left to right, starting with feed products or raw materials on the left, and ending with finished products on the right. Other information found on a PFD includes variables, pump capacities, heat exchangers, equipment symbols, equipment designations, major process piping, and control instruments.

Symbology charts are used along with PFDs and/or P&IDs to show the major pieces of equipment, piping, temperatures, pressures at critical points, and the flow of the process. The use of symbology allows for the standardization of information on industrial drawings. Each industrial drawing has similar lines and symbols that represent the various components. These lines and symbols (with some subtle changes) are used all over the world.

PIPING AND INSTRUMENTATION DIAGRAMS (P&IDs)

Piping and Instrument Diagrams (P&IDs), which are sometimes referred to as Process and Instrument Drawings, are similar to Process Flow Diagrams (PFDs). However, P&IDs show more detailed process construction information, such as equipment, piping, flow arrows, materials of construction, and insulation. Additional information on a P&ID includes equipment numbers, piping specifications, and instrumentation. Figure 2-3 shows an example of a P&ID.

For a process technician, a vital part of a P&ID is the instrumentation information. This information gives the technician a firm understanding of how a product flows through the process and how it can be monitored and controlled. P&IDs are also critical during maintenance tasks, troubleshooting, modifications, and upgrades.

P&IDs use ISA (formerly known as the Instrumentation, Systems and Automation Society) standard symbols, (ISA and instrumentation tag numbers are discussed later in this chapter.) Process technicians should be able to recognize these symbols and any special lettering conventions used on a P&ID. In addition, technicians must also be able to interpret process flows, and instrument and equipment designations.

FIGURE 2-3 Sample Piping and Instrument Diagram (P&ID)

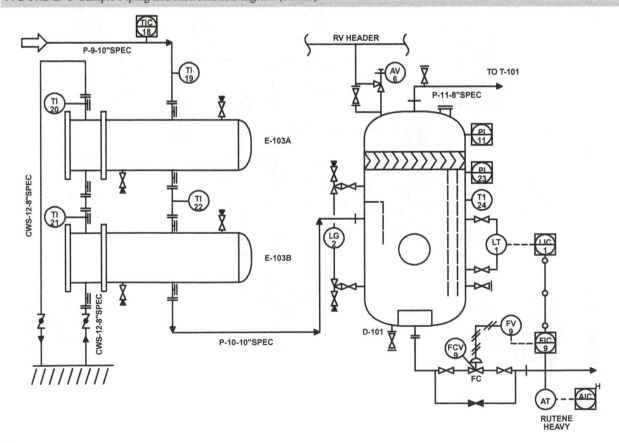

UTILITY FLOW DIAGRAMS (UFDs)

Utility Flow Diagrams (UFDs) provide process technicians a PFD-type view of the utilities used for a process. UFDs represent the way utilities connect to the process equipment, along with the piping and main instrumentation used to operate those utilities.

Typical utilities shown on a UFD include:

- Steam
- Condensate
- Cooling water
- Instrument air
- Plant air
- Nitrogen
- Fuel gas

Figure 2-4 shows an example of a Utility Flow Diagram (UFD).

FIGURE 2-4 Sample Utility Flow Diagram (UFD)

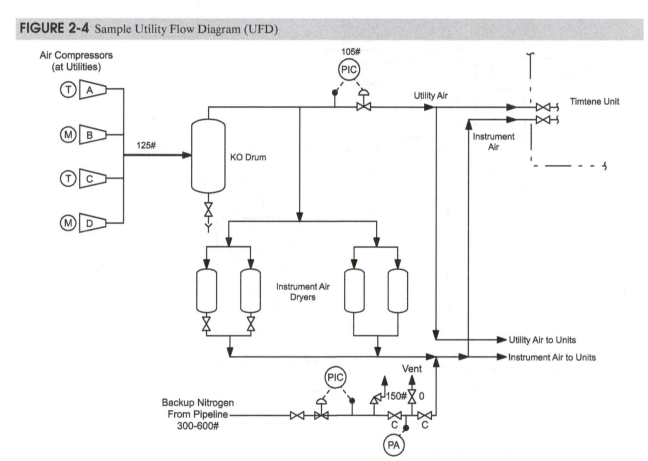

ELECTRICAL DIAGRAMS

Many processes rely on electricity, so it is important for process technicians to understand electrical systems and how they work. **Electrical diagrams** help process technicians understand power transmission and how it relates to the process. A firm understanding of these relationships is critical when performing lockout/tagout procedures (i.e., control of hazardous energy) and monitoring various electrical measurements.

Electrical diagrams show the various electrical components and their relationships, for example:

- Switches used to stop, start, or change the flow of electricity in a circuit
- Power sources provide by transmission lines, generators, or batteries
- Loads (the components that actually use the power)
- Coils or wire used to increase the voltage of a current
- Inductors (coils of wire that generate a magnetic field and are used to create a brief current in the opposite direction of the original current) that can be used for surge protection

- Transformers (device that takes electricity of one voltage and changes it into another voltage)
- Resistors (coils of wire used to provide resistance in a circuit)
- Contacts used to join two or more electrical components.

There are different types of electrical diagrams. However, the most common types of electrical diagrams are wiring diagrams and schematics. Figure 2-5 shows an example of a schematic.

FIGURE 2-5 Electrical Diagram (Schematic)

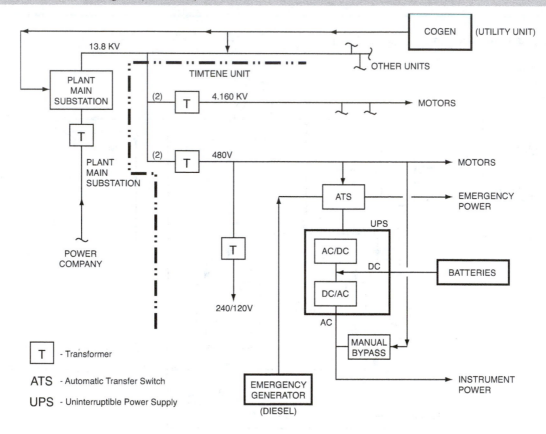

SCHEMATICS

Schematics show the direction of current flow in a circuit, typically beginning at the power source. Process technicians use schematics to visualize how current flows between two or more circuits. Schematics also help electricians detect potential trouble spots in a circuit.

ISOMETRICS

Isometric drawings show objects like equipment as they would appear to the viewer. In other words, they are like a 3D drawing that appears to come off the page. Isometric drawings may also contain cutaway views to show the inner workings of an object. Figure 2-6 shows an example of an isometric drawing.

Isometric drawings show the three sides of the object that can be seen, with the object appearing at a 30-degree angle with respect to the viewer. Isometrics are typically used during new unit construction or unit revisions. These drawings can prove useful to new process technicians as they learn to identify equipment and understand its inner workings.

FIGURE 2-6 Isometric Drawing

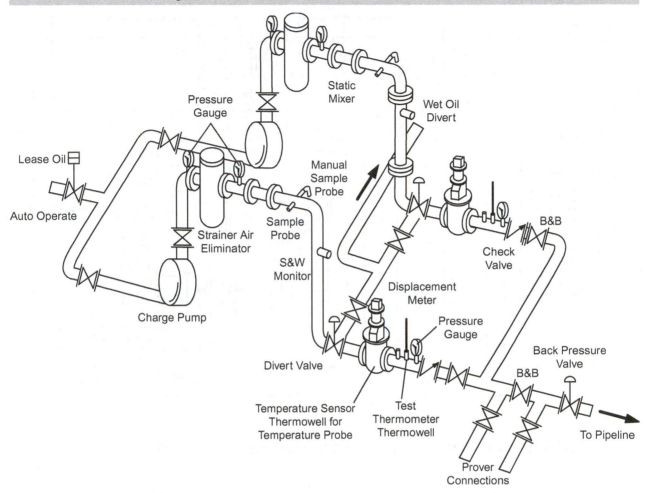

PLOT PLAN

Plot plans show the layout and dimensions of equipment, units, and buildings. They are drawn to scale so that everything is of the correct relative size. For example, plot plans show the location of machinery (e.g., pumps and heat exchangers) in an equipment room. On a larger scale, a plot plan shows the location and dimensions of process units, buildings, roads, and other site constructions such as fences. A site plot plan also shows elevations and grades of the ground surface. Figure 2-7 shows an example of a plot plan.

OTHER DRAWINGS

Along with the drawings mentioned in the previous sections, process technicians might also encounter other types of drawings such as elevation diagrams, equipment location diagrams, loop diagrams, and logic diagrams.

Elevation diagrams show the relationship of equipment to ground level and other structures.

Equipment location diagrams show the relationship of units and equipment to a facility's boundaries.

Loop diagrams show all components and connections between instrumentation and the control room.

Logic diagrams show the sequential steps within the computer or safety system.

FIGURE 2-7 Plot Plan

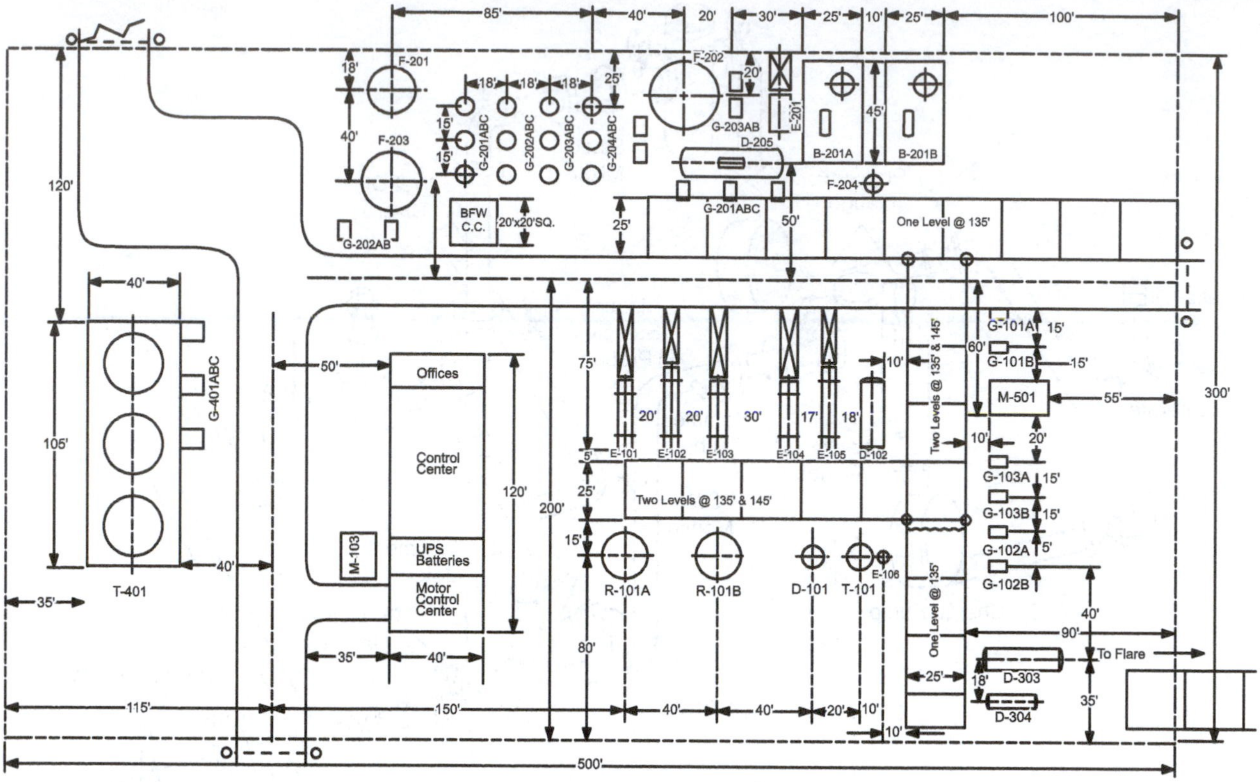

Common Information Contained on Process Drawings

LEGEND

A **legend** is a section of a drawing that explains or defines the information or symbols contained within the drawing (like a legend on a map). Legends include information such as abbreviations, numbers, symbols, and tolerances. Figure 2-8 shows an example of a legend.

TITLE BLOCK

The **title block** is a section of a drawing (typically located in the bottom right corner) that contains the drawing title, drawing number, revision number, sheet number, company or process unit, and approval signatures and dates. Figure 2-9 shows an example of a title block.

APPLICATION BLOCK

An **application block** is the main part of a drawing that contains symbols and defines elements such as relative position, types of materials, equipment descriptions, and functions. Figure 2-10 shows an example of an application block.

Symbols

Symbols are simple illustrations used to identify types of equipment. A set of common symbols has been developed to represent actual equipment, piping, instrumentation, and other components. While some symbols may differ from facility to facility, many are universal, with only subtle differences. It is critical that process technicians recognize and understand these symbols. Figure 2-11 shows examples of a few equipment symbols.

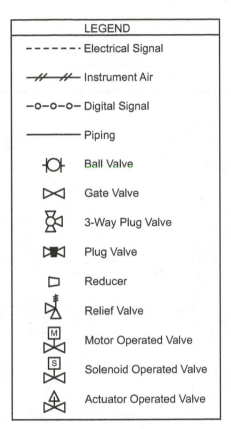

```
                    LEGEND
- - - - - - -   Electrical Signal

  //    //      Instrument Air

-o-o-o-         Digital Signal

                Piping

   ⊶          Ball Valve

   ⋈            Gate Valve

   ⛣            3-Way Plug Valve

   ⋈            Plug Valve

   ⬠            Reducer

   ⋈            Relief Valve

  [M]
   ⋈            Motor Operated Valve

  [S]
   ⋈            Solenoid Operated Valve

  [A]
   ⋈            Actuator Operated Valve
```

FIGURE 2-8 Example of a P&ID Legend

FIGURE 2-9 Example of a P&ID Title Block

ANYCorp	ANYCorp Construction Company			
Contract No.21609	Texas City, Texas			
	Project No. 2447			
ANYCorp	Drawn S. Turnbough		Date 1/06	
	Chk. By M. Collins		App. By A. Toupard	
Piping & Instrumentation Diagram	Scale None		Unit 85	
Primary Amine Unit	AFE No.		Chc. No. 6848-56	
Fuel Gas Amine Absorber	Dwg. No. F-52-4-0505			Rev 2

PIPING SYMBOLS

Pipes are long, hollow tubes used to transport fluids throughout a process facility. Many symbols are associated with piping. While standards exist, symbols can vary slightly from facility to facility. Figure 2-12 show some examples of piping symbols.

COMPRESSOR SYMBOLS

Compressors are used to increase the pressure on gases and vapors. To locate compressors on a P&ID, process technicians must be familiar with the different types of compressors and their symbols.

Figure 2-13 shows examples of centrifugal compressor symbols, and Figure 2-14 shows examples of positive displacement compressor symbols. These symbols may vary from facility to facility.

FIGURE 2-10 Example of a P&ID Application Block

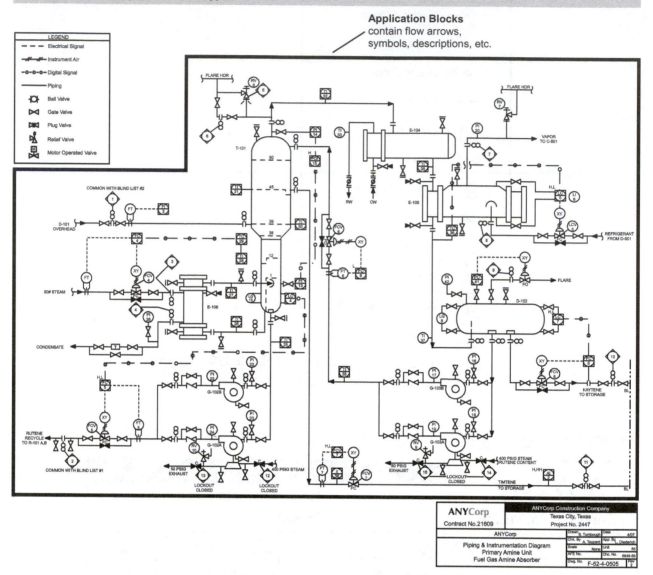

Application Blocks
contain flow arrows,
symbols, descriptions, etc.

FIGURE 2-11 Examples of Symbols Used in Process Diagrams

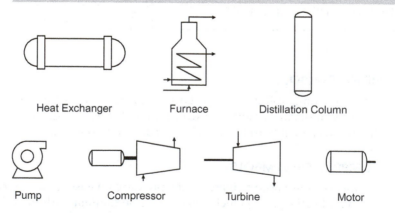

Heat Exchanger   Furnace   Distillation Column

Pump   Compressor   Turbine   Motor

ACTUATOR SYMBOLS

Actuators are devices that convert electrical control signals to physical actions.

Figure 2-15 shows examples of actuator symbols. However, these symbols may vary from facility to facility.

FIGURE 2-12 Examples of Piping Symbols Used in Process Diagrams

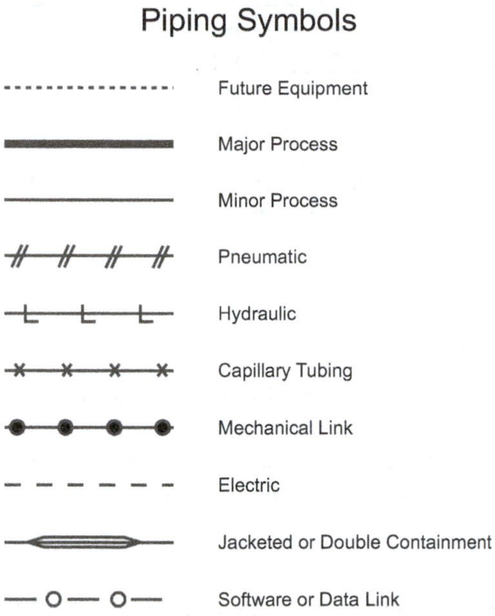

Piping Symbols

- - - - - - - - - - - - - - -	Future Equipment
━━━━━━━━━━━	Major Process
─────────────	Minor Process
─//──//──//──//─	Pneumatic
─L──L──L─	Hydraulic
─x──x──x──x─	Capillary Tubing
●──●──●──●─	Mechanical Link
─ ─ ─ ─ ─ ─	Electric
──◁▷──	Jacketed or Double Containment
─○──○─	Software or Data Link

FIGURE 2-13 Centrifugal Compressor Symbols

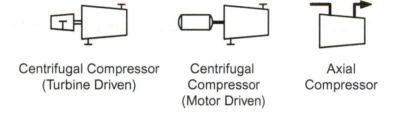

Centrifugal Compressor
(Turbine Driven)

Centrifugal
Compressor
(Motor Driven)

Axial
Compressor

FIGURE 2-14 Positive Displacement Compressor Symbols

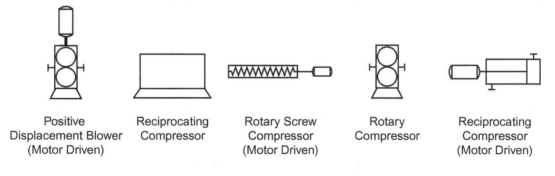

Positive
Displacement Blower
(Motor Driven)

Reciprocating
Compressor

Rotary Screw
Compressor
(Motor Driven)

Rotary
Compressor

Reciprocating
Compressor
(Motor Driven)

FIGURE 2-15 Examples of Actuator Symbols

Actuator Symbols

Hand

Pneumatic

Motor

Hydraulic

Solenoid

FIGURE 2-16 Common Cooling Tower Symbols

Induced Draft
Cross-Flow

Natural Draft
Counter-Flow

Forced Draft
Cooling Tower

COOLING TOWER SYMBOLS

Cooling towers are used to lower the temperature of water using latent heat of evaporation and sensible heat loss. To locate cooling towers accurately on a P&ID, process technicians must be familiar with the different cooling tower symbols.

Figure 2-16 shows some of the symbols used to indicate cooling towers.

VALVE SYMBOLS

Valves are used to control the flow of fluids through a pipe. Many types of valves are used in the process industries. Each valve type has a unique symbol used to identify it on a process drawing. Table 2-1 shows some examples of different valve symbols.

TABLE 2-1   Examples of Valve Symbols Used in Process and Instrument Drawings (P&IDs)

Valve Type	Symbol	Valve Type	Symbol
Ball valve		Needle valve	
Butterfly valve		Non-return valve	NR
Check valve		Piston valve	
Actuator-operated valve		Plug valve	
Gate valve		Relief valve	
Globe valve		Stop-check valve	
Hand valve		Three-way valve	
Motor-operated valve			

PUMP SYMBOLS

Pumps are used to move liquid materials through piping systems. The process industries use many different types of pumps. Each pump type has a unique symbol that appears on P&IDs. Table 2-2 shows some examples of pump symbols.

TABLE 2-2   Examples of Pump Symbols

Pump Type	Symbol	Pump Type	Symbol
Centrifugal pump		Rotary lobe pump	
Positive displacement pump		Turbine driven equipment	
Electric motor driven pump		Electric motor driven variable positive displacement pump	
Piston pump		Vertical centrifugal or positive displacement pump	

HEAT EXCHANGER SYMBOLS

Heat exchangers are used to transfer heat from one substance to another without the two substances physically contacting each other. The symbols shown in Figure 2-17 are examples of heat exchanger symbols that a process technician might encounter.

FIGURE 2-17 Examples of Heat Exchanger Symbols

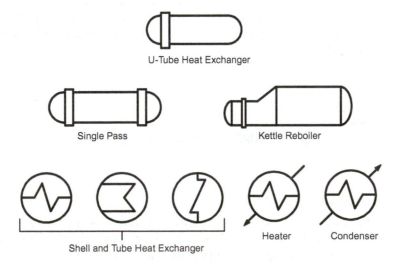

U-Tube Heat Exchanger

Single Pass   Kettle Reboiler

Shell and Tube Heat Exchanger   Heater   Condenser

ELECTRICAL EQUIPMENT AND MOTOR SYMBOLS

Electrical equipment can be used for a variety of functions. Each equipment type has a unique symbol that is used to identify it on process drawings. Table 2-3 shows examples of electrical equipment symbols and their descriptions.

TABLE 2-3   Electrical Equipment and Motor Symbols

Symbol	Name	Description
	Transducer	A device that converts one type of energy to another, such as electrical to pneumatic
	Motor driven	A symbol that indicates a piece of equipment is motor driven (either AC or DC)
	Current transformer	A device that can provide circuit control and current measurement
	Transformer	A device that can either step up or step down the voltage of AC electricity
	Electrical signal	A signal that indicates voltage or current
	Potential transformer	A device that monitors power line voltages for power metering
	Inductor	An electronic component consisting of a coil of wire
Motor	Motor	A motor (either AC or DC) that converts electrical energy into mechanical energy
	Outdoor meter device	A meter used to monitor electricity, such as a voltmeter (used to measure voltage) or ammeter (used to measure current)

TURBINE SYMBOLS

Turbines are used to produce the power necessary to drive equipment. To locate turbines accurately on a P&ID, process technicians must be familiar with turbine symbols. These symbols may vary from facility to facility. Figure 2-18 shows an example of a turbine symbol.

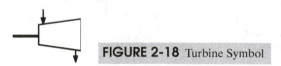

FIGURE 2-18 Turbine Symbol

FURNACE SYMBOL

Furnaces are devices that are used to produce heat required for processes. In order to accurately locate furnaces on a P&ID, process technicians must be familiar with furnace symbols. Figure 2-19 provides an example of a furnace symbol although these symbols may vary from facility to facility.

FIGURE 2-19 Furnace Symbol

BOILER SYMBOL

Boilers are devices used to produce the steam required for various parts of a process. To locate boilers accurately on a P&ID, process technicians must be familiar with boiler symbols. Figure 2-20 provides an example of a boiler symbol, although these symbols may vary from facility to facility.

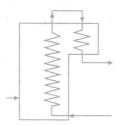

FIGURE 2-20 Boiler Symbol

REACTOR AND DISTILLATION COLUMN SYMBOLS

Reactors are vessels in which chemical reactions are initiated, controlled, and sustained. Distillation columns are devices used to separate the components of a liquid mixture by partially vaporizing the lighter components of the liquid and then recovering both the vapors and the bottoms. To locate columns and reactors accurately on a P&ID, process technicians must be familiar with distillation column (tower) and reactor symbols. These symbols may vary from facility to facility. Figure 2-21 provides an example of a tower symbol, and Figure 2-22 provides an example of a distillation column symbol.

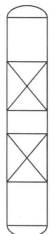

FIGURE 2-21 Reactor Symbol

FIGURE 2-22 Distillation Column Symbol

VESSEL SYMBOLS

Pressure vessels are pressure-rated containers in which materials are processed, treated, or stored. To locate vessels accurately on a P&ID, process technicians must be familiar with vessel symbols. Figure 2-23 shows examples of some of the vessel symbols that can be found on a P&ID.

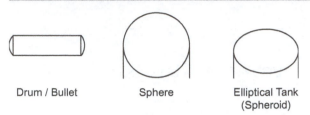

FIGURE 2-23 Common Pressure Vessel Symbols

Drum / Bullet   Sphere   Elliptical Tank
                                       (Spheroid)

INSTRUMENTATION SYMBOLS

Instruments are devices used to measure and control process data (e.g. indicate flow, temperature, level, pressure, or analytical data). Instrumentation symbols are used to identify instrumentation throughout a facility. Instrumentation symbols may or may not look like the physical device they represent (e.g., a 7/16 inch diameter circle, called a balloon, is commonly used to represent any number of functionally different instruments). Figure 2-24 shows an example of an instrumentation balloon, and Figure 2-25 explains what each of the letters in the instrument balloon represents.

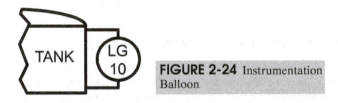

FIGURE 2-24 Instrumentation Balloon

FIGURE 2-25 Instrument Symbol Interpretation Key

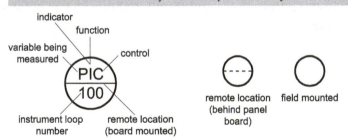

The only difference from one balloon to another is its unique alphanumeric tag number. Considering the complexity of many control systems, this schematic approach works very well. The tag number is the primary key to defining the functionality of the instrument, whereas slight modifications of the balloon depict where the instrument is physically located.

A typical legend (Figure 2-26) shows how the various instrument balloons are represented for a particular drawing. Balloons are further explained by the letters and numbers found on the tag.

Instrument tag numbers identify the measured variable, the function of the specific instrument, and the loop number. Instrument tag numbers give process technicians an indication of what that instrument is monitoring or controlling. An ISA instrument tag number (shown in Figure 2-27) is described with both letters and numbers and should be unique because most process facilities use a global database to identify devices.

FIGURE 2-26 Legend With Instrument Symbols

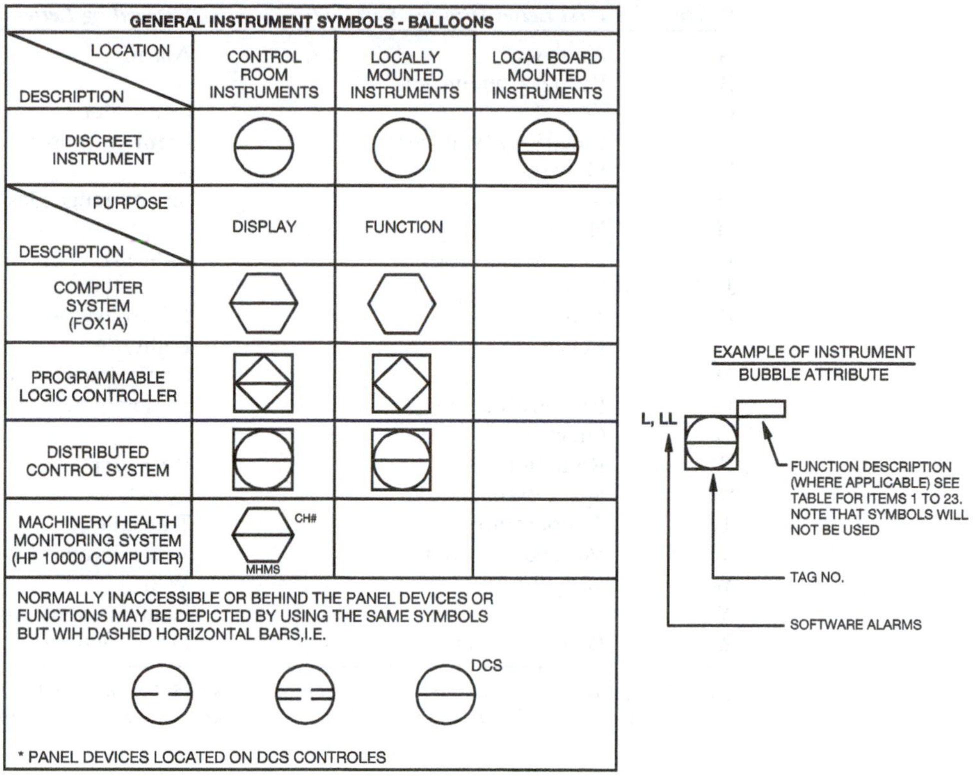

GENERAL INSTRUMENT SYMBOLS - BALLOONS			
LOCATION / DESCRIPTION	CONTROL ROOM INSTRUMENTS	LOCALLY MOUNTED INSTRUMENTS	LOCAL BOARD MOUNTED INSTRUMENTS
DISCREET INSTRUMENT	⊖	◯	⊖
PURPOSE / DESCRIPTION	DISPLAY	FUNCTION	
COMPUTER SYSTEM (FOX1A)	⬡	⬡	
PROGRAMMABLE LOGIC CONTROLLER	◇	◇	
DISTRIBUTED CONTROL SYSTEM	⊕	⊕	
MACHINERY HEALTH MONITORING SYSTEM (HP 10000 COMPUTER)	CH# ⬡ MHMS		

NORMALLY INACCESSIBLE OR BEHIND THE PANEL DEVICES OR FUNCTIONS MAY BE DEPICTED BY USING THE SAME SYMBOLS BUT WIH DASHED HORIZONTAL BARS,I.E.

⊖   ⊖   ⊖ DCS

* PANEL DEVICES LOCATED ON DCS CONTROLES

EXAMPLE OF INSTRUMENT BUBBLE ATTRIBUTE

L, LL

FUNCTION DESCRIPTION (WHERE APPLICABLE) SEE TABLE FOR ITEMS 1 TO 23. NOTE THAT SYMBOLS WILL NOT BE USED

TAG NO.

SOFTWARE ALARMS

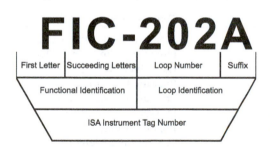

FIC-202A

First Letter	Succeeding Letters	Loop Number	Suffix
Functional Identification		Loop Identification	
ISA Instrument Tag Number			

FIGURE 2-27 Sample ISA Instrument Tag Number

The first letter identifies the measured or initiating variable, and the subsequent or succeeding letters describe the function of the instrument. For example, in Figure 2-27, F stands for "flow," I for "indicate," and C for "controller." Thus, this instrument is a flow indicating controller, a controller that controls flow and has an indicator on its faceplate. The ISA functional identification table shown in Table 2-4 lists examples of first letter identifiers with possible modifiers and succeeding letters. Table 2-5 shows examples of instrument tag letters and their interpretations based on the information contained in Table 2-4.

Equipment Standards

With the development of process drawings and equipment standards, a system of symbols has been created to describe how both equipment and its associated instrumentation are interconnected. While various engineering and chemical companies have created symbol systems unique to their facilities or companies, other groups have

TABLE 2-4   Sample ISA Functional Identification Labels

Letter	First Letter (Process Variable)	Succeeding Letters (Type of Instrument)
A	Analysis	Alarm
B	Burner Combustion	—
C	—	Controller
E	Electricity (voltage)	Element/sensor
F	Flow rate	—
G	—	Site or gauge glass/monitor
H	Hand	—
I	Current (electric)	Indicator
J	Power	—
K	Time	Control station
L	Level	Light
O	—	Orifice/restriction
P	Pressure/vacuum	Sample point
Q	Quantity	—
R	Radiation	Recorder
S	Speed/frequency	Switch
T	Temperature	Transmitter
V	Voltage/Vibration	Valve/damper/louver
W	Weight/force	Well
Y	Event	Relay/compute/convert
Z	Position/dimension	Driver/actuator

The information found in this table was taken from the ISAS5. 1 standard table, but it is subject to change. Thus, the most current ISA table should always be consulted for verification purposes.

TABLE 2-5   Sample Instrument Tag Letters with Functional Identifications

Letters	Functional Interpretation
FIC	Flow ratio Indicating Controller
FRC	Flow Recording Controller
LI	Level Indicator
PC	Pressure Controller
PIC	Pressure Indicating Controller
PT	Pressure Transmitter
PY	Pressure Relay
TE	Temperature Element
TT	Temperature Transmitter

formed organizations or societies to address issues and standardize symbols across the various process industries, for example:

- **"ISA"** (International Society of Automation) is a global, non-profit technical society that develops standards for automation, instrumentation, control, and measurement. For instrumentation, the ISA is the dominant source for instrumentation symbology under standard 5.1 (see Figure 2-28).

  The ISA standard S5.1 is comprised of both specific symbols and a coded system built on the letters of the alphabet that depicts functionality. Although many, if not most, large companies have moved toward adopting the ISA 5.1 standard in its entirety, other preferred symbols may be kept in their inventory. Anyone who uses a drawing should not assume that the ISA standard is used. All symbols, standard and nonstandard alike, should be identified in the legend of each drawing.

- **The American National Standards Institute (ANSI)** oversees and coordinates the voluntary standards in the United States. ANSI accreditation is used as a baseline

ISA
67 Alexander Drive
Research Triangle Park, NC 27709
www.isa.org

FIGURE 2-28 ISA logo
Courtesy of ISA

or "backbone" for standardization in various industries. ANSI develops and approves norms and guidelines that affect many business sectors while coordinating U.S. standards with international standards. This coordination allows American products to be used competitively worldwide.

- **The American Petroleum Institute (API)** is a trade association that represents the oil and gas industry. API speaks on behalf of the petroleum industry to the public and the various government branches. API sponsors and researches economic analyses and provides statistical indications to the public. API is a leader in the development of the petroleum and petrochemical industries for equipment and operating standards. Currently API maintains over 500 standards and recommendations. API has a certification program for the inspection of industry equipment. API also has various education programs, including seminars, workshops, and conferences available to industry for ongoing education.
- **The American Society of Mechanical Engineers (ASME)** specifies requirements and standards for pressure vessels, piping, and their fabrication.
- **The National Electric Code (NEC)** specifies electrical cable sizing requirements and installation practices. NEC was established by the National Fire Protection Agency (NFPA).
- **The Occupational Safety and Health Administration (OSHA)**, a U.S. government agency, was created to establish and enforce workplace safety and health standards, conduct workplace inspections, propose penalties for noncompliance, and investigate serious workplace incidents.

Summary

Many different types of drawings are used within the process industries. Each drawing type represents different aspects of the process and various levels of detail. Studying combinations of these drawings provides a more complete picture of the processes at a facility.

Process drawings provide process technicians with visual descriptions and explanations of processes, their flows, equipment, and other important items in a facility. Process facilities use process drawings to assist with operations, modifications, training, and maintenance. The information contained within process drawings includes a legend, title block, and application block.

Examples of process drawings include block flow diagrams (BFDs), process flow diagrams (PFDs), piping and instrumentation diagrams (P&IDs), and plot plans.

Block Flow Diagrams (BFDs) are the simplest drawings used in the process industry. While providing a general overview of the process, they contain few specifics. Block flow diagrams include the feed, product location, intermediate streams, recycle, and storage.

Process Flow Diagrams (PFDs) are basic drawings that use symbols and direction arrows to show the primary flow of a product through a process. PFDs describe the actual process including flow rate, temperature, pressure, pump capacities, heat exchangers, equipment symbols, equipment designations, reactor catalyst data, cooling water flows, and symbol charts.

Piping and Instrument Diagrams (P&IDs) are similar to Process Flow Diagrams, but show more detailed process information, such as equipment numbers, piping specifications, instrumentation, and other detailed information.

Plot plans are scale drawings that show the layout of equipment, units, and buildings. They are drawn to scale so that everything is of the correct relative size and shows proper dimensions.

Symbols are figures used to represent types of equipment. Each specific piece of equipment has a unique symbol. Different industry organizations develop and publish symbology standards to improve consistency among process drawings.

Checking Your Knowledge

1. Define the following terms:
   a. Application block
   b. Legend
   c. Process drawing
   d. Symbol
   e. Title block
   f. Block Flow Diagram (BFD)
   g. Electrical diagrams
   h. Elevation diagrams
   i. Isometric
   j. Loop diagram
   k. Piping and Instrument Diagram (P&ID)
   l. Plot plan
   m. Process Flow Diagram (PFD)
   n. Schematic
   o. Utility Flow Diagram (UFD)

2. Which drawing provides a general overview of the process and contains few specifics?
   a. PFD
   b. BFD
   c. P&ID
   d. UFD

3. Process Flow Diagrams are typically drawn from the _____.
   a. right to left
   b. left to right

4. Which of the following items are located on a Process Flow Diagram? (Select all that apply.)
   a. Pump capacities
   b. Equipment symbols
   c. Equipment designations

5. *(True or False)* P&IDs show less detailed process construction information, including materials of construction, insulation, and equipment details.

6. *(True or False)* UFDs provide process technicians a P&ID view of the instrumentation used for a process.

7. Which of the following are included in a process drawing? (*Select all that apply.*)
   a. Legend
   b. Title block
   c. Application block
   d. Symbols

8. *(True or False)* A balloon is an instrumentation symbol that is commonly used to represent the function of different instruments.

9. On the following ISA tag, what does the first letter "F" stand for?
   a. frequency
   b. flow
   c. force
   d. function

10. Which society develops standards for automation, instrumentation, control, and measurement symbols?
    a. API
    b. NFPA
    c. ISA
    d. ASME

11. Complete the following chart by writing three to five sentences about each drawing and how they are used.

Drawing Type	*Description and Use*
Block Flow Diagram (BFD)	
Process Flow Diagram (PFD)	
Piping and Instrument Diagram (P&ID)	
Plot plan	

Student Activities

1. Write a paper on one or more instrumentation standards (e.g., ANSI, API, ASME, and ISA).
2. Work with a classmate to draw a flow sequence of a simple process and describe the basic steps associated with the process.
3. Identify all the symbols in Figure 2-2.

CHAPTER 3

Piping, Tubing, Hoses, and Fittings

Objectives

After completing this chapter, you will be able to:

- Identify and explain the purpose of piping, tubing, hoses, and fittings in the process industries.
- Explain pressure and temperature limits of piping, tubing, hoses, and fittings.
- List and describe various fittings used for piping, tubing, and hoses.
- Discuss the uses, advantages, and constraints of the materials of construction.
- Discuss different schedules for piping thickness and ratings on flanges for required service.
- Discuss selection and sizing criteria as related to pressure, temperature, flow, and corrosiveness of fluids.
- Identify types of connections.
- Describe the use of sealant compounds.
- Identify and describe equipment tests.
- Identify and describe plugs and caps.
- Describe the use of gaskets.
- Explain the purpose of heat tracing, insulation, and jacketing.
- Explain the purpose, types, and operation of steam traps.
- Describe the process technician's responsibilities regarding the selection, maintenance, and repair of piping, tubing, hoses, and fittings.
- Identify typical problems associated with piping, tubing, hoses, and fittings.

Key Terms

Alloys—compounds composed of two or more metals that are mixed together in a molten solution.

Blinds—solid plates or covers that are installed between pipe flanges to prevent the flow of fluids and isolate equipment or piping sections when repairs are being performed; typically made of metal.

Corrosion—deterioration of a metal by a chemical reaction (e.g., iron rusting).

Expansion loop—segment of pipe that allows for expansion and contraction during temperature changes.

Gasket—flexible material used to seal components together so they are air- or watertight.

Insulation—any substance that prevents the passage of heat, light, electricity, or sound from one medium to another.

Jacketed pipes—pipes that have a pipe-within-a-pipe design so that hot or cold fluids can be circulated around the process fluid without the two fluids coming into direct contact with each other.

Pipe clamp—piping support that protects piping, tubing, and hoses from vibration and shock; also a ring-type device used to temporarily stop a pipe leak.

Pipe hanger—piping support that suspends pipes from the ceiling or other pipes.

Pipe shoe—piping support that supports pipes from underneath.

Schedule—piping reference number that pertains to the wall thickness, inside diameter (ID), outside diameter (OD), and specific weight per foot of pipe.

Steam trap—device used to remove condensate from the steam system or piping.

Stress corrosion—type of corrosion that results in the formation of stress cracks.

Tensile strength—the pull stress, in force per unit area, required to break a given specimen.

Introduction

Piping, tubing, hoses, and fittings are the most prevalent pieces of equipment in the process industries. Some estimates state that piping accounts for 30 to 40 percent of the initial investment when building a new facility.

In most plants you will see large segments of pipe going from one location to another. These pipes carry chemicals and other materials into and out of various processes and equipment.

When building a process facility, it is important to select proper construction materials and connectors because some materials and connectors are inappropriate for certain processes, pressures, or temperatures. Improper material selection and improper operation can lead to leaks, wasted product, and/or hazardous conditions.

Piping, Tubing, Hoses, and Blinds

Pipes are long, hollow cylinders through which fluids are transmitted. In the process industries, piping is used to transport fluids throughout a facility and to different parts of a process. In some instances, piping may even be used to carry materials or products outside a facility and across the long distances. An example of this type of piping is the 800-mile-long Trans Alaska Pipeline.

Piping comes in many different sizes and materials. It can be rigid (e.g., metal pipe) or flexible (e.g., tubing and hoses). It can be high pressure (e.g., steam) or low pressure

Did You Know?

The Trans Alaska Pipeline was designed and constructed to move oil from the North Slope of Alaska to the northern most ice-free port in Valdez, Alaska.

The pipeline itself is 800 miles long and 48 inches in diameter. It crosses three mountain ranges and more than 800 rivers and streams

The construction of the pipeline, which began in March 1975 and took three years and $8 billion to complete, was the largest privately funded construction project at that time.

Since its construction, more than 14 billion barrels of oil have moved through the Alaska pipeline.

(e.g., instrument air). It can be permanent (e.g., welded joints) or temporary (e.g., screw ends), and it can be used to carry fluids over long or short distances.

TUBING

Tubing is small diameter hose or pipe used to transport process fluids. Unlike piping, tubing is used when fluid flow rates are relatively small because the diameter of tubing is usually less than 1 inch.

Tubing can be made from a variety of materials, including copper, synthetics, plastics, and alloys, and it can be used in a variety of applications (e.g., instrument air lines, heat exchanger tubing, metered chemical additions, sample systems, and sensing equipment).

HOSES

Hoses are flexible tubes that carry fluids. Hoses may be used in a variety of applications, including steam lines for cleaning or flushing, plant air for air-driven equipment, water for washing, and nitrogen or other gas services. Unlike piping, which is more permanent, hoses are typically used to make temporary connections. For example, if a piece of pipe is damaged or working improperly, it is common for a piece of hose to be installed temporarily until a replacement pipe can be created (see Figure 3-1).

Because of the pressure, temperature, and the corrosive nature of some fluids, the material the hose is made of must be designed to withstand the service for which it is being used. For example, utility hoses are used to deliver various utilities like air, water, nitrogen, and steam. These types of hoses can be made of a variety of materials including plastic, rubber, fiber, metal, or a combination.

FIGURE 3-1 Hose Used for Temporary Pipe Repair

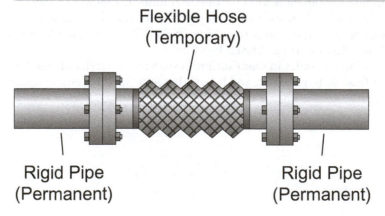

Flexible Hose
(Temporary)

Rigid Pipe
(Permanent)

Rigid Pipe
(Permanent)

One unique feature of utility hoses is the style of their connectors. Each connector is designed to connect to a specific service or utility (e.g., air hoses can be connected only to air lines and water hoses can be connected only to water lines). This prevents someone from accidentally connecting a particular hose to an incorrect service line (e.g., connecting an air hose to a steam line).

Steam hoses are flexible tubes that are used to transport steam through the piping system. Because of the heat and pressure associated with steam, steam hoses are typically constructed of a braided metal fiber core with a rubber outer coating. This durable construction allows for higher temperatures and pressures (e.g., 75 pounds per square inch gauge [psig], the pressure of the fluid above the pressure of the atmosphere).

Steel-braided hoses are flexible rubber hoses with a woven metal fiber outer or inner casing. They are normally used to connect piping to a transport vehicle, and for steam pressure above 250 psi. Steel-braided hoses use quick-lock connectors that use a flange or union lever to lock the hose in place. The covering of the hose reinforces the strength and protects the rubber from the wear of long-term use. The metal fiber casing strengthens and protects the hose for long-term use and has specific pressure, temperature, and chemical ratings. Figure 3-2 shows an example of a braided hose with a quick-lock connector.

FIGURE 3-2 Braided Hose with Quick-Lock Connector

Color-coded flex hoses (shown in Figure 3-3) are flexible hoses that are color coded to ensure that the proper hose is used for a particular service (e.g., water hoses may be one color, while steam hoses are another color). Color coding helps process technicians know which service the hose is providing and reduces the risk of error or injury. The colors that are used for each type of hose varies from company to company.

FIGURE 3-3 Flex Hose

LIMITATIONS OF HOSES AND FITTINGS

All hoses have a maximum pressure rating at a specific temperature. Pressure ratings decrease with increasing temperature. The maximum hose pressure should not be exceeded because of the danger of rupture, which could cause personal injury or environmental hazards. Before using a hose, inspect the hose for any damage and find the inspection date to ensure that the hose is still within the allowed inspection period.

When the temperature, pressure, or chemical limitations of a hose are about to be reached or exceeded, it is not uncommon to replace the hose with temporary piping. Temporary piping is used for the same purpose as temporary services; however, the replacement piping material must have the ability to withstand higher temperatures and pressures, and be compatible with process chemicals.

BLINDS

Blinds or "blanks" are solid plates or covers that are installed between pipe flanges to prevent the flow of fluids and to isolate equipment or piping sections when repairs are being performed. Blinds are normally made of metal and have properties that are appropriate for the type of service in which they are being used. For example, if the service is highly corrosive, then the blind will be made of a **corrosion** (deterioration of a metal by a chemical reaction) resistant material.

Two common types of blinds that process technicians may encounter are paddle blinds (see Figure 3-4) and spectacle blinds (see Figure 3-5).

FIGURE 3-4 Paddle Blinds

FIGURE 3-5 Spectacle Blind

Both paddle and spectacle blinds are installed inside a flange. Paddle blinds (see Figure 3-6), have a handle that sticks out of the flange (see Figure 3-6). This handle provides a quick visual that informs process technicians that the pipe has been blocked with a blind. Spectacle blinds, like the one in Figure 3-7, on the other hand, have a disk sticking out of the flange (either solid or with a hole in it) instead of a handle. This disk indicates to a process technician that the flange contains a blind and indicates if the flow is open or blocked (that is, if the side with the hole in it is visible from the outside, then the technician knows the pipe is blocked. If the solid side is visible, then the technician knows the pipe is ready for service).

Figure 3-8 shows an example of a paddle blind installed in a flange. Because of the hazards associated with the blinding process, blank/blind checklists are frequently used

Paddle Blind

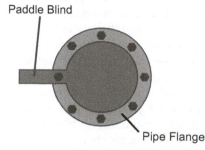

Pipe Flange

FIGURE 3-6 Paddle Blind Installed

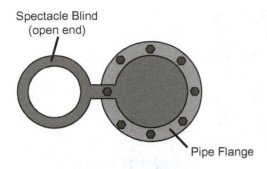

Spectacle Blind
(open end)

Pipe Flange

FIGURE 3-7 Spectacle Blind Installed

FIGURE 3-8 Pipe Flange Blind Installed

Courtesy of Brazosport College

to ensure complete isolation during repair, and to ensure that all blinds are removed before normal operations are resumed.

PIPING SUPPORTS

As we mentioned earlier, it is not uncommon to find extremely long runs of pipe in the process industries. However, for these long pipe runs to function correctly, they must be properly supported. Figure 3-9 through Figure 3-11 show a few of the supports that are available.

Pipe shoes are piping supports that support pipes from underneath. **Pipe hangers** are piping supports that suspend pipes from the ceiling or other pipes. **Pipe clamps** are piping supports that protect piping, tubing, and hoses from vibration and shock.

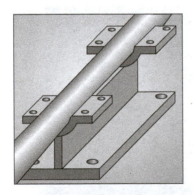

FIGURE 3-9 Pipe Shoes

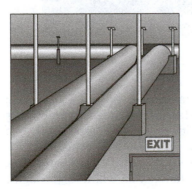

EXIT

FIGURE 3-10 Pipe Hangers

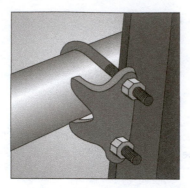

FIGURE 3-11 Pipe Clamp

Piping supports hold piping in place to prevent movement or damage from the weight of the piping run. They also allow for thermal and mechanical expansion and contraction. Without proper support, the piping and the equipment associated with these supports can easily become damaged. For example, if a line is improperly supported, it could throw a pump out of alignment, causing seal, coupling, and impeller damage.

EXPANSION LOOPS AND JOINTS

An **expansion loop** (also referred to as an expansion joint) is a segment of pipe that allows for expansion and contraction during temperature changes. Figure 3-12 shows an example of an expansion loop.

The temperature inside a pipe can change dramatically during normal operations. For example, 650 psig superheated steam is approximately 725 degrees Fahrenheit (°F). As this steam cools it reduces to 450°F and the pipe contracts. Expansion joints allow for this type of thermal expansion and contraction and minimize its impact. Without an expansion joint, the pipe could rupture, damage the foundation, or break the supports. The pipe could also become misaligned due to expansion and contraction, thereby damaging the seals on pumps and turbines. These types of damages can be very expensive and could cause serious physical and environmental hazards.

FIGURE 3-12 Expansion Loops Allow for Thermal Expansion

Materials of Construction

Industrial piping, tubing, hoses, and fittings can be made of many different materials like carbon steel, alloy steel, stainless steel, exotic metals, glass, plastic, or clay. However, the most common piping material is carbon steel because it is appropriate for a wide range of temperatures and is relatively economical.

When designing piping and valve systems, the engineer or designer must be familiar with the process and the substances that will flow through the pipes. Specifically, they must know the temperature of the substance; its viscosity; how much pressure it exerts; and its flammability, corrosivity, and reactivity. Failure to take these factors into account could have serious effects on the safety and performance of the plant once it is operational. For example, some metals become brittle at extremely low temperatures. Table 3-1 contains a list of common construction materials and their characteristics.

TABLE 3-1 Common Construction Materials and Their Characteristics

Construction Material	Temperature Ranges	Description
Carbon steel	−20°F to 800°F	The most commonly used construction material because of its flexibility, weldability, strength, and relatively low cost.
Stainless steel	−150°F to 1400 + °F	Less brittle than carbon steel at extremely low temperatures; appropriate for use in high-temperature applications; more corrosion-resistant than carbon steel, especially in acid service; susceptible to stress corrosion cracking.
Brass, bronze, and copper	−50°F to 450°F	Corrosion-resistant at low or moderate temperatures; excellent heat transfer properties.
Alloys (e.g., hastelloy and inconel)	Varies with the alloy	May be used in high-temperature (above 800°F) applications (e.g., furnace tubes) and highly corrosive service; high cost.
Plastics	Varies with the plastic	Used for low-pressure applications and corrosive services; easy to install, low in cost, and lightweight.

CARBON STEEL AND ITS ALLOYS

Carbon steel is the most common material used for construction of piping systems where corrosion is not a problem. The benefits of carbon steel, compared to stainless steel or zirconium, are its availability, affordability, and increased **tensile strength** (the pull stress, in force per unit area, required to break a given specimen. The greater the tensile strength, the thinner you can make the pipe walls).

While some construction materials are pure metals, others may be substances called alloys. **Alloys** are compounds composed of two or more metals that are mixed together in a molten solution (e.g., bronze is an alloy of copper and tin). Alloys improve the properties of single-component metals and provide special characteristics that are not present in the original metals.

Carbon steel is routinely used in the construction of piping in which organic chemicals as well as neutral or basic aqueous solutions at moderate temperatures flow. In instances where corrosion is an issue, the walls of the pipe are lined with a protective coating (e.g., Teflon, glass, and polymers like butyl rubber) or are made extra thick, or an alloying material is added to change the properties of the carbon steel. The result is a more corrosion-resistant and higher strength material. The most common alloying elements are nickel, manganese, chromium, and molybdenum.

STAINLESS STEELS

Stainless steel is an iron-carbon alloy that is commonly used in the process industries because it resists heat and corrosion, such as rust. Because of these properties, stainless steel is often used in pump shafts, valves, gears, and vessels.

The use of stainless steel does have some drawbacks, however. For example, stainless steel is significantly more expensive than carbon steel, and it is subject to **stress corrosion**, a type of corrosion that results in the formation of stress cracks. This type of corrosion may occur when handling chlorides (e.g., sea water) or caustics (e.g., sodium hydroxide).

There are several grades of stainless steel. The grade of the steel is determined by the amount of chromium and other elements present (e.g., nickel, molybdenum, silicon, and manganese). Stainless steel alloys can be grouped into three categories: martensitic, ferritic, and austenitic. This categorization is based on their composition, molecular structure, and physical properties.

Martensitic stainless steel (e.g., SS 410) is a small category of magnetic steels that typically contain 12% chromium, a moderate level of carbon, and a very low level of nickel. Martensitic steel can be hardened with heat treatment. The result is a very hard, high-strength alloy.

Ferritic stainless steel (e.g., SS 430) is a type of steel that has a low carbon content and contains chromium as the main alloying element. Ferritic steels are very resistant to oxidation and cannot be hardened by heat treatment.

Austenitic stainless steel (e.g., SS 304 and SS 316) is a group of chromium-nickel steels that are nonmagnetic, are very corrosion resistant, and cannot be hardened by heat treatment. Table 3-2 lists some of the more common grades of stainless steel, their types, compositions, properties, and uses.

TABLE 3-2   Common Stainless Steel Grades and Properties

Type	*Grade*	*% Chromium*	*% Nickel*	*% Other*	*Properties/Uses*
Martensitic	SS 410	11.5–13.5%	—	—	Very hard; magnetic; used to create turbine blades and valve trim.
Ferritic	SS 304	18–20%	8–12%	1% Silicon (Si)	Corrosion resistant; magnetic; used in process piping.
Austenitic	SS 316	16–18%	10–14%	2–3% Molybdenum (Mo)	Nonmagnetic; improved corrosion resistance over SS 304; used in process piping.

OTHER METALS

Medium and high alloys are generally proprietary groups of alloys that have better corrosion resistance than stainless steels. Table 3-3 lists examples of different types of alloys and nonferrous metals and their descriptions.

TABLE 3-3   Common Alloys and Other Metals

Medium Alloys	*High Alloys*	*Nonferrous Metals*
Description:	**Description:**	**Description:**
Moderate corrosion and temperature resistance.	High corrosion and temperature resistance.	Extremely high corrosion and temperature resistance.
Examples:	**Examples:**	**Examples:**
• Durimet 20	• Hastelloy B-2	• Titanium
• Carpenter 20	• Chlorimet 2, and 3	• Zirconium
• Incoloy 825	• Hastelloy C-276, C-4	• Tantalum (composed of
• Hastelloy G-3	• Inconel 600	nickel, aluminum, and copper)

NONMETALS

A variety of nonmetallic piping is used in corrosive service (e.g., acid or high pH service). Organic and inorganic materials are used because they provide superior corrosion resistance.

TABLE 3-4 Examples of Common Nonmetallic Piping

Substance	Description
Clay	Different varieties are used to construct settling basins and waste treatment ponds; an inexpensive, economical material.
Elastomeric membranes	Membranes that have the elastic properties of rubber and are made from rubber, epoxy, or other resins; these membranes are often used to separate the vessel wall (usually carbon steel) from the process.
Glass	Resistant to most acids except hydrofluoric acid and hot, concentrated phosphoric acids; brittle and subject to thermal shock; impact, abrasion, and thermal shock resistance are added properties for a nucleated crystalline ceramic-metal form of glass that is similar to, or better than, the corrosion resistance of conventional glass.
Plastics	Good for applications that have normal temperatures (not extremely high or extremely low), are not corrosive, and do not contain high pressures.
	Plastics include polyvinyl chloride (PVC), chlorinated PVC, high-density polyethylene, polypropylene, and polybutylene.
Porcelain and stoneware	Acid resistant, but brittle; brick-lined construction is used for many severely corrosive conditions.

Pipe Selection and Sizing Criteria

To determine the proper piping material and size, one must consider the pressure, temperature, flow, and corrosiveness of the process fluid.

When working with corrosive fluids it is important to use gasket materials that are appropriate for corrosive fluids, because corrosion occurs much more quickly when the pressure and temperature of the fluid is high.

When working with high temperatures, it is important to apply insulation to prevent burns and conserve heating or cooling energy. When working with high pressures, it is important to ensure that the connection fittings are appropriate for the pressure ratings of the host piping systems. Ratings and sizes for pipe flanges range from 150 to 2500 psi. For pipe sizes up to 3" and 150 psi pressure, the flange has four bolts. The number of flange bolts increases in steps as the pipe size increases and as the pressure rating increases.

CAUTION: Never use any chemical in a pipe not rated to manage that process!

PIPE THICKNESS AND RATINGS SERVICE

Piping comes in many different dimensions and ratings. To standardize these dimensions and ratings, the American National Standards Institute (ANSI), (sponsored by the American Society for Testing Materials (ASTM) and the American Society of Mechanical Engineers (ASME),) published a set of guidelines that industries follow. The dimensions and characteristics specified in the ANSI guide are referred to as a pipe's schedule.

Schedule is a piping reference number that pertains to the wall thickness, inside diameter (ID), outside diameter (OD), and specific weight per foot of pipe. To distinguish pipe sizes from actual measured diameters, the terms iron pipe size (IPS) or nominal pipe size (NPS) are usually used. In pipes larger than 12 inches, the pipe is known by its actual outside diameter (OD). For most pipe sizes, however, the outside diameter (OD) remains relatively constant. What changes is the variations in wall thickness, or the inside diameter (ID). To distinguish different weights of pipe, three long-standing

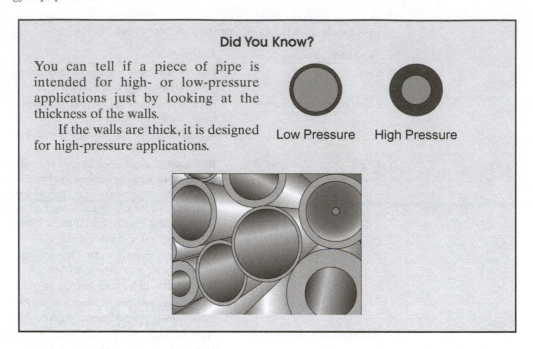

designations are used: standard wall (STD), extra strong wall (XS), and double extra strong wall (XXS) (also referred to as extra heavy wall [XH], and double extra heavy wall [XXH].

Pipe sizes and schedules are assigned to piping to help piping systems designers specify safe installations. Schedule numbers range from 5S (thinnest wall) to 160S (thickest wall). For example, a normal 1″ schedule 5S, stainless steel pipe (Note: the "S" in "5S" stands for "stainless"), has an outside diameter of 1.315″, an inside diameter of 1.185″, and a wall thickness of 0.065″. A normal 1″ schedule 160, carbon steel pipe has a 0.358″ wall thickness. Figure 3-13 illustrates the relative increase in the wall thickness of various schedules of pipe.

FIGURE 3-13 Pipe Schedule Reference Chart

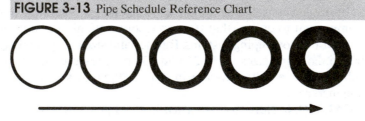

Increasing Wall Thickness (Schedule)

The schedule number defines the wall thickness of the pipe and the maximum pressure rating at a specific temperature. Ratings are typical for flange connections and some fittings. A raised face flange contains a gasket. Flanges with 300 PSI and 600 PSI rating have the same basic dimensions. What is different, however, is that for 300 PSI and below, the gasket thickness is 1/16″. For 600 PSI and above the thickness is 1/4″.

Connecting Methods

Pipes and valves can be connected together in a variety of ways. They can be screwed (threaded), flanged, or bonded (e.g., welded, glued, soldered, or brazed). Figure 3-14 shows examples of each of these connection methods.

The factor that determines which connection type is the most appropriate is the purpose of the pipe. For example, if the pipe is used in low-pressure water service, a threaded joint might be appropriate because it is cheaper and easier to install than

FIGURE 3-14 Examples of Screwed, Flanged, Welded, and Bonded Connections

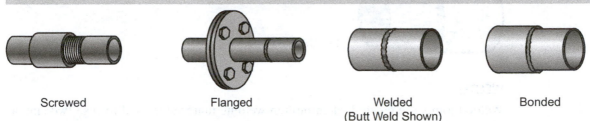

Screwed   Flanged   Welded   Bonded
                                (Butt Weld Shown)

welded or flanged joints. However, if the pipe is used in high pressure, flammable, or corrosive service, a welded joint would be a better option because threaded joints and flange joints with gaskets are more likely to leak (this is why welded joints are usually the connection method of choice for critical service piping).

THREADED

Threaded connections involve the joining together of two pipes through a series of tapered threads similar to the ones shown in Figure 3-15. In a threaded connection, the pipe is cut with "male" threads and the connector is cut with "female" threads so the two can be joined together. When these threads are cut, they are generally cut with precision to ensure the two pieces fit together tightly to avoid leaks. However, this tightness can make it difficult to connect the two pieces together. Because of this, threading compound or Teflon® tape is often employed to lubricate the joints, facilitate the connection, and seal against leaks.

Threaded connections are more prone to leak because the connections are weaker (i.e., more easily broken due to vibration or external force). Because of this, threaded connections are typically used for low-pressure, nonflammable, and nontoxic service.

FIGURE 3-15 Threaded Connection

FLANGED

Flanged connections, like the one shown in Figure 3-16, are typically used in instances where the piping may need to be disconnected from another pipe or a piece of equipment.

In a flanged connection, two mating plates are joined together with bolts. Between the two mating plates is a **gasket** (a flexible material used to seal components together so they are air- or watertight). As the bolts are tightened, the gasket is compressed between the two plates. This compression increases the tightness of the seal and prevents leakage.

FIGURE 3-16 Flanged Connection

BONDED

Bonded pipe joints, like the one show in Figure 3-17, may be glued, brazed, or soldered. Gluing involves fusing joints together with glue. Soldering or brazing involves fusing joints together with molten metal.

Glued joints are typically found on plastic lines and pipes; for example, lines made of polyvinyl chloride (PVC) or high-density polyethylene (HDPE). Soldered joints are found on metal pipes.

FIGURE 3-17 Bonded Joint

WELDED

Welded joints are created when molten welding material is used to fuse two similar pieces of metal together. A butt weld is used to connect two pipes of the same diameter. If one pipe is small enough to fit snugly inside the other, a socket weld is used. Figure 3-18 shows examples of butt and socket welds.

FIGURE 3-18 Butt and Socket Welds

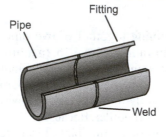
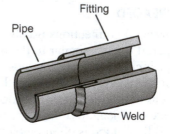

Butt Weld   Socket Weld

TEMPORARY CONNECTIONS
Flexible Hose Connections

Flexible or flex hose connections are normally used to provide a temporary connection. Flex hose connections should always be cleaned properly. Otherwise, the mixing of chemicals could create a condition where chemicals react with one another, thus causing a potential spill, leak, fire, or explosion. When using flex hose connections, it is essential to check the pressures, temperatures, and corrosiveness of chemicals to make sure the flex hose connection is appropriate for the conditions in which it will be used.

Fitting Types

Fittings are piping system components used to connect together two or more pieces of pipe. There are many different types of fittings in the process industries. Figure 3-19 shows examples of some of the most common fittings. Table 3-5 contains a description of the fittings shown in Figure 3-19.

FIGURE 3-19 Examples of Some of the Most Common Fittings

Flange   Bushing   Plug   Cap   Union   Nipple

Coupling   Bell Reducer   90° Elbow   45° Elbow   Tee   Y Strainer   Cross

TABLE 3-5   Descriptions of Fittings Shown in Figure 3-19

Fitting Name	Description
Flange	Consists of a mating plate used to join two pieces of pipe or a valve to a piece of pipe.
Bushing	Threaded on both the internal and external surfaces; used to join two pipes of differing sizes.
Plug	Used to close the end of a piping run; similar to caps, except that the threading is male and plugs a female-threaded fitting; can be glued, threaded, or welded into place.
Cap	Used to cap (seal) the open end of a pipe; may be threaded, glued, or welded.
Union	Joins two sections of threaded pipe but allows them to be disconnected without cutting or disturbing the position of the pipe.
Nipple	A short length of pipe (usually less than 6 inches) with threads on both ends; can be used to extend the length of a pipe or make temporary connections for maintenance purposes.
Coupling	Used to connect shafts, pipe, tubing sections, or hoses together.
Bell reducer	Used to connect two pipes of different diameters (large gauge to smaller gauge); both ends are female, with one end being larger than the other.
Elbow	A 45° or 90° angle that is used to change the direction of flow.
Tee	A T-shape that allows the splitting or joining of flows; normally the openings join at a 90° angle, but they may have a lesser angle.
Strainer	Contains a fine mesh that allows fluid to flow through while holding back solid particles; located before process equipment.
Cross	A crossshape that allows four pipes to be connected together at 90° angles.

In addition to the fittings shown in Figure 3-19, there are also other types of fittings. Table 3-6 provides descriptions of these other types of fittings.

TABLE 3-6   Additional Fitting Descriptions

Fitting Name	Description
Compression fitting	Tubing connection that is sealed by a slip-on tapered sleeve and that presses against the joining tubing when pressure is applied by a threaded connection that fits over the sleeve.
Utility connection	Type of quick connecting fitting.
Chicago pneumatic (CP) coupling	Type of coupling with two prongs and a sealing gasket used primarily to connect air or water supplies to a hose; also called a crow's foot.
Flared fitting	Fitting composed of tubing, copper, or stainless of which the ends are flared in a bell shape and sealed with a ferrule and nut.
Nitrogen hose coupling	Prevents the accidental connection to the wrong utility; unique for nitrogen service.
Quick connect	Special tubing and piping connections that are sealed by spring-loaded seals so they can be connected and disconnected easily.
Steam hose fitting	Special steam hose connectors that screw together (as opposed to quick connects) so high-pressure steam cannot blow them apart; also referred to as Boss fittings.

GASKETS

Gaskets are used between all flange connections to seal the connection. The types of gaskets include spiral wound, tang, and ring. Gasket size, material, and rating must match the application.

FIGURE 3-20 Gaskets

Piping gasket material must be compatible with the substance inside the pipe. Otherwise, the gasket could corrode, fail, and leak process fluid. Gasket materials include Teflon, Garlock, rubber, and cork. Figure 3-20 shows examples of gaskets.

Gasket ratings are grouped into classes. Common classes are 150, 300, 400, 600, 900, 1500, and 2500. The higher-numbered classes have higher pressure and temperature ratings. The service pressure for each class varies with the temperature of the application and the size of the pipe.

Flange bolts should always be tightened to maintain the correct placement of the gasket. The gasket should always be properly sized and should never be folded or crimped. There should always be even, but not excessive, pressure on all of the bolts to prevent leakage. If the bolts are tightened excessively, the gasket can be damaged and the seal will be compromised. Gaskets should always cover the flange face and be flush with the outer surface of the flange face, but they should never restrict the inside diameter of the pipe.

SEALANT COMPOUNDS

Sealants and tapes are used to lubricate and seal the threads on a threaded pipe. Fresh sealant and tape should be used every time a threaded pipe is connected or reconnected. Process technicians should also inspect pipe connections routinely for leakage because leakage signals that the sealant compound or tape was not applied correctly. Table 3-7 lists some examples of sealant materials.

When dismantling a pipe section where sealant or tape is used, be sure to remove the old sealant or tape before applying any new sealant material. Failure to remove the old sealant or tape before applying the new can cause leaks. If sealant or tape is applied improperly, it can bunch up on the threading and prevent a proper seal or connection.

TABLE 3-7 Types of Sealing Compounds

Sealant	*Description*
Sealant compound (pipe dope)	Paste applied to the thread of a pipe to seal the thread connections.
Teflon tape	A thin, white tape made of Teflon®, used to lubricate and seal male pipe threads (see Figure 3-21)

FIGURE 3-21 Teflon Tape

Leak Testing

From time to time, it is necessary to perform tests in order to check pipes for leaks or structural problems. The most common types of tests are hydrostatic and pneumatic.

Hydrostatic testing, also known as tightness testing, uses a hydraulic pump to increase the pressure on a water-filled system in order to test for leaks. Because water is a noncompressible fluid, a decrease in pressure caused by a leak can be quickly detected. Once the leak is identified, the test pressure is released, repairs are made, and a retest is performed.

One disadvantage of hydrostatic testing is that the ambient temperature must be within a certain range (i.e., the temperature outside cannot be too low). Before conducting a test of this type, technicians should always check ANSI standards to determine temperature limitations for all of the materials used in the system test.

In cases where water or other liquids cannot be used, pneumatic testing is employed. In pneumatic testing, the system is isolated and then air is introduced and compressed to a desired pressure. The pressure is then monitored for leaks.

A disadvantage of pneumatic testing is that it takes longer to depressurize and repressurize the system should the test be unsuccessful. Also, in high-pressure, large-volume systems, hazards can be created by a possible rupture.

When conducting leak tests, it is important to realize that pressure is being applied to the piping structure and that risks are associated with over pressurization. Thus, it is necessary to refer to the ANSI Code for Pressure Piping for various pressure tolerances of the pipes and tubing prior to conducting the test.

Piping Protection

All piping components and systems are designed to operate within three specified limits: temperature, pressure, and appropriate process chemicals. ASME, ANSI, or other standards organizations rate all piping and fittings for these three limitations. The rating assigned to a given pipe is based on a pressure/temperature curve that, if exceeded, could result in a rupture.

Thermal relief valves are commonly used to keep piping within the safety zone of these curves. In some cases, the equipment automatic shutdown systems are used to keep the pressures and temperatures within the curve.

TRACING

Heat tracing is a coil of heated wire or tubing that is adhered to or wrapped around a pipe to increase the temperature of the process fluid, reduce fluid viscosity, and facilitate flow. Heat tracing can be very important in cold climates because frozen or

---

Did You Know?

(Pipe Pig)

There are pigs in the pipeline!

Not real pigs, mind you, but pipe pigs.

Pipe pigs are mechanical devices that clean and scrape residue from the inner walls of a pipe. Some pigs, called smart pigs, use sensors, television cameras, and isotopes to detect corrosion.

(Cross Section of a Pig in a Pipe)

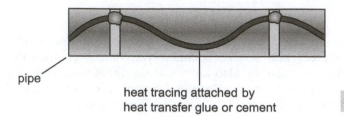

pipe

heat tracing attached by
heat transfer glue or cement

FIGURE 3-22 Heat Tracing

extremely viscous process fluids can damage piping and equipment and can result in a unit-upset condition. Figure 3-22 shows an example of heat tracing.

In extremely cold climates, heat tracing may not be sufficient to prevent freezing. In these instances, freezing is prevented by removing process fluids from the piping and equipment. Some systems, such as safety showers, eye baths, and fire protection, are protected by small bleed valves that allow dripping, which prevents freezing inside the pipe.

Steam heat tracing is a constant flow of steam through a small tube that runs beside the pipe to keep the pipe warm, thus reducing viscosity and preventing the materials within the pipe from freezing. Engineering and maintenance standards and the amount of steam supplied determine the steam pressure rating and flow.

Electrical heat tracing involves an electrical wire that is wrapped around a pipe and is controlled by a thermostat. The thermostat cycles on and off at preset temperatures to keep the fluid in the pipe warm.

INSULATION

Insulation is any substance that prevents the passage of heat, light, electricity, or sound from one medium to another. In the process industries, pipe insulation is primarily used to maintain the temperature of process fluids and refrigeration systems, and to protect workers from thermal burns that could occur if a worker were to come in contact with piping that is extremely hot or extremely cold. Protecting the process fluid from temperature changes due to ambient conditions (i.e., temperatures outside the pipe) is desirable because it reduces energy cost for the process unit. Figure 3-23 shows an example of a pipe surrounded with insulation.

Insulation can be made of many materials, including:

- Glass
- Kaowool (ceramic fiber)
- Mineral wool
- Fiberglass
- Insulation brick
- Wrap
- Foam

FIGURE 3-23 Pipe Surrounded with Insulation

When selecting insulation, it is important to consider the fluid type, fluid flow rate, the distance from the starting point of the fluid to the end point, weather conditions, protective metal walls, and the need for personal protection.

JACKETING

Jacketed pipes have a pipe-within-a-pipe design so hot or cold fluids can be circulated around the process fluid without the two fluids coming into direct contact with each other. In other words, a jacketed pipe operates like a heat exchanger. Figure 3-24 shows an example of jacketed pipe.

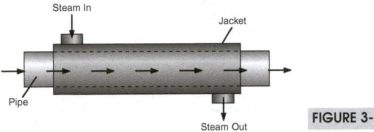

FIGURE 3-24 Jacketed Pipe

STEAM TRAPS

Steam traps are devices used to remove condensate from a steam system or piping. When steam cools, condensate forms, so steam traps were invented to catch and remove the condensate and prevent any steam from escaping. Figure 3-25 shows examples of different types of steam traps.

Steam traps may vary in design, but they use three basic principles: density differences between steam and condensate, steam and temperature, and steam and velocity. The best design to use depends on the operating system and the application.

FIGURE 3-25 Examples of Steam Traps

Bellows Thermostatic

Inverted Bucket

Potential Hazards

In some process facilities, process technicians are responsible for the care and inspection of piping, tubing, hoses, and fittings. If the process technician has any doubt about whether the piping, tubing, hoses, and fittings can be used for a particular job, he or she should consult with a supervisor.

Many hazards are associated with piping, tubing, hoses, and fittings. Table 3-8 lists some of these hazards. Figure 3-26 shows examples of corroded and eroded pipes.

TABLE 3-8 Hazards and Possible Impacts

Improper Operation	Possible Impacts			
	Individual	*Equipment*	*Production*	*Environment*
Leaks	• Leaks carry the potential danger of fire or explosion • Contact with or breathing chemicals may cause burns, acute or chronic illnesses, or death • Leaks may cause slipping hazards	• Leaks carry the potential danger of fire or explosion • Potential damage to equipment can range from minimal to complete loss • Other problems may include reduced suction or discharge to pumps or compressors, resulting in damage	• Impaired or totally shut down • Quality could be affected due to reductions in addition of required materials	• Potential for problems is high; leaks can affect ground water and air • Agency-reportable quantities could be exceeded resulting in fines and other business consequences
Corrosion		• Pipes can leak under insulation	• Impurities in the process can affect reactions, quality, etc.	• Hidden leaks under insulation can break through insulation, causing an environmental spill
Erosion	• Exposure to chemicals	• Pipes can leak	• Loss of material due to leakage, and downtime due to repairs	• Potential exposure to chemicals
Cross-connection	• If nitrogen and air lines are cross-connected, individuals could die from asphyxiation (e.g., if nitrogen is inadvertently being used to power air-driven equipment in a confined space, the nitrogen exhaust from the tool could replace breathing air and result in death. • If drinking water and cooling water lines are cross-connected, personnel could drink water filled with the chemicals used to control algae and prevent corrosion.	• If compressed air is in a line where a nitrogen purge is specified and flammables are present, an explosion could result in the line. • If drinking water is cross-connected to cooling water lines, the water will not be adequately treated to control bacteria and corrosion.		

FIGURE 3-26 Pipe Hazards

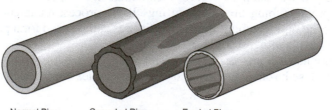

Normal Pipe   Corroded Pipe   Eroded Pipe

Process Technician's Role in Operation and Maintenance

When working with piping, tubing, hoses, and fittings, process technicians should always conduct monitoring and maintenance activities to ensure that the piping and equipment are functioning properly. When performing operations and maintenance tasks, process technicians should always look at, listen to, and feel for the items listed in Table 3-9.

TABLE 3-9 Process Technician's Role in Operation and Maintenance

Look	*Listen*	*Feel*
• For leaks in abnormal locations • Inspect fittings • Check gauges	• For hissing noises that are not normally present.	• For temperatures within acceptable ranges (using care not to touch hot pipes) • Vibrations signifying problems with upstream equipment

Piping Symbols

Piping and Instrumentation Diagrams (P&IDs) are drawings that show the equipment, piping, and instrumentation contained in a process in a facility. Contained on a P&ID are many different symbols. Process technicians must be able to identify these symbols and interpret what they mean. Figure 3-27 shows some of the more commonly used symbols for piping.

It is important to note that, while there are standards, it is possible that symbols can vary slightly from plant to plant.

Piping Symbols

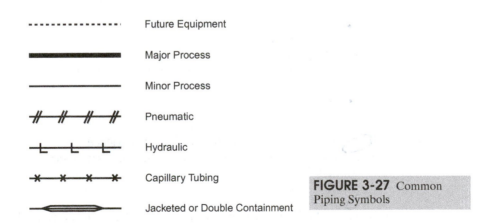

FIGURE 3-27 Common Piping Symbols

Summary

Piping, tubing, hoses, and fittings are some of the most prevalent pieces of equipment in the process industries. In any facility, it is not unusual to see large segments of pipe going from one location to another. Pipes are connected with tubing, hoses, and fittings.

Industrial piping, tubing, hoses, and fittings can be made of many different materials, including carbon steel, stainless steel, alloy steel, iron, exotic metals, flexible membranes, and plastics. The most common type of piping is carbon steel because it is appropriate for a wide range of temperatures and is relatively economical.

The selection and sizing criteria of piping material is critical, and such decisions must consider pressure, temperature, flow, and corrosiveness of the process fluid. Corrosion occurs much more quickly as the pressure and temperature of a process increases. All gasket materials must be appropriate for the chemical properties of the fluid, and operating temperature determines if insulation is required.

Piping connections may be welded, flanged, threaded, or bonded. The factor that determines which connection type is the most appropriate is the purpose of the pipe. Sealant compounds are used to lubricate and seal the connections on threaded pipe. Gaskets are used to seal the connections on flanged pipe. Flared fittings and compression fittings are commonly used for tubing connections. These are mechanical seals and do not normally require separate sealing materials.

All of the material used within the process industries have pressure, temperature, and process chemical property limitations. These limitations are important to the safety and performance of the unit.

Equipment testing is used to check pipes for structural problems. The most common type of equipment tests are hydrotesting, tightness testing, and pneumatic testing.

As a process technician, you must constantly be aware of your surroundings and must analyze the entire system in order to completely understand the functionality of the unit. A part of this understanding includes using the proper types of piping, tubing, hoses, and fittings for the job.

Identifying the need for maintenance and repair is a vital role of the process technician. While the process technician may not perform the repairs, in-depth familiarity with the equipment is necessary to ensure safe operation and maintenance.

Checking Your Knowledge

1. Define the following terms:
   - a. Alloys
   - b. Schedule
   - c. Fitting
   - d. Gasket
   - e. Expansion loop
   - f. Insulation
   - g. Corrosion
   - h. Tensile strength
   - i. Sealant
   - j. Quick connects
   - k. Piping
   - l. Hose
   - m. Tubing
   - n. Jacketed pipes

2. Which of the following is the primary material for piping?
   - a. Glass
   - b. Metal
   - c. Plastic-lined material
   - d. Solid plastic

3. A type of support that suspends pipe from the ceiling is a(n):
   - a. pipe clamp
   - b. pipe hanger
   - c. pipe shoe
   - d. expansion loop

4. (*True* or *False*) Caps may be threaded, glued, or welded.

5. Carbon steel is used in which of the following temperature ranges?
   - a. −50°F to 450°F
   - b. −150°F to 1400+°F
   - c. −20°F to 800°F

6. Which of the following is *NOT* a material used for piping insulation?
   - a. Elastomeric membrane
   - b. Fiberglass
   - c. Asbestos
   - d. Foam glass

7. Cracking shows that the sealant compound or tape:
   - a. was not removed before new sealant/tape was applied.
   - b. is bunched up on the threading of the pipe.
   - c. was not applied correctly.
   - d. clogs the pipe.

8. Give the proper name for each of the following fittings:

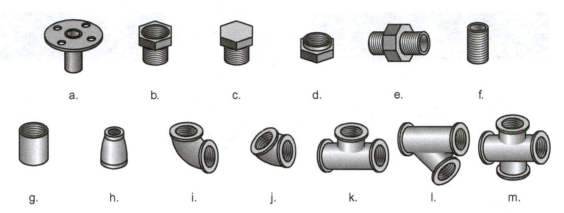

a. b. c. d. e. f.

g. h. i. j. k. l. m.

9. Which of the following is a white tape like substance used to lubricate the thread connections of a male pipe thread?
   a. Pipe dope
   b. Oil
   c. Sealant compound
   d. Teflon® tape
10. Which of the following tests are used to check for leaks in pipe (select all that apply)?
    a. Hydrostatic testing
    b. Pneumatic testing
    c. Temperature testing
    d. Hypertesting
11. To prevent the liquid contents in a pipe from freezing, engineers often apply coils of heated wire or tubing called:
    a. insulation.
    b. heat shielding.
    c. heat tracing.
    d. steam tubes.

Match the type of pipe fitting to the description:

Type of Pipe	Description
Plug	a. Used to cap (seal) the open end of a pipe; may be threaded, glued, or welded.
Cap	b. Threaded on both the internal and external surfaces; used to join two pipes of differing sizes.
Bushing	c. Consists of a mating plate used to join two pieces of pipe or a valve to a piece of pipe.
Flange	d. Used to close the end of a piping run; the threading is male and plugs a female-threaded fitting; can be glued, threaded, or welded into place.

Student Activities

1. Examine the various types of piping in and around your home (e.g., water, sewage, and natural gas). Identify the various piping materials and the characteristics of the service (e.g., high pressure, low temperature, corrosive).
2. Research a type of piping and connection assigned by your instructor and prepare a two-page report.
3. Given various piping, tubing, hoses, and fittings in a lab setting, identify the different types of connections and practice connecting them (e.g., connect a steam hose to a boss fitting or connect a water hose to a crow's foot).

CHAPTER 4

Valves

Objectives

After completing this chapter, you will be able to:

- Describe the purpose of valves in the process industry.
- Identify and explain the use of common valve types.
- Identify the components of valves.
- Explain the purpose of each valve component.
- Describe safety and environmental concerns associated with valves.
- Describe the process technician's role in valve operation and maintenance.

Key Terms

Ball check valve—a valve used to control the flow of heavy fluids; it is available in horizontal, vertical, and angle designs.

Ball valve—a type of valve that uses a flow control element shaped like a hollowed-out ball, attached to an external handle, to increase or decrease flow.

Block valve—used to block flow to and from equipment and piping systems (e.g., during outage or maintenance). Block valves differ from other types of valves in that they should not be used to throttle flow.

Bonnet—the portion of a valve body through which the stem leaves the body and contains the stem packing. The bonnet is a bell-shaped dome mounted on the body of a valve.

Butterfly valve—a type of valve that uses a disc-shaped flow control element to increase or decrease flow.

Check valve—a type of valve that allows flow in only one direction and is used to prevent reversal of flow in a pipe.

Control valve—a valve that automatically controls the increase or decrease of fluid flow through a pipe by remote operation.

Damper—a movable plate that regulates the flow of air, draft, or flue gases.

Diaphragm valve—a type of valve that uses a flexible, chemical-resistant, rubber-type diaphragm to control flow.

Gate valve—a positive shutoff valve that utilizes a gate or guillotine that, when moved between two seats, causes a tight shutoff.

Globe valve—a type of valve that uses a plug and seat to regulate the flow of fluid through the valve body; the plug is shaped like a sphere or globe.

Hand wheel—the mechanism that raises and lowers a valve stem to allow or restrict the flow of fluid through the valve.

Lift check valve—a valve that has built-in globe valve bodies and is available in horizontal and vertical designs.

Louvers—moveable, slanted slats that are used to adjust the flow of air.

Manual valve—a hand-operated valve that is opened or closed using a hand wheel or lever.

Multi-port valve—a type of valve used to split or redirect a single flow into multiple directions.

Needle valve—a type of globe valve that controls small flows using a long, tapered plug that passes through a circular hole in a plate or pipe.

Non-rising stem valve—a valve stem that remains in place when the valve is opened or closed.

Packing—a substance such as Teflon® or graphite-coated material that is used inside the packing gland to keep leakage from occurring around the valve stem.

Plug valve—a type of valve that uses a flow control element shaped like a hollowed-out plug, attached to an external handle, to increase or decrease flow.

Relief valve—a safety device designed to open slowly if the pressure of a liquid in a closed space, such as a vessel or a pipe, exceeds a preset level.

Rising stem valve—a valve stem that rises out of the valve when the valve is opened.

Safety valve—a safety device designed to open quickly if the pressure of a gas in a closed vessel exceeds a preset level.

Straight-through diaphragm valve—a valve that contains a flexible diaphragm that extends across the valve opening.

Swing check valve—a valve used to control the direction of flow and prevent contamination or damage to equipment caused by backflow.

Throttling—a condition in which a valve is partially opened or closed to restrict or regulate fluid flow rates.

Valve body—the lower exterior portion of the valve; contains the fluids flowing through the valve.

Valve disc—the section of a valve that attaches to the stem and that can fully or partially block the fluid flowing through the valve.

Valve knocker—a device used to facilitate the movement (opening or closing) of a valve.

Valve seat—a section of a valve designed to maintain a leak-tight seal when the valve is shut.

Valve stem—a long slender shaft that attaches to the flow control element in a valve.

Weir diaphragm valve—a valve that contains a plate-like device that functions as a seat for the diaphragm.

Introduction

Valves are piping system components used to control or stop the flow of fluids through a pipe. Valves allow process technicians to achieve a desired flow rate or pressure which may, in turn, be used to achieve a desired parameter, such as tank level. Valve classifications are based on the design of the internal components, how the valve is operated, or the function of the valve.

When determining the type of valve to be used in a process, engineers must take into consideration the function the valve will perform, the fluid that will be passing through the system, and the pressure of the fluid.

Valves are made of many different materials, including cast iron, brass, carbon steel, steel alloys, bronze, plastic, and glass. The materials used in the construction of a valve are based on the intended use, particularly the fluid that will be passing through the system. Valve ratings are based on the temperature and pressure of the fluid in the process.

Valve Components

All valves contain a closure device attached to a stem which passes through the valve body. To prevent leakage, the stem attaches to a ring seal that opens or closes the valve and seals in the fluid as the stem rotates. A **hand wheel** is a mechanism that raises and lowers a valve stem to allow or restrict the flow of fluid through the valve.

Packing is a substance such as Teflon® or graphite-coated material that is used inside the packing gland to keep leakage from occurring around the valve stem. The packing sealing material is located within a packing gland inside the valve bonnet (the top of the valve). This packing is held tightly in place by bolts or a cap nut.

It is important to tighten the packing periodically to prevent leakage. However, over tightening can result in over compression against the stem of the valve. This over

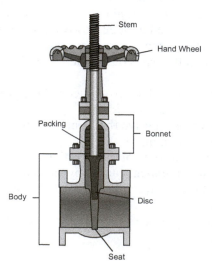

FIGURE 4-1 Basic Components of a Gate Valve

compression causes excessive wear and the loss of packing material. Over time, this can cause the valve to leak. Thus, it is essential that potential expansion and contraction of equipment due to temperature changes be considered when tightening valve components.

The **valve body** is the lower exterior portion of the valve that contains the fluids flowing through the valve. The **bonnet** is the portion of a valve body through which the stem leaves the body and contains the stem packing. The bonnet is a bell-shaped dome mounted on the body of a valve. In nonrising stem valves, the stem does not exit the valve body.

The **valve stem** is the long slender shaft that attaches to the flow control element in a valve. Gate valves are classified as either rising stem, or non rising stem. In a **rising stem valve**, the stem rises out of the valve when the valve is opened. Rising stem valves are used when it is important to know if the valve is opened or closed. In a **non-rising stem valve**, the valve stem remains in place when the valve is opened or closed. Because the stem is threaded into the gate, the gate travels up or down as the handwheel is rotated. Because the stem does not indicate the position of the valve (i.e., whether it is opened or closed), many nonrising stem valves have a pointer or indicator threaded onto the upper end of the stem to indicate the position of the gate. Figures 4-2 and 4-3 show examples of rising and nonrising stem valves.

The **valve disc**, is the section of a valve that attaches to the stem and can fully or partially block the fluid flowing through the valve. Disc design can vary depending on the type of valve and the type of blockage required. The **valve seat** is a section of a valve designed to maintain a leak-tight seal when the valve is shut. The seat is located in the interior and near the bottom of the valve.

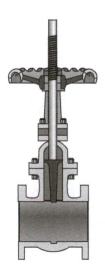

FIGURE 4-2 Rising Stem Valve

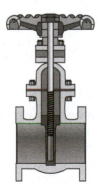

FIGURE 4-3 Nonrising Stem Valve

Types of Valves

Many different types of valves are used in the process industries. The most common valve types include:

- Gate valves
- Globe valves
- Check valves
- Ball valves
- Plug valves
- Butterfly valves
- Diaphragm valves
- Relief and safety valves
- Multiport valve

Process technicians must be familiar with each of the different valve types and their maintenance and operating characteristics.

GATE VALVES

Gate valves are positive shutoff valves that utilize a gate or guillotine which, when moved between two seats, causes a tight shutoff. Gate valves are the most common type of valve in the process industries. They are designed for on/off service and are not intended for **throttling** (a condition in which a valve is partially opened or closed to restrict or regulate the fluid flow rates).

Figure 4-4 shows a cutaway of an actual gate valve. Figure 4-5 and Figure 4-6 show a cutaway of a gate valve with the gate open and closed respectively.

In a gate valve a wedge-shaped disc or gate is lowered into the body of the valve with a hand wheel. The stem is either passed through a wheel that is fastened to the valve body (rising stem), or through an opening fixed to the valve body, with the wheel fixed to the top of the stem (non-rising stem). The stem is sealed by packing and can be tightened with bolts or cap nuts (packing collars).

The body of the valve contains two seat rings that seal around the gate. When the gate is fully closed, flow is completely blocked. Because of the design, gate valves should not be used for throttling, since throttling can cause disc chatter, metal erosion, and seat damage which can prevent the valve from sealing properly.

FIGURE 4-4 Cutaway Photo of a Closed Gate Valve

Courtesy of Bayport Technical Training Center

GLOBE VALVES

Globe valves are valves that use a sphere or globe-shaped plug and seat to regulate the flow of fluid through the valve body. These types of valves are designed to regulate flow in one direction and are usually used in throttling service. Globe valves are very common in the process industries. Figure 4-7 and Figure 4-8 show cutaways of a globe valve with the flow opened and closed respectively.

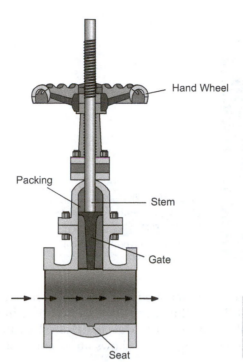

Hand Wheel

Packing

Stem

Gate

Seat

FIGURE 4-5 Open Gate Valve Showing Flow (Notice the Gate in the Upper Valve Body)

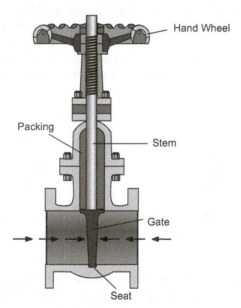

Hand Wheel

Packing

Stem

Gate

Seat

FIGURE 4-6 Closed Gate Valve Showing Flow (Notice the Gate Position in the Valve)

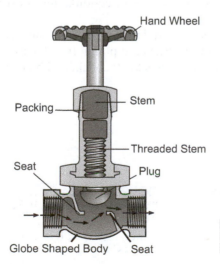

Hand Wheel

Stem

Packing

Threaded Stem

Seat

Plug

Globe Shaped Body

Seat

FIGURE 4-7 Opened Globe Valve with Ball-Shaped Plug

FIGURE 4-8 Cutaway Photo of an Open Globe Valve with a Cylindrical-Shaped Plug

Courtesy of Bayport Technical Training Center

In a globe valve, fluid flow is increased or decreased by raising or lowering the plug or flow control element. These flow control elements come in a variety of shapes, including ball, cylindrical, and needle. The seats in these valves are designed to accommodate the shape of the plug. Figure 4-9 shows an example of these different plug shapes.

A **needle valve** is a type of globe valve that controls small flows using a long, tapered plug that passes through a circular hole in a plate or pipe. Some needle valves have a three-way design that allows them to be used directly in the process piping line. These valves are used for metering or sampling flows (for example, metering a specific additive flow to a tank or other vessel, or sampling material from a process product line).

FIGURE 4-9 Globe Valve Plug Designs

Ball-Shaped   Cylindrical-Shaped   Needle-Shaped

CHECK VALVES

Check valves are devices that prevent backflow (reverse flow in piping when the flow stops). By eliminating backflow, these valves prevent equipment damage and contamination of the process. All check valves are directional and must be placed in the line correctly.

Check valves, which are also referred to as nonreturn valves, are primarily installed in pump discharge lines and other discharge piping (see Figure 4-10). The most common types of check valves are swing check, lift check, and ball check.

FIGURE 4-10 Cutaway Photo of a Swing Check Valve Allowing Flow

Courtesy of Bayport Technical Training Center

Swing check valves are used to control the direction of flow and prevent contamination or damage to equipment caused by backflow. These valves usually have an arrow stamped on the body signifying the direction of flow through them.

Swing check valves contain discs and seats that are set at an angle. As the fluid moves forward through the valve, the valve disc is lifted (forward flow). If the flow stops or changes direction (backward flow), gravity or back pressure forces the valve closed to prevent backflow.

Figures 4-11 and 4-12 show an example of a swing check valve in the open and closed positions, respectively.

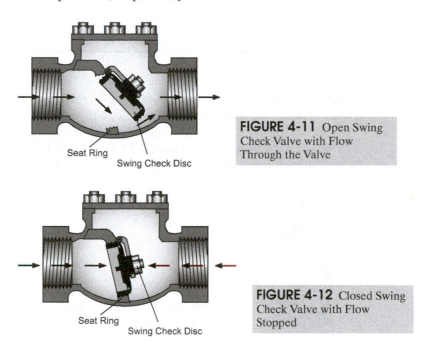

Seat Ring

Swing Check Disc

FIGURE 4-11 Open Swing Check Valve with Flow Through the Valve

Seat Ring

Swing Check Disc

FIGURE 4-12 Closed Swing Check Valve with Flow Stopped

Lift check valves have built-in globe valve bodies and are available in horizontal and vertical designs. Lift check valves have a disc-shaped flow control element that lifts up as fluid flows through (forward flow) and drops back onto the valve seat if the fluid flow stops or attempts to flow back (backward flow).

Figures 4-13 and 4-14 show examples of a lift check valve in the open and closed position, respectively.

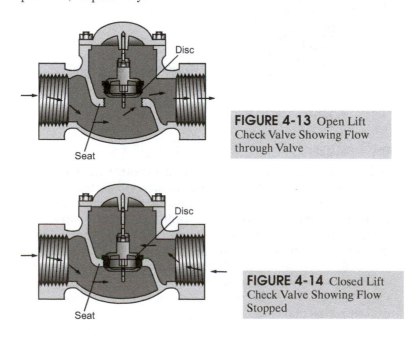

Disc

Seat

FIGURE 4-13 Open Lift Check Valve Showing Flow through Valve

Disc

Seat

FIGURE 4-14 Closed Lift Check Valve Showing Flow Stopped

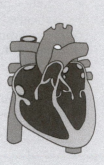

Did You Know?

The human heart has valves that function like check valves.

Without these valves, blood would not circulate properly.

Ball check valves are used to control the flow of heavy fluids and are available in horizontal, vertical, and angle designs. Ball check valves have a ball or sphere-shaped flow control element that responds to gravity or back pressure by moving back onto the seat, thus preventing backflow. Figure 4-15 and Figure 4-16 show examples of a ball check valve in both the open and closed positions, respectively.

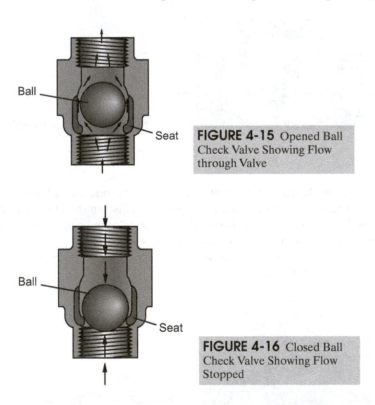

Ball

Seat

FIGURE 4-15 Opened Ball Check Valve Showing Flow through Valve

Ball

Seat

FIGURE 4-16 Closed Ball Check Valve Showing Flow Stopped

BALL VALVES

A **ball valve** is a device that uses a flow control element shaped like a hollowed-out ball, attached to an external handle, to increase or decrease flow. Ball valves are quick opening and are used where a low pressure drop is required. Ball valves are referred to as quarter-turn valves. In other words, turning the valve's stem a quarter of a turn brings it to a fully open or fully closed position (in comparison to other valves, such as gate valves, which require multiple turns to fully open or fully close).

Ball valves use a sphere held against a cup-shaped seat to control flow. The seat has a circular opening with a smaller diameter than the ball. Stops are provided on the valve handle to indicate open and closed positions. (Figures 4-17, 4-18, and 4-19 show examples of a ball valve.

FIGURE 4-17 Cutaway Photo of a Ball Valve in the Open Position

Courtesy of Bayport Technical Training Center

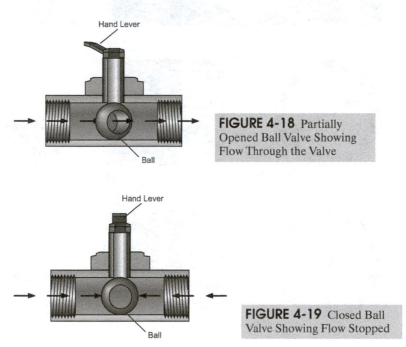

FIGURE 4-18 Partially Opened Ball Valve Showing Flow Through the Valve

FIGURE 4-19 Closed Ball Valve Showing Flow Stopped

Ball valves are typically used for on/off service. When a ball valve is opened, the hollowed-out portion of the ball (sometimes referred to as the port) lines up perfectly with the inner diameter of the pipe. When a ball valve is closed, the port aligns with the wall of the pipe.

The hole in most ball valves is smaller in diameter than the internal diameter of the pipe. Some ball valves, however, are designed to have an opening that is the same diameter as the internal diameter of the pipe. These are called full port ball valves.

PLUG VALVES

A **plug valve** uses a flow control element shaped like a hollowed-out plug, attached to an external handle, to increase or decrease flow. Plug valves are quick-opening valves that are similar to ball valves.

Like ball valves, plug valves are quarter-turn valves that use a hollowed-out, flow control element to control flow. In plug valves, the opening for the flow is machined through a tapered cylindrical plug. The opening is rotated perpendicular (at a 90° angle) to the direction of flow to close.

Plug valves are designed for on/off service and are well suited for certain types of applications such as low pressure, slurry, lubrication, and fuel gas. Figures 4-20, 4-21 and 4-22 show examples of plug valves.

BUTTERFLY VALVES

A **butterfly valve** uses a disc-shaped flow control element to increase or decrease flow. Like ball valves, butterfly valves can be fully opened or closed by turning the valve handle one-quarter of a turn.

FIGURE 4-20 Cutaway Photo of a Closed Plug Valve

Courtesy of Bayport Technical Training Center

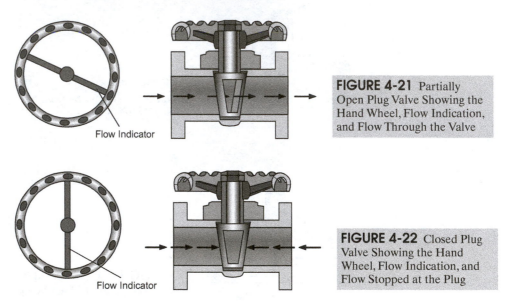

Flow Indicator

FIGURE 4-21 Partially Open Plug Valve Showing the Hand Wheel, Flow Indication, and Flow Through the Valve

Flow Indicator

FIGURE 4-22 Closed Plug Valve Showing the Hand Wheel, Flow Indication, and Flow Stopped at the Plug

Butterfly valves have a similar construction to dampers, except the interior body of the butterfly valve is shaped to fit the disc. This results in a tighter seal when the valve is closed. Because of the way they are designed, butterfly valves are best suited for low-temperature, low-pressure applications such as cooling water systems.

Figures 4-23, 4-24, and 4-25 show cutaways of a butterfly valve in both the open and closed positions.

Unlike many other valves, butterfly valves can be used for throttling and on/off service. However, the throttling capabilities of a butterfly valve are not uniform or exact (for example, opening the valve halfway may provide a flow that is near maximum).

DIAPHRAGM VALVES

Diaphragm valves use flexible, chemical resistant, rubber-type diaphragms to control flow. Their flexibility and material composition make them different from other flow

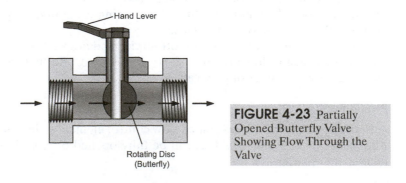

Hand Lever

FIGURE 4-23 Partially Opened Butterfly Valve Showing Flow Through the Valve

Rotating Disc (Butterfly)

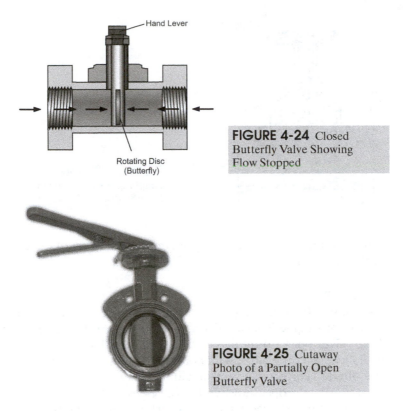

FIGURE 4-24 Closed Butterfly Valve Showing Flow Stopped

FIGURE 4-25 Cutaway Photo of a Partially Open Butterfly Valve

control elements. In this type of valve, the diaphragm seals the parts above it (e.g., the plunger) from the process fluid.

Figures 4-26 and 4-27 show examples of a diaphragm weir valve in both the open and closed position, respectively.

Weir diaphragm valves contain a plate-like device that functions as a seat for the diaphragm. The flow must go over the top of the weir and lift the diaphragm to exit.

Straight-through diaphragm valves contain a flexible diaphragm that extends across the valve opening. There is very little pressure drop across this type of valve when open.

Because of their unique design, diaphragm valves work well with process substances that are exceptionally sticky, viscous, or corrosive. However, they are not adequate for applications with high pressures or excessive temperatures. Diaphragm valves are easy to clean, but leaks can occur if the diaphragm fails.

RELIEF AND SAFETY VALVES

Relief and safety valves are used to protect equipment and personnel from the hazards of overpressurization. Both of these valves are designed to open and discharge when the pressure in a line, vessel, or other equipment exceeds a preset threshold. The difference,

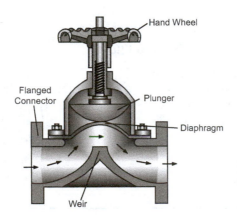

FIGURE 4-26 Opened Diaphragm Valve Displaying Flow Through Valve

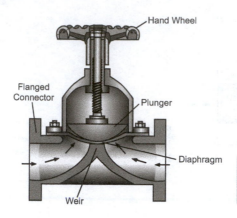

FIGURE 4-27 Closed Diaphragm Valve Showing Flow Stopped in Valve

however, is the type of service they are intended for (liquid versus gas), and the speed at which they open.

Relief valves

Relief valves (shown in Figure 4-28) are safety devices designed to open slowly if the pressure of a liquid in a closed space, such as a vessel or a pipe, exceeds a preset level. These valves open slower and with less volume than safety valves because liquids are virtually noncompressible. With noncompressible substances, a small release is all that is required to correct overpressurization.

Because they open slowly, relief valves are not good for gas service. However, they are good for pressurized liquid service. These valves can be found in liquid drain lines in which the contents are vaporized and sent to a flare system, or found on the top of tanks that are intended to operate liquid full. For example, there is a relief valve on the top of most residential hot water heaters.

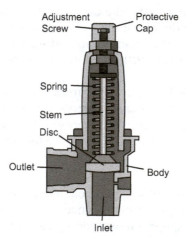

FIGURE 4-28 Relief Valve Components

Did You Know?

Process technicians often use the slang term *pop valve* to refer to safety valves because they open quickly, or pop off, once the pressure threshold has been exceeded.

In a relief valve, a flow control disc is held in place by a spring. Once the pressure in the system exceeds the threshold of the spring, the valve is forced open (proportional to the increase in pressure) and liquid is allowed to escape into a containment receptacle, flare, or other safety system. As the pressure drops below the threshold, the spring forces the flow control element back into the seat, thereby resetting the valve.

Safety valves

Safety valves are safety devices designed to open quickly if the pressure of a gas in a closed vessel exceeds a preset level. A few examples of safety valve applications include steam lines, compressor discharge lines, and overhead tower vapor lines. Safety valves are designed to open and vent excess pressure to the flare header or through a large exhaust port into the atmosphere (depending on the substance being vented).

Safety valves contain a spring that is used to set the opening pressure and two seats. One seat is exposed to the pressure source. When it is opened, this seat exposes the other seat area to the pressure source, causing the valve to "pop" open quickly. Figures 4-29 and 4-30 show examples of safety valves in both labeled and cutaway views, respectively.

To protect equipment and piping, safety valves do not close until the system pressure falls beneath the design specification. Once activated, some safety valves will reseat themselves, while others must be taken to a shop to be manually reset by a quailfied technician.

Safety valves differ from relief valves because they:

- have a different handle device
- are used with gas service, not liquid
- have a faster response time
- have a larger exhaust port

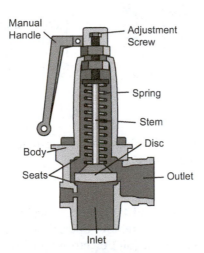

FIGURE 4-29 Safety Valve with Manual Handle Labeled

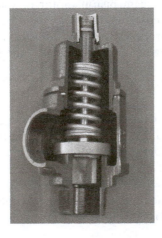

FIGURE 4-30 Cutaway Photo of a Safety Valve

Courtesy of Bayport Technical Training Center

BLOCK VALVES

Block valves are devices used to block flow to and from equipment and piping systems (e.g., during an outage or maintenance). Block valves differ from other types of valves because they should not be used to throttle flow. They can be rising stem, ball, or plug.

MULTIPORT VALVES

A **multiport valve** is used to split or redirect a single flow into multiple flows. Multiport valves can contain three-, four-, and six-way openings. These types of valves normally have globe valve-type closures that allow the flow to be directed two ways in the three-way valve, three ways in the four-way valve, and so forth. An example of a multiport valve is a three-way valve used to redirect the process flow from a vessel to a flare header.

In Figure 4-31, flow is entering the three-way valve though the inlet. The normal direction has the flow going through the left outlet (to the vessel). When we redirect the flow, the flow exits to the flare through the outlet on the right. Since the flow can be routed by the valve to more than one location, another common use for these valves is on the inlet and outlet of heat exchangers to facilitate backwashing (reversing flow in order to clean the exchanger of debris or other unwanted material).

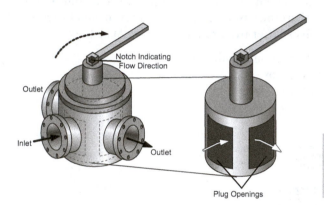

FIGURE 4-31 Multiport Valves are used to Split or Redirect a Single Flow into Multiple Directions

AUTOMATIC VALVES

Automatic valves are automated on/off valves that are designed to start and stop flow during normal and emergency operations.

CONTROL VALVES

Control valves (shown in Figures 4-32 and 4-33) automatically control the increase or decrease of fluid flow through a pipe by remote operation. The operation of control valves can be conducted using pneumatic (air), electronic, or hydraulic signals. Control valves have actuators that can be diaphragm, piston, motor, or solenoid in design.

Control valves are mechanically controlled and they can be operated automatically or manually. Automatically controlled valves are adjusted by the control instrument operator. Manually controlled valves are controlled by the process technician in the assigned area.

HAND MANUAL VALVES

A **manual valve** is a hand-operated valve that is opened or closed using a hand wheel or lever. Two types of manual valves discussed are valve operators with knockers and valve operators with louvers and dampers.

Valve knockers

A **valve knocker** is a device used to facilitate the movement (opening or closing) of a valve. The purpose of valve knockers is to help operate the wheel when valves are in

FIGURE 4-32 Cutaway Photo of an Open Control Valve

Courtesy of Design Assistance Corporation (DAC)

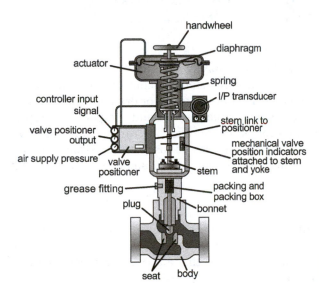

FIGURE 4-33 Internal Elements of a Control Valve

hard-to-reach places or high-pressure service locations. Valve knockers contain a heavy cam (called a "knocker") attached to a wheel that slams against a cam attached to the valve stem. The knocking action on the cam attached to the stem raises the closure device off the seat so the valve can be opened. The knocking action causes small movements of the valve that can be used to either open or close the valve. Valves with knockers are usually set into the line in pairs, and at an angle, and they are always welded.

Louvers and Dampers

Louvers and dampers are adjustable plates that are used to regulate the air flow (draft) or off-gas flows in furnaces, boilers, and air fin heat exchangers.

Louvers are moveable, slanted slats that are used to adjust the flow of air. Axles and gears attach to each plate, allowing these plates to open or close together and control gaseous flows. Louvers are commonly used to control air supply to furnaces, boilers, and fin-fan heat exchangers. Figure 4-34 shows an example of a louver.

Dampers are movable plates that regulate the flow of air, draft, or flue gases. Dampers consist of a metal plate attached to an axle that allows the plate to turn perpendicular (at right angles) to the flow to open or close. They primarily control the flow of stack gases in furnaces. Figure 4-35 shows an example of a damper.

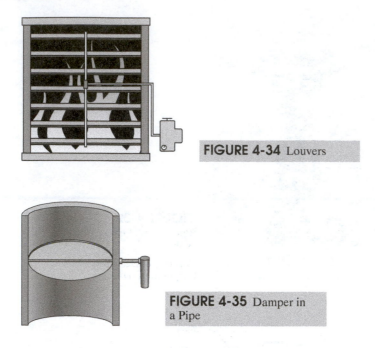

FIGURE 4-34 Louvers

FIGURE 4-35 Damper in a Pipe

Louvers and dampers may be operated by hand, by a switch, or by an electric motor. They may also be connected to a control loop and manipulated as part of a system design to control the amount of air, or draft, through a furnace or boiler.

It is important to know that louvers and dampers cannot provide a total shutoff of flow; some leak-by can occur. Additionally, care must be taken when opening and closing louvers and dampers. They can be easily damaged and may stick in an open or closed position, which can cause damage to the equipment associated with the damper system.

Potential Problems

Process technicians are commonly required to troubleshoot valve operations and determine the best solution to return the valve to normal operating conditions. The most common valve conditions that a process technician may need to investigate are plugging and leakage.

Valves can become plugged when they are depressurized. When depressurizing the valve, the process can form a solid in the valve due to abnormalities of process fluids flowing through the valve. If the valve is a drain valve, solids can collect in the valve (the lowest point in the system), and over time these solids can build up and plug the valve.

Another problem with valves is leakage. Valve leakage can occur either across the valve (within the pipe) or into the atmosphere. Atmospheric leakage normally occurs through the valve stem packing or flanges. Causes of leakage include corrosion (internal or external), erosion (usually internal), vibration and wear (external), and damage from an external source (e.g., impact from something striking the pipe). Because valve leakage can result in fire, explosion, personnel exposure, and environmental impacts, it is important to resolve leaks as quickly as possible.

Hazards Associated with Improper Valve Operation

Many hazards are associated with piping and valves. Table 4-1 lists some of those hazards and their impacts.

PROPER BODY POSITION

An operator should never lean over the valve handle when operating the valve. If the valve bonnet were to fail under high pressure, the bonnet could fly up and strike the process technician in the face.

TABLE 4-1 Safety and Environmental Concerns Associated with Valves

Improper Operation	*Possible Impacts*			
	Individual	*Equipment*	*Production*	*Environment*
Throttling a valve that is not designed for throttling		Valve damage to the point that it will not seat and stop flow, even when closed	Off spec product due to improper flows	Valve leakage could cause a vessel to overflow to the atmosphere
Use of excessive force when opening or closing a valve	Slipping and falling, as well as muscle and back strain	Damage to the valve seat, the packing, or the valve stem. This causes leakage and makes the valve difficult to open or close.	Off spec product due to improper flows	Valve leakage could cause a vessel to over flow to the atmosphere
Failure to clean and lubricate valve stems	Possible injuries as a result of a valve wrench (required because the valve is difficult to open) slipping off of the valve handle	Valve stem seizure or thread damage. This makes the valve difficult to open and close. The valve handle or wheel could break or the stem could shear.		
Improperly closing a valve on a high-pressure line	Possible injury due to equipment over pressurization	Equipment damage (e.g., overpressurizing a pump)		Possible leak to the environment
Failure to wear proper protective equipment when operating valves in high-temperature, high-pressure, hazardous, or corrosive service	Burns or chemical exposure and other serious injuries			

When opening a bleed or vent valve, never stand in front of the opening. Instead, operate the hand wheel with your body away from the valve opening (see Figure 4-36). This will prevent potential injury and exposure.

To prevent bruises, abrasions, and "knuckle busting" from slipping wheels or levers, always use proper tools and body positioning to adjust a valve that is stuck. Improper body positioning can cause slipping, falling, or back strain due to overexertion or hyperextension.

When working with valves, always wear appropriate gloves, use a valve wrench to reduce the effort required, and maintain proper body position to prevent injury.

OPERATING OVERHEAD VALVES

Valves are sometimes located in positions that can create safety hazards. Whenever possible, valves should be operated at a point above or level with the valve. This helps prevent injury from downward-flowing fluid that may be present as a result of valve ruptures or failures. Fluids flowing through valves may be hazardous and could cause serious injuries.

Elevated valves (e.g., valves in pipe racks) may have a chain operator extension that allows the technician to operate the valve from ground level. Care should be taken when manipulating a chain-operating device to ensure the chain does not become dislodged from its cradle around the handwheel or get tangled in clothing. Figure 4-37 shows an example of a chain-operated valve.

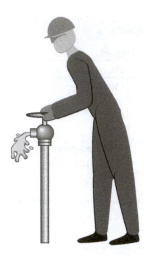

FIGURE 4-36 Proper Body Position when Opening Bleed or Vent Valves

FIGURE 4-37 Chain-Operated Valve

SAFE FOOTING

When adjusting a valve, the optimum operating position is usually at ground level. However, if ground operation is not possible, the process technician should at least attempt to obtain solid footing by firmly planting both feet on the surface being used for elevation (e.g., ladder, scaffold, or secured platform). To prevent slipping or falling, ensure the platform is stable and has adequate traction (i.e., it is not slippery).

When adjusting a valve, use a pushing motion instead of a pulling motion. A pushing motion allows you to react more effectively if you begin to slip.

USE OF VALVE WRENCHES AND LEVERS

Always use the proper valve wrench to adjust a valve. See Figure 4-38 for examples of valve wrenches. Make sure that the valve wrench is the proper size and in good condition. Also, be sure to check for pinch points before using the valve wrench or lever, and ensure that the hands, or any other parts of the body, do not enter those pinch points.

Always use the correct tool for the job! Never improvise with the closest tool at hand (e.g., never use a pair of pliers instead of a valve wrench). Using the wrong tool can cause injury or damage to the valve, hand wheel, and process technician.

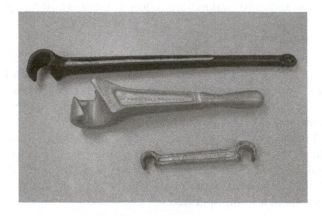

FIGURE 4-38 Examples of Valve Wrenches

When using a valve wrench or lever, always wear leather gloves to protect against abrasions, burns, or handles that could cause cuts. Once you have put on proper protective equipment, establish a firm grip on the valve handle or wrench, secure your footing, and then position your body to prevent a strain or to avoid striking an object. Use your body weight instead of back muscles when adjusting valves that are difficult to open or close. Seek help in operating particularly stubborn valves.

Process Technician's Role in Operation and Maintenance

Depending on company policy, a process technician may be responsible for incidental (minor) maintenance and repair of valves. Minor maintenance may include tightening valve packing or replacing a small valve. The most critical repairs are performed by maintenance personnel with specialized expertise.

When monitoring valves, process technicians are generally responsible for inspecting for leaks, greasing and lubricating stems, and monitoring the operation of valves.

INSPECTING FOR LEAKS AND CORROSION

Inspections are performed to check for leaks and signs of external corrosion. The most obvious places to look for leaks and corrosion include areas around the packing, bonnets, gaskets, or threads.

GREASING AND LUBRICATING

Valves must be greased and lubricated regularly for proper operation. When greasing and lubricating, process technicians should typically grease the valve stems, lubricate additional components, adjust the packing, and inspect the packing for leaks.

MONITORING

When monitoring valves, technicians must always remember to look at, listen to, and feel for the items listed in Table 4-2.

Failure to perform proper maintenance and monitoring could impact the process and result in equipment damage.

TABLE 4-2 Process Technician's Role in Operation and Maintenance

Look	*Listen*	*Feel*
• Check valves to make sure there are no leaks	• Listen for abnormal noises (e.g., check valve "chatter")	• To make sure the valve is not being overly tightened
• Check valve for excessive wear		
• Check to make sure the valve stems are properly lubricated		

Summary

Valves, which control the flow of materials through piping, are one of the most prevalent pieces of equipment in the process industries. While there are many different types of valves in the process industries, the most common types of valves are ball, butterfly, check, diaphragm, gate, globe, plug, and relief. These valves can be manually or automatically controlled.

Ball and plug valves are both quarter-turn valves that use a hollowed-out ball or plug to increase or decrease flow. Both of these valve types are primarily used for on/off service.

Butterfly valves are also quarter-turn valves, but they use a disc-shaped element to control flow. Butterfly valves can be used for throttling. However, the throttling capabilities of butterfly valves are not linear (e.g., opening the valve halfway may provide a flow that is near maximum).

Check valves are regulating devices designed to prevent backflow. The most common types of check valves are swing, lift, and ball.

Diaphragm valves use a flexible, chemical-resistant, rubber-type diaphragm to control flow instead of a typical flow control element. Because of their design, diaphragm valves are good for low-pressure service of viscous or corrosive materials.

Gate valves are the most common type of valve in the process industries. These valves use a metal gate to block the flow of fluids through the valve. They are intended for on/off service and should not be used for throttling.

Globe valves, which are also common in the process industries, use a plug to block the flow of fluid through a valve. These valves can come in a variety of shapes and are designed for throttling.

Relief and safety valves are safety devices designed to open if the pressure of a fluid exceeds a preset threshold. Relief valves are primarily used in liquid service, while safety valves are used for gases.

It is important to operate valves properly. Excessive wear and damage to the valves may be caused by throttling valves that are not designed for throttling, using excessive force to open or close a valve, or failing to clean and lubricate valve stems. With time, this damage may cause the valve to leak, seize up, or fail to open or close completely.

Improperly closing a valve on a high-pressure line (e.g., blocking or "dead heading" a pump) or operating a steam-filled valve without proper protective gear can cause equipment damage or personal injury.

When making rounds, process technicians should always inspect valves for leaks and perform proper maintenance and lubrication procedures to prevent damage or excessive wear. Technicians should also listen for abnormal noises or "valve chatter" that could be an indication of improper throttling.

Checking Your Knowledge

1 Define the following terms:

a. Ball valve	l. Plug valve
b. Block valve	m. Relief valve
c. Butterfly valve	n. Safety valve
d. Check valve	o. Bonnet
e. Control valve	p. Damper
f. Diaphragm valve	q. Hand wheel
g. Gate valve	r. Louver
h. Globe valve	s. Packing
i. Multiport valve	t. Throttling
j. Manual valve	u. Valve seat
k. Needle valve	v. Weir

2. *(True or False)* Gate valves are the most common type of valve in the process industry.
3. Globe valves use a _____ to seal off the flow.
    a. cylindrical plug
    b. needle plug
    c. ball plug
4. Check valves are used to prevent _____ in piping when the flow stops.
    a. overflows
    b. continuous flows
    c. trickle flows
    d. reverse flows
5. Which of the following are types of check valves (*select all that apply*)?
    a. Swing
    b. Lift
    c. Ball
    d. Gate
6. *(True or False)* A ball valve opens completely with a quarter turn of the wheel or lever.

7. A _____ valve is a quick-opening valve that is easier to operate than ball valves because of the lower surface area.
   a. check
   b. gate
   c. multiport
   d. plug

8. *(True or False)* The two types of diaphragm valves are weir diaphragm valve and straight-through diaphragm valve.

9. *(True or False)* Relief valves are designed to open quickly to release pressure when a liquid exceeds its designed level.

10. *(True or False)* Safety valves operate quickly to prevent the overpressurization of gases, which can cause equipment damage or injury.

11. What can happen to a valve over time if the valve is used for throttling, even though it is not designed for throttling? *Chattering*

12. Why should a process technician refrain from using excessive force when opening or closing a valve? *slip or fall*

13. Why should process technicians clean and lubricate valve stems regularly? *for proper operation*

Match the valve type with its description.

Valve Type	Description
Ball	a. Designed to open if the pressure of a gas exceeds a preset threshold.
Butterfly	
Check	b. Uses a disc-shaped flow control element to increase or decrease flow.
Diaphragm	
Gate	c. Uses a hollowed-out plug to increase or decrease flow.
Globe	d. Uses a rubber-type diaphragm to control flow.
Plug	e. Uses a metal gate to block the flow of fluids.
Relief	f. Uses a hollowed-out ball to increase or decrease flow.
Safety	g. Designed to open if the pressure of a liquid exceeds a preset level.
	h. Uses a spherical or globe-shaped plug to block fluid flow.
	i. Used to prevent accidental backflow.

14. What type of valve is used to split or redirect a single flow into different directions?
   a. Butterfly
   b. Check
   c. Multiport
   d. Gate

Student Activities

1. Given a hand valve (globe or gate), dismantle the valve and label all the separate components.
2. Using a simulator or process skid unit, identify all the valve types in use.
3. Given a P&ID from Chapter 2 (or one that your instructor has provided), label the valves and identify their types and the direction of flow.
4. Look around your house and identify at least five valves (e.g., the valve under your kitchen sink). Tell where each valve is located and try to identify what type of valve you think it might be (e.g., ball, gate, or globe valve).
5. Given a drawing of a valve, identify all of the components (e.g., stem, seat, and flow control element).

Tanks and Vessels

Objectives

After completing this chapter, you will be able to:

- Explain the purpose of tanks and vessels in the process industries.
- Identify the major tank and vessel types and explain the purpose of each.
- Identify the components of tanks and vessels and explain the purpose of each.
- Describe safety and environmental hazards associated with tanks and vessels.
- Identify typical procedures associated with tank and vessel operation and maintenance.
- Describe the process technician's role in tank and vessel operation and maintenance.
- Identify potential problems associated with tanks and vessels.

Key Terms

Agitator/mixer—a device used to mix the contents inside a tank or vessel.

Articulated drain—a hinged drain that is attached to a floating roof. Its purpose is to allow drainage as the roof raises and lowers.

Baffle—a metal plate placed inside a vessel or tank and used to alter the flow of chemicals, facilitate mixing, or cause turbulent flow.

Barge—a flat-bottomed boat used to transport fluids (e.g., oil or liquefied petroleum gas) and solids (e.g., grain or coal) across shallow bodies of water such as rivers and canals.

Blanketing—the process of introducing an inert gas, usually nitrogen, into the vapor space above the liquid in a tank to prevent air leakage into the tank. Often referred to as a "nitrogen blanket."

Bullet vessel—cylindrically shaped container used to store contents at moderate to high pressures.

Containment wall—a wall used to protect people and the environment against tank failures, fires, runoff, and spills.

Dike—a wall (earthen, shell, or concrete) built around a piece of equipment to contain any liquids in the event of an equipment rupture or leak.

Firewall—an earthen bank or concrete wall built around a storage tank to contain the contents in the event of a spill or rupture.

Fixed roof tank—a container that has a roof permanently attached to the top. The roof can be cone-shaped, flat, or cylindrical.

Floating roof tank—a container with a roof that floats on the surface of a stored liquid to minimize vapor space.

Foam chamber—a reservoir on the side of a tank designed to contain chemical foam that can be used to extinguish fires within the tank.

Gauge hatch—an opening on the roof of a tank that is used to check tank levels and obtain samples of the product or chemical.

Grounding—connecting an object to the earth using copper wire and a grounding rod to provide a path for the electricity to dissipate harmlessly into the ground.

Hemispheroid vessel—a pressurized container used to store material with a vapor pressure slightly greater than atmospheric pressure (i.e., .5 psi to 15 psi). The walls of a hemispheroid tank are cylindrically shaped and the top is rounded.

Hopper car—a type of railcar designed to transport solids like plastics and grain.

Level indicator—a gauge placed on a vessel to denote the height of the liquid level within the vessel.

Lining—a coating applied to the interior wall of a vessel to prevent corrosion or product contamination.

Manway—an opening (usually 24 inches in diameter) in a vessel or tank that permits entry for inspection or repair.

Mist eliminator—a device in a tank designed to collect droplets of mist from gas and is composed of mesh, vanes, or fibers.

Platform—a strategically located structure designed to provide access to instrumentation and to allow the performance of maintenance and operational tasks. Platforms are generally accessed by ladders.

Pressure relief device—a component that discharges or relieves pressure in a vessel in order to prevent overpressurization.

Scale—dissolved solids deposited on the inside surfaces of equipment.

Ship—a large seagoing vessel used to transport cargo across large bodies of water.

Skirt—support structure attached to the bottom of freestanding vessels or tanks.

Spherical vessel—spherically shaped container designed to distribute the pressure evenly over every square inch of the container. Spherical vessels are often used to store volatile or pressurized materials at 15 pounds per square inch (psi) and above.

Sump—the lowest section at the base of a tank. The sump allows materials to be removed from the bottom of the tank and the tank to be completely emptied.

Tank breather vent—a type of vacuum relief device that maintains the pressure in the tank at a specific level, allowing air flow in and out of the tank.

Tank car—a type of railcar designed to transport liquids in bulk.

Tank truck—a special container designed to transport fluids in bulk.

Underground storage tank—a container and its respective piping system in which a minimum of 10 percent of the combined process fluids are stored underground.

Vacuum breaker—a safety device used to remove vacuum by adding pressure to a vessel. This device can help to prevent implosion and backflow.

Vacuum relief device—a safety device that prevents pressure in a tank from dropping below normal atmospheric pressure. This device functions to prevent tank implosion (sudden inward collapse).

Vapor recovery system—process used to capture and recover vapors. Vapors are captured by methods such as chilling or scrubbing. The vapors are then purified and are either returned to the process, moved to storage, or incinerated.

Vortex—cyclonelike rotation of a fluid.

Vortex breaker—a metal plate, or similar device, that prevents a vortex from being created as liquid is drawn out of the tank.

Weir—a flat or notched dam that functions as a barrier to flow.

Introduction

Vessels are a vital part of the process industries. Without these types of equipment, the process industries would be unable to create and store products in large amounts. By using vessels to create products in large quantities, companies can improve both cost effectiveness and efficiency.

In addition to long-term storage, vessels are also used to provide intermediate storage between processing steps and to provide residence time so reactions can be completed or to provide time for material to settle.

Vessels vary greatly in design (e.g., size and shape) based on the requirements of the process. Factors that affect vessel design may include pressure requirements (e.g., high and low), product storage (liquid, gas, or solid), temperature requirements (e.g., insulation), corrosion factor, and volume.

The most common vessels are tanks, drums, cylinders, hoppers, and bins. However, there are other types of specialized vessels (e.g. reactors, towers, and columns). These are covered in other chapters of this textbook.

Types of Tanks and Vessels

Tanks and vessels come in many different shapes and sizes, and many factors affect their design and manufacture. For example, pressure, temperature, and chemical properties are key factors that affect wall thickness, materials of construction, and shape.

TANKS

Tanks are containers in which atmospheric pressure is maintained (i.e., they are neither pressurized nor placed under a vacuum; they are at the same pressure as the air around them). Most tanks are cylindrical in shape and equipped with either a fixed or floating roof.

Tanks are usually made of steel plates that are welded together in large sections and do not seal as tightly as vessels. As a result, tanks are appropriate only for substances that do not contain toxic vapors or high vapor pressure liquids.

Fixed roof tanks are containers that have a roof permanently attached to the top. The roof can be cone-shaped, flat, or cylindrical.

Floating roof tanks, like the one shown in Figure 5-1, have a roof that floats on the surface of a stored liquid to minimize vapor space. This decreased vapor space makes

FIGURE 5-1 Example of a Floating Roof Tank with an Articulated Drain

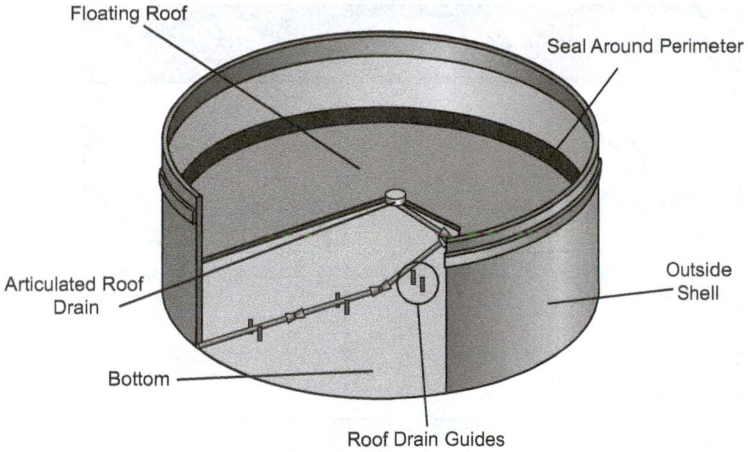

them ideal for storing volatile liquids. Floating roofs may be external or internal, with a standard fixed cone roof above the floating roof.

All floating roofs have a roof that floats on top of the liquid stored inside the tank. They also have a stairway from the outside top edge of the roof to the top of the floating roof. The stairway is essential for inspection of the roof and annual maintenance activities. The removal of water from the external floating roof is accomplished with a system of drains.

An **articulated drain** is a hinged drain that is attached to a floating roof. Its purpose is to allow drainage as the roof raises and lowers. Figure 5-1 shows an example of a floating roof tank with an articulated drain.

An **underground storage tank** is a container and its respective piping system in which a minimum of 10 percent of the combined process fluids are stored underground. Because an underground storage tank is installed in the ground, this type of vessel must be built to structurally withstand the pressure exerted by the soil around the tank. Often, integral ribs are included in the tank design to provide this additional strength.

Underground storage tanks, like the one shown in Figure 5-2, must be watertight to prevent groundwater from contaminating the tank contents and to prevent leakage of tank contents into the soil and groundwater surrounding the tank. All underground storage tanks must have protection against spills, overfill, and corrosion incorporated into their design. Underground storage tanks that store certain hazardous materials are regulated under the Comprehensive Environmental Response, Compensation, and Liability Act (CERCLA). These tanks must have secondary containment systems with interstitial monitoring (devices that indicate the presence of a leak in the space between the first and the second wall).

VESSELS

Vessels are enclosed containers in which the pressure is maintained at a level that is higher than atmospheric pressure. The most common types of vessels are spheroid, bullet, and hemispheroid. Figure 5-3 shows examples of spherical, bullet, and hemispheroid (domed) vessels.

Spherical vessels are spherically shaped containers designed to distribute the pressure evenly over every square inch of the container. Spherical vessels are often used to store volatile or pressurized materials at 15 pounds per square inch (psi) and above.

Bullet vessels are cylindrically shaped containers used to store contents at moderate to high pressures. The rounded ends of a bullet vessel help distribute pressure more evenly than in a nonrounded vessel.

FIGURE 5-2 Underground Storage Tank System

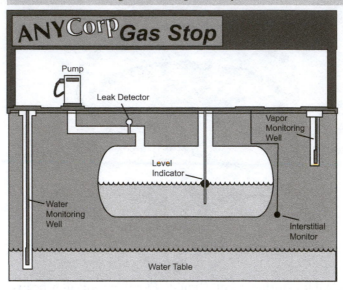

FIGURE 5-3 Sphere, Bullet, and Hemispheroid Vessels

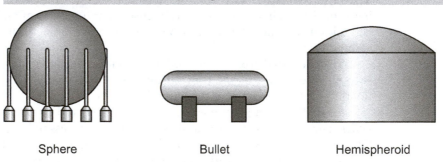

Sphere   Bullet   Hemispheroid

Hemispheroid vessels are pressurized containers used to store material with a vapor pressure slightly greater than atmospheric pressure (i.e., .5 psi to 15 psi). The walls of hemispheroid vessels are cylindrically shaped and the top is rounded.

MOBILE CONTAINERS

Process industry products are transported from manufacturing sites to customers in a variety of mobile tanks and vessels. For example, tank trucks and tank cars move products on land, while barges and ships transport products across water.

Did You Know?

Spherical vessels are much stronger than nonspherical vessels.

That is why they are used for extremely high pressure applications.

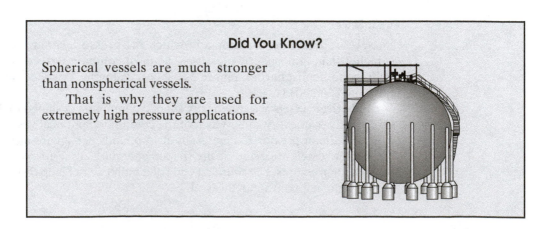

A **tank car** (see Figure 5-4) is a type of railcar designed to transport liquids in bulk. Tank cars may be insulated to maintain their loads at specific temperatures, or pressurized to maintain their loads at set pressures. Tank cars may also be divided into compartments in order to carry several different products in one car.

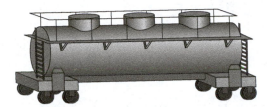

FIGURE 5-4 Tank Car Used to Transport Liquid Cargo by Rail

Hopper cars (see Figure 5-5) are another type of railcar designed to transport solids like plastics and grain.

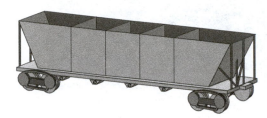

FIGURE 5-5 Hopper Car Used to Transport Solids by Rail

Tank trucks (see Figure 5-6) are special containers designed to transport fluids in bulk over roadways. Tank trucks are similar to tank cars because they can be insulated, pressurized, and compartmentalized.

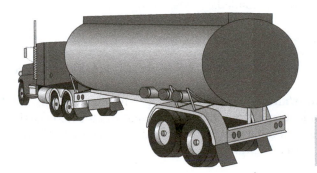

FIGURE 5-6 Tank Truck Used to Transport Fluid Cargo Over Roadways

Ships are large seagoing vessels used to transport cargo across oceans and other large bodies of water. The most common types of ships used by the process industries are container ships, tankers, or supertankers. Container ships transport truck-size storage containers loaded with products, and tankers transport bulk liquids like oil and chemicals. Figure 5-7 shows an example of a container ship, and Figure 5-8 shows an example of a tanker.

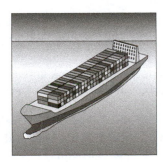

FIGURE 5-7 Example of a Container Ship

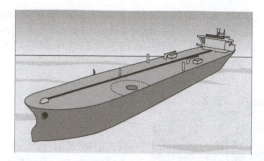

FIGURE 5-8 Example of a Tanker

FIGURE 5-9 Barge Used to Transport Heavy Goods

A **barge** (see Figure 5-9) is a flat-bottomed boat used to transport fluids (e.g., oil or liquefied petroleum gas) and solids (e.g., grain or coal) across shallow bodies of water such as rivers and canals. These vessels can be pushed or pulled by tugboats.

Both barges and tankers are compartmentalized to stabilize the load. Each compartment must be filled and unloaded in a prescribed sequence to keep the vessel level while it is loaded or unloaded.

Common Components of Vessels

Vessels, including tanks and drums, have many components. Process technicians should be familiar with these components.

An **agitator/mixer** is a device used to mix the contents inside a tank or vessel. Process technicians operate mixers using motors located outside the vessels. Pumps can be used to mix the liquid by recirculating the content. Figure 5-10 shows an example of a tank with an agitator.

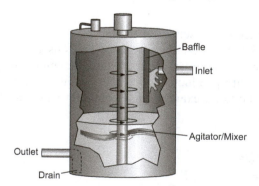

Baffle

Inlet

Agitator/Mixer

Outlet

Drain

FIGURE 5-10 Tank with Agitator Mixing Liquids

A **baffle** is a metal plate placed inside a vessel or tank and used to alter the flow of chemicals, facilitate mixing, or cause turbulent flow. Baffles can be vertical or horizontal. Figure 5-11 shows an example of a baffle.

A **sump** (also called a boot) is the lowest section at the base of a tank. The sump allows materials to be removed from the bottom of the tank and the tank to be completely emptied. Figure 5-12 shows an example of a sump.

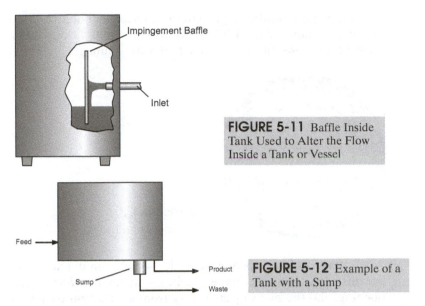

FIGURE 5-11 Baffle Inside Tank Used to Alter the Flow Inside a Tank or Vessel

FIGURE 5-12 Example of a Tank with a Sump

A **foam chamber** (see Figure 5-13) is a reservoir on the side of a tank designed to contain chemical foam that can be used to extinguish fires within the tank.

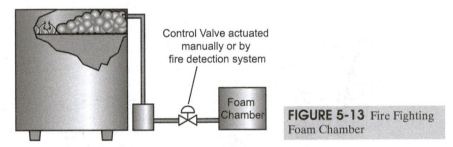

FIGURE 5-13 Fire Fighting Foam Chamber

A **gauge hatch** (see Figure 5-14) is an opening on the roof of a tank that is used to check tank levels and obtain samples of the product or chemical.

FIGURE 5-14 Gauge Hatch

A **level indicator** (see Figure 5-15) is a gauge placed on a vessel to denote the height of the liquid level within the vessel.

FIGURE 5-15 Example of a Sight Glass Level Indicator on a Small Tank

A **lining** is a coating applied to the interior wall of a vessel to prevent corrosion or product contamination. Linings are usually made of polymers, elastomers, or glass. Figure 5-16 shows an example of a tank with a lining.

FIGURE 5-16 Cross Section of a Tank with a Lining

A **manway** is an opening (usually 24 inches in diameter) in a vessel or tank that permits entry for inspection or repair. The number of manways, the size of the manway opening, and the manway style vary from application to application. Most manways have circular flanged openings located on the side of a tank for entry. Figure 5-17 shows an example of a manway.

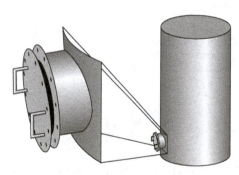

FIGURE 5-17 Example of a Manway with Bolts and Nuts Removed in Preparation for Entry

A **mist eliminator** is a device in a tank designed to collect droplets of mist from gas and is composed of mesh, vanes, or fibers. The droplets coalesce (merge) to form larger droplets until they are too large and heavy to remain on the mesh; at this point, they drop back into the tank. Figure 5-18 shows an example of a mist eliminator.

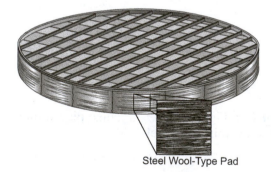

Steel Wool-Type Pad

FIGURE 5-18 Mist Eliminator

A **platform** is a strategically located structure designed to provide access to instrumentation and to allow the performance of maintenance and operational tasks. Platforms are generally accessed by ladders. Their use and construction is regulated by the Occupational Safety and Health Administration (OSHA). Figure 5-19 shows an example of a platform installed on the side of a storage tank.

PRESSURE OR VACUUM RELIEF

Storage tanks are built to hold large volumes of liquid, but they cannot withstand any pressure or vacuum. Unfortunately, many catastrophic tank farm fires and oil spills

Platform

FIGURE 5-19 Platform Attached to a Storage Tank

have been caused by overpressurizing or collapsing a tank. Devices have been developed to prevent tank damage and to prevent these hazards from occurring.

A **pressure relief device** (see Figure 5-20) is a component that discharges or relieves pressure in the event of overpressurization. Pressure relief devices include relief valves, safety valves, and rupture discs. These components open when a set pressure is reached thus prevent bursting and rupture of the tank and piping.

FIGURE 5-20 Pressure Relief Device (Valve) Used to Prevent Overpressurization

Courtesy of Design Assistance Corporation (DAC)

A **vacuum relief device** (see Figure 5-21) is a safety device that prevents the pressure in a tank from dropping below normal atmospheric pressure. This prevents tank implosion (sudden inward collapse).

FIGURE 5-21 Vacuum Relief Device Used as Air Pressure Regulator

Courtesy of Design Assistance Corporation (DAC)

Another type of vacuum relief device is a **tank breather vent** (also referred to as a conservation vent). Breather vents maintain the pressure in the tank at a specific level, thus allowing air flow in and out of the tank. Figure 5-22 shows an example of a tank breather vent.

Tank breather vents are installed on the outer roof of a storage tank (storage tanks that have no internal floating roof). These relief devices conserve tank vapors while protecting against overpressurization and vacuum. Inside a breather vent is a pallet (a guided disk that opens and closes in response to a pressure differential). This disk permits the intake of air or the escape of vent vapors, as needed. This process keeps the tank within safe working pressures. These pallets are weight-loaded to correspond to set pressures for each tank. Tank breather valves have the pressure and vacuum pallets mounted side by side on the tank.

FIGURE 5-22 Breather Vent (Conservation Vent)

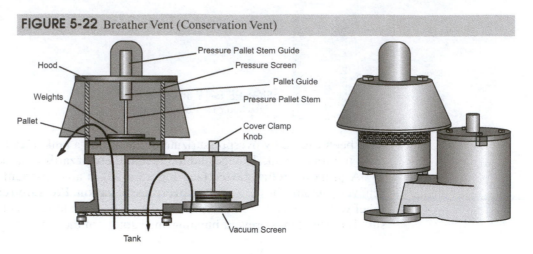

A **vacuum breaker** is a safety device used to remove vacuum by adding pressure to a vessel. This device can help to prevent implosion and backflow.

Auxiliary Equipment

In addition to the components previously mentioned, tanks and vessels also contain additional auxiliary equipment. This equipment includes blanketing, corrosion monitoring, fire protection, heating and cooling, insulation and tracing, secondary containment, and vapor recovery systems.

BLANKETING SYSTEMS

Blanketing is the process of introducing an inert gas into the vapor space above the liquid in a tank to prevent air leakage into the tank. This layer of gas is referred to as a nitrogen blanket. Blanketing a tank reduces the amount of oxygen present and decreases the risk of fire and explosion by preventing air (oxygen) from reacting with tank contents. Nitrogen blankets also reduce the risk of tank implosion (collapse) by decreasing the amount of vacuum created when a tank is emptied. Figure 5-23 shows an example of a tank blanketed with an inert gas.

FIGURE 5-23 Tank Blanketed with Inert Gas

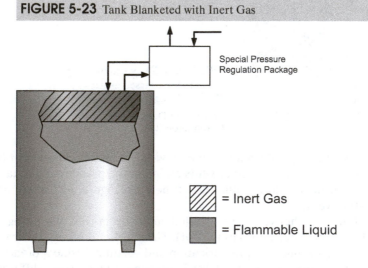

CORROSION MONITORING SYSTEMS

Corrosion in tanks can cause serious problems, so it must be controlled as much as possible. This corrosion control is often handled by corrosion monitoring systems. An example of a corrosion monitoring system is shown in Figure 5-24.

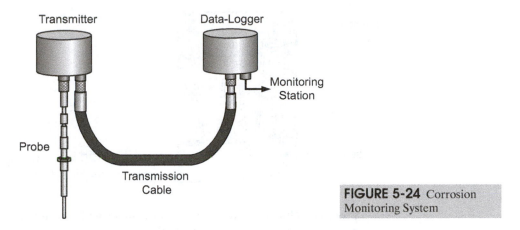

FIGURE 5-24 Corrosion Monitoring System

Typical corrosion monitoring systems consist of corrosion probes, transmitters, communication lines, and receivers. The probes are attached to the vessel and are connected to the transmitters. The transmitters send corrosion data to a receiver via a communication cable. The data can then be downloaded for analysis by authorized testing personnel (individuals who are qualified to test for metal thickness and symptoms of corrosion).

FIRE PROTECTION SYSTEMS

Fire protection can be provided through water sprinkler systems, which keep supports and shells cool, or by chemical foam systems, which extinguish the fire by removing air.

GROUNDING SYSTEMS

Grounding involves connecting an object to the earth using copper wire and a grounding rod to provide a path for the electricity to dissipate harmlessly into the ground. Grounding is installed on vessels and tanks in order to redirect any static electricity that might build up on the surfaces of the container. This prevents the ignition of flammable vapors by sparks that are generated when static electricity is discharged. Grounding is also used on electrical equipment to prevent electrical shock.

HEATING AND COOLING

Heating and cooling can be applied to vessels or tanks by a variety of methods. For example, water may be added to heat or cool the contents of a vessel or tank. This may be accomplished by direct injection of steam or refrigerated water into the vessel or tank. Additionally, heating media (such as steam or hot oil) and cooling media (such as refrigerated water or ammonia) may be circulated through coils or jackets attached to the vessel or tank to adjust the temperature of the contents.

INSULATION AND TRACING

Insulation and tracing (either steam or electrical) are provided to prevent tank contents from freezing or to keep hot lines flowing. Insulation is also used as a safety measure to

---

Did You Know?

Static electricity can ignite gasoline vapors. Static electricity can be generated by:

- Filling a portable gas can in the bed of a pickup truck.
- Getting in and out of your car while the car's gas tank is being filled.

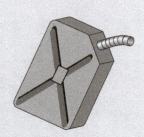

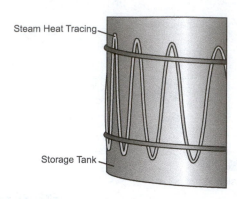

FIGURE 5-25 Steam Tracing System on the Side of a Tank

keep process technicians from coming in contact with hot lines. Figure 5-25 shows an example of heat tracing on a storage tank.

Some pipes contain fluids that will freeze if they are not kept warm. This freezing may or may not have anything to do with the temperature outside. Instead, the freezing may be attributed to the chemical properties of the fluid inside. For example, some fluids freeze at very high temperatures (e.g., some tars freeze at 200°F!). Without adequate heat tracing, these fluids would be unable to pass properly through the piping system.

SECONDARY CONTAINMENT SYSTEMS

A secondary containment system is built around tanks or vessels to contain the vessel's contents in the event of a leak or spill. A containment system, which is usually made from earth or concrete, can take the form of a dike, firewall, or containment wall. A **dike** is a wall (earthen, shell, or concrete) built around a piece of equipment to contain any liquids in the event of an equipment rupture or leak. A **firewall** is an earthen bank or concrete wall built around a storage tank to contain the contents in the event of a spill or rupture. A **containment wall** (see Figure 5-26) is a wall used to protect people and the environment against tank failures, fires, runoff, and spills.

The main purposes of all the various containment systems are to:

- Minimize safety risks by containing chemical spills in a small area and thus trap contaminants before they can spread to other areas
- Protect the soil and the environment from contaminants
- Protect humans from potential hazards (e.g., fire or chemical release)
- Contain wastewater and contaminated rainwater until it can be drained into a proper sewage line
- Facilitate cleanup after a release of hazardous material
- Protect the environment and people against tank failures

In the event of rainfall accumulation within the containment wall, the process technician must open a valve and drain the liquid to prevent the tank from floating. However, water inside the containment wall must be tested for possible pollutants before it can be drained into the process sewer.

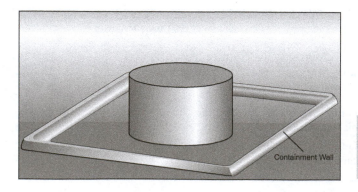

FIGURE 5-26 Containment Wall Surrounding a Tank to Protect Against Tank Failures, Fires, Runoff, and Spills

VAPOR RECOVERY SYSTEM

A **vapor recovery system** is the process used to capture and recover vapors. Vapors are captured by methods such as chilling or scrubbing. The vapors are then purified and either returned to the process, stored, or incinerated.

OTHER COMPONENTS

A **skirt** is the support structure attached to the bottom of freestanding vessels or tanks. Figure 5-27 shows an example of a tank skirt during construction and prior to the tank being placed on the skirt.

FIGURE 5-27 Skirt that Surrounds the Bottom of a Tank

A **vortex breaker** (see Figure 5-28) is a metal plate or similar device that prevents a **vortex** (the cyclonelike rotation of a fluid) from being created as liquid is drawn out of the tank. This breaker prevents gas or vapor from creating the phenomenon known as cavitation, which can damage pumps over time. Vortex breakers help maintain stable levels in tanks and proper suction levels for pumps.

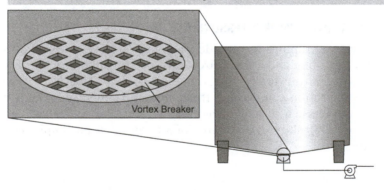

FIGURE 5-28 Vortex Breaker Example on Bottom of a Tank

A **weir** is a flat or notched dam that functions as a barrier to flow. A weir can also be used to separate two substances (e.g., gasoline and water) from a mixture. Figure 5-29 shows an example of a vessel (decanter) with a weir. In this example, anything that is above the weir flows over. Because gasoline is lighter than water, gasoline is the substance that goes over the weir and is separated out.

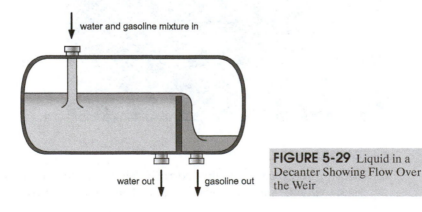

FIGURE 5-29 Liquid in a Decanter Showing Flow Over the Weir

Potential Problems with Tanks and Vessels

As with any piece of process equipment, various problems may arise with tanks and vessels. Potential problems include corrosion, scale buildup, over- and underpressurization, operating upsets, shutdowns, and malfunctions during startup.

CORROSION

Corrosion occurs when metal is deteriorated through a chemical reaction (e.g., iron rusting). Corrosion can occur for a variety of reasons, whether it is from direct contact with chemicals (e.g., acids, bases, and salts), temperature extremes, oxidation, or water exposure. Corrosion can have a catastrophic effect on the integrity of pipelines, vessels, and equipment.

Corrosion causes thinning, pitting or cracking (e.g., chloride stress corrosion) of vessel walls, which can lead to leaks or cause the integrity of the vessel to be compromised. Compromised vessel integrity can result in tank rupture or collapse. Where corrosion is a possibility, an additional metal thickness (corrosion allowance) is calculated into the vessel design.

Metal corrosion must be checked at regular intervals. While vessel wall thickness checks may be made while the vessel is in operation, more accurate thickness measurements are made when a vessel is empty and clean.

SCALE BUILDUP

Scale consists of dissolved solids deposited on the inside surfaces of equipment. Scale normally does not affect vessel integrity, but it may affect the quality or purity of the vessel's contents. Samples should be taken to determine if scaling has occurred and if the scales have altered the purity or quality of the vessel's contents in anyway.

OVER- AND UNDERPRESSURIZATION

Tanks can rupture or explode when they are overpressurized and implode (collapse inward) when underpressurized. Overpressurization can occur if too much material is pumped into a tank and the pressure relief valve fails to work properly. Underpressurization can occur if the tank vent is plugged while a tank is being emptied. This results in a vacuum being created within the vessel, which may result in implosion. Figure 5-30 shows an example of a railcar that has imploded as a result of vacuum pressure.

FIGURE 5-30 Imploded Railcar

(Image is from Department of Energy - http://www.hanford.gov/RL/?page=525&parent=506)

OPERATING UPSETS

Upsets within a process can allow corrosive chemicals to enter equipment that was not designed to handle such chemicals. This can lead to rapid corrosion and vessel damage.

Did You Know?

Special attention should be given when "steaming out" or applying steam to low-pressure tanks.

Rapid cooling from a rain shower can cause the internal pressure to drop and result in a tank collapse.

Changes or deviations (excursions) in operating temperatures to extreme levels can also accelerate corrosion rates and damage vessels by overheating.

SHUTDOWNS AND STARTUPS

Following shutdowns, vessels should be checked to remove any foreign materials that may have been left during vessel maintenance. During vessel closure, gaskets must be properly installed and bolts properly tightened. On vessels that are operated at high temperatures, flange bolts can become loose as the vessel heats up. These bolts must be tightened while the tank is heating up.

Typical Procedures

Startup, shutdown, lockout/tagout (a procedure used in industry to isolate energy sources from a piece of equipment), and emergency procedures vary depending on the specific process facility. Each process technician is required to be familiar with his or her site-specific procedures. Discussed below, however, are a few typical procedure examples.

STARTUP

The following are common steps that must be followed when returning a storage tank to service. These can vary from facility to facility, however, so process technicians should always follow procedures that are specific for their unit.

- Inspect for leaks by hydrotesting the tank.
- Inspect the interior of the tank for cleanliness.
- Check the gauging system.
- Connect swing line cables (if applicable).
- Check to see that all manway heads are in place.
- Remove blinds and check to see that valves are bolted properly.
- Close and seal water draw valves (if applicable).
- Check to be sure that all necessary walkways are in place.
- Open roof drains on external floating roof tanks.
- Check to see that mixers are in place (if applicable).
- Check to see if valves are in alignment with the venting and vacuum protection devices.

The following startup activities may be initiated once the checklist above is completed.

1. Slowly begin product flow into the tank by gradually opening the appropriate tank valve.
2. Observe the tank at the prescribed times listed on the Returning Tank to Service Form to ensure that no leaks are present.

SHUTDOWN

The following are common steps that must be performed during shutdown:

1. Confirm that all of the liquid has been removed from the tank or vessel.
2. Remove all of the harmful vapors from the tank or vessel if personnel will be entering the container (refer to site-specific procedures as required).
3. Refer to site-specific procedures if personnel will not be entering the tank or vessel.
4. Open the manway hatches and place an electrically grounded air mover in one of the hatches to remove harmful vapors if a person is going to enter the tank or vessel.
5. Monitor the lower explosive limit (LEL), the minimum percentage of fuel required for combustion, and do not let any personnel inside the tank or vessel until an acceptable LEL reading is obtained.
6. No one should enter the tank or vessel until all the above have been completed and a trained and certified attendant is present.

EMERGENCY

Each process facility has specific emergency procedures. Each process technician is responsible for knowing and being able to execute the emergency procedures for his or her operating unit.

Hazards Associated with Improper Operation

The improper operation of a tank or a vessel can create unsafe or hazardous situations. Table 5-1 lists some of these scenarios.

TABLE 5-1   Hazards Associated with Improper Tank and Vessel Operation

Improper Operation	Possible Effects			
	Individual	*Equipment*	*Production*	*Environment*
Overfilling	Exposure to hazardous chemicals; possible injury	Damage to tank and equipment, especially floating roof tanks	Added cost for cleanup; lost product or raw material	Spill, possible fire, vapor release
Putting wrong or off-spec material in the storage tank	Potential exposure to hazardous chemicals; possible injury	Potential damage to equipment due to undesirable chemical reactions from incompatible chemical mixtures	Added cost to remove material, clean tank, and rerun material	Possible spills when removing material and cleaning tank, or unwanted chemical reactions
Misalignment of blanket system	Potential exposure to hazardous chemicals; possible injury	Loss of blanket; collapse of tank due to vacuum	Loss of production due to reduced storage	Possible vapor release
Misalignment of pump systems	Potential exposure to hazardous chemicals; possible injury	Damaged pump	Contaminate other tanks	Possible spill
Pulling a vacuum on a tank while emptying	Exposure to hazardous chemicals; possible injury	Collapse of tank due to vacuum	Loss of production due to reduced storage	Possible vapor release
Overpressurization	Exposure to hazardous chemicals; possible injury	Possible rupture of vessel	Loss of production due to reduced storage	Possible vapor release

Process Technician's Role in Operation and Maintenance

MONITORING AND MAINTENANCE ACTIVITIES

Process technicians are often responsible for the routine monitoring of tanks and vessels. These activities are usually performed during scheduled rounds when a technician is specifically performing tank and vessel equipment checks. However, many of the activities listed below should become part of a technician's regular routine when traveling throughout the facility to perform other tasks. Technicians must always remember to look, listen, feel, and smell for the items listed in Table 5-2. Failure to perform proper monitoring can affect the process and result in equipment damage or an environmental incident.

TABLE 5-2   Process Technician's Role in Operation and Maintenance

Look	*Listen*	*Feel*	*Smell*
• Monitor levels • Check firewalls, sumps, and drains • Check auxiliary equipment associated with the tank • Check to ensure that the drain remains closed • Visually inspect for leaks (especially if associated with abnormal odor) • Check sewer valves • Use level gauges and sight glasses to monitor level • Monitor leakage, level, and pressure • Inspect for corrosion	• Listen for abnormal noise	• Inspect for abnormal heat on vessels and piping • Check for excessive vibration on pumps and mixers	• Be aware of abnormal odors that can indicate leakage • Use sniffers to detect gas leaks and vapors

SPECIAL PROCEDURES

Tanks and vessels must be taken out of service to prepare for turnarounds or large-scale maintenance projects. To prepare a vessel for maintenance, the vessel must be isolated from all types of hazardous energy, including electrical, mechanical, hydraulic, pneumatic, chemical, or thermal. Process technicians are often involved in the isolation of vessels and use their company's lockout/tagout procedures to perform this isolation.

Once vessels are removed from service and isolated, vessel entry is often necessary to perform various maintenance tasks or repairs. Process technicians should follow their company's confined-space entry procedures to safely enter and work inside tanks and vessels. Process technicians should realize the hazards of confined-space entry. For example, looking inside a manway on a tank may not appear to be very hazardous, but it could be deadly if it's in a low-oxygen environment.

Vessel and Reactor Symbols

To locate tanks and vessels accurately on a Piping and Instrumentation Diagram (P&ID), process technicians should be familiar with the symbols that represent the different types of tanks and vessels.

Figure 5-31 shows some examples of tank and vessel symbols. It is important to note, however, that the appearance of these symbols may vary from facility to facility.

FIGURE 5-31 Tank and Vessel Symbols

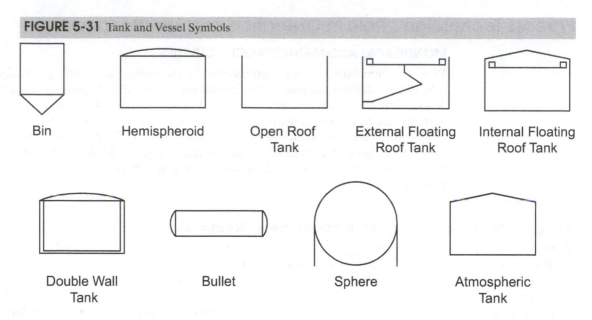

Bin Hemispheroid Open Roof Tank External Floating Roof Tank Internal Floating Roof Tank

Double Wall Tank Bullet Sphere Atmospheric Tank

Summary

Vessels are used to store raw materials and additives, intermediate products, final products, and waste materials. Underground storage tanks are used by the process industries to store materials beneath the surface of the soil. Tank trucks, tank cars, and hopper cars move products on land, while barges and ships transport products across water.

Process technicians must recognize and understand vessel components, including floating roofs, articulated drains, foam chambers, sumps, mixers, manways, vapor recovery systems, vortex breakers, baffles, weirs, boots, mist eliminators, and vane separators.

Most vessels have a variety of auxiliary systems (e.g., fire protection and secondary containment) incorporated into their designs to enhance safe operation and prevent environmental incidents such as leaks and spills. Process technicians must always remember to look, listen, feel, and smell when monitoring and maintaining vessels and must also be aware of the hazards of improper vessel operations.

Checking Your Knowledge

1. Define the following terms:
   a. Articulated drain
   b. Baffle
   c. Blanket
   d. Boot
   e. Containment wall
   f. Floating roof
   g. Gauge hatch
   h. Manway
   i. Mist eliminator
   j. Mixer
   k. Sump
   l. Vortex breaker
   m. Weir

2. *(True or False)* Tanks are good for storing substances with toxic vapors.

3. Which of the following vessels would be the most appropriate choice for storing volatile substances under extreme pressure?
   a. Spherical
   b. Bullet
   c. Hemispheroid

4. *(True or False)* Underground storage tanks are built to structurally withstand the pressure of the soil around the tank.

5. Which of the following types of storage containers are used to transport liquid across roadways?
   a. Tank car
   b. Tank truck
   c. Container ship
   d. Tanker

6. A _____ is a metal plate or similar device placed inside a cylindrical or cone-shaped vessel that prevents a cyclonelike rotation from being created as the fluid is drawn out of the tank.
    a. sump
    b. manhole
    c. vane separator
    d. vortex breaker

7. A _____ is commonly referred to as a dam, barrier, or baffle normally used for the measurement of fluid flows or to maintain a given depth of fluid.
    a. decanter
    b. sump system
    c. weir
    d. tank breather valve

8. Air is prevented from leaking into a tank by a _____.
    a. tank breather vent
    b. level valve
    c. pressure relief device
    d. blanketing system

9. *(True or False)* Heating and cooling systems on tanks or vessels are used to cool or warm the contents inside a tank.

10. List three hazards that dikes and containment walls protect against.

11. List at least three details that a process technician should look, listen, feel, and smell for during normal tank and vessel monitoring and maintenance.

12. List at least three hazards and/or effects associated with improper tank operations, and explain the effects these hazards might have on individuals, equipment, production, and the environment.

Student Activities

1. On the cross-section of a storage tank below, identify the labeled components.

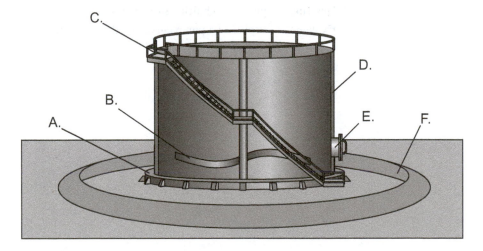

2. Research and write a one- to two-page paper on a tank implosion or explosion.

3. Given a piping and instrumentation diagram (P&ID), identify all of the tanks and vessels.

Pumps

Objectives

After completing this chapter, you will be able to:

- Explain the purpose of pumps in the process industry.
- Identify common pump types and describe the operating principles of each.
- Identify pump components and explain the purpose of each.
- Describe the operating principles of pumps.
- Identify potential problems associated with pumps.
- Describe safety and environmental hazards associated with pumps.
- Identify typical procedures associated with pumps.
- Describe the process technician's role in pump operation and maintenance.
- Explain the purpose of a pump curve and demonstrate its use.

Key Terms

Axial pump—a dynamic pump that uses a propeller or row of blades to propel liquids along the shaft.

Canned pump—a seal-less pump that ensures zero emissions. It is often used on EPA-regulated liquids.

Cavitation—a condition inside a pump wherein the liquid being pumped partly vaporizes due to variables such as temperature and pressure drop.

Centrifugal force—the force that causes something to move outward from the center of rotation.

Centrifugal pump—a pump that uses an impeller on a rotating shaft to generate pressure and move liquids.

Dead head—the maximum pressure (head) that occurs at zero flow.

Diaphragm pump—a mechanically or air-driven pump that consists of two flexible diaphragms connected by a common shaft.

Discharge check valve—prevents liquid from back-flowing into the pump during the suction stroke.

Double-acting piston pump—a type of piston pump that pumps by reciprocating motion on every stroke (one on one side of the piston, and one on the other side).

Dynamic head—the amount of "push" a pump must have in order to overcome the pressure of the liquid.

Dynamic pump—converts the spinning motion of a blade or impeller into dynamic pressure to move liquids.

Gear pump—a pump that rotates two gears with teeth in opposing directions, allowing the liquid to enter the space between the teeth of each gear in order to move the liquid around the casing to the outlet.

Head pressure—the pressure of a liquid exerted on a system; the amount of head pressure is determined by the height of the liquid.

Impeller—a vaned device that rapidly spins a liquid in order to generate centrifugal force.

Liquid head—the pressure developed from the pumped liquid passing through the volute.

Lobe pump—consists of two or three lobes; liquid is trapped between the rotating lobes and is subsequently moved through the pump.

Magnetic (mag) drive pump—use magnetic fields to transmit torque to an impeller.

Mechanical seals—seals that typically contain two flat faces (one that rotates, and one that is stationary), that are in constant contact with one another in order to prevent leaks.

Multistage centrifugal pump—uses two or more impellers on a single shaft. These are generally used in high-volume, high-pressure applications, such as boiler feed water pumps.

Net positive suction head (NPSH)—the liquid pressure that exists at the suction end of a pump. If the NPSH is insufficient, the pump will cavitate.

Piston pump—a type of reciprocating pump that is driven by either a motor or direct-acting steam. Piston pumps use cylinders mounted on bearings within a casing.

Plunger pump—displaces liquid using a piston and is generally used for moving water and pulp. A plunger pump maintains a constant speed and torque.

Positive displacement pump—a pump that uses pistons, diaphragms, gears, or screws to deliver a constant volume with each stroke.

Priming—the process of filling the suction of a pump with liquid to remove any vapors that might be present.

Pump—a mechanical device that transfers energy to move materials through piping systems.

Pump curve—a specification that describes the capacity, speed, horsepower, and head needed for correct pump operations.

Reciprocating pump—a positive displacement pump that uses the inward stroke of a piston or diaphragm to draw liquid into a chamber, followed by a subsequent outward stroke to positively displace that liquid.

Rotary pump—a positive displacement pump that moves liquids by rotating a screw or a set of lobes, gears, or vanes.

Screw pump—a rotary pump that displaces liquid with a screw. The pump is designed for use with a variety of liquids and viscosities and is designed to accommodate a wide range of pressures and flows.

Seal—a device that holds lubricants and process fluids in place while keeping out foreign materials.

Seal flush—a small flow (slip stream) of pump discharge or externally supplied liquid that is routed to the pump's mechanical seal. This acts as a barrier liquid between the two faces of the seal to reduce friction and remove heat.

Single-acting piston pump—a type of piston pump that pumps by reciprocating motion on every other stroke.

Static head—the amount of pressure the liquid is placing on the pump.

Stuffing box seals—the seals around a moving shaft or stem that contains packing material designed to prevent the escape of process liquids.

Suction check valve—prevents liquid from back-flowing during the discharge pump stroke.

Suction head—the pressure required to force liquids into a pump.

Vane pump—a rotary pump having either flexible or rigid vanes designed to displace liquid. Vane pumps are used for low-viscosity liquids that may run dry for short periods without causing damage.

Vanes—raised ribs on the impeller of a centrifugal pump designed to accelerate a liquid during impeller rotation.

Venturi—a device consisting of a converging section, a throat, and a diverging section.

Viscosity—the degree to which a liquid resists flow under applied force (e.g., molasses has a higher viscosity than water at the same temperature).

Volute—a widened spiral casing in the discharge section of a centrifugal pump designed to convert liquid speed to pressure without shock.

Wear ring—a ring that allows the impeller and casing suction head to seal tightly together without wearing each other out. Wear rings are close-running, non-contacting replaceable pressure breakdown devices located between the impeller and casing of a centrifugal pump.

Introduction

Pumps are mechanical devices that transfer energy to move materials through piping systems. Pumps are used in many applications. including filling or emptying tanks, wells, pits, and trenches; supplying water to boilers; providing circulation for systems; supplying fire control water; lubricating equipment; and drawing samples.

The two main categories of pumps are dynamic and positive displacement. Dynamic pumps use impellers to generate centrifugal force, which is then converted to dynamic pressure to move liquids. Positive displacement pumps use pistons, lobes, gears, or vanes to move or push liquids. Dynamic pumps tend to be used more often than positive displacement pumps because they are less expensive, easier to operate, and require less space and maintenance.

When working with pumps, process technicians should always conduct monitoring and maintenance activities to ensure that pumps are sealed; properly lubricated; and not overpressurizing, overheating, leaking, or cavitating.

Selection of Pumps

To select the appropriate pump for a job, several factors must be considered. One factor is the viscosity of the liquid. **Viscosity** is the degree to which a liquid resists flow under applied force. Some liquids are less viscous, so they flow more easily than others (i.e., they are less viscous). The pump must be able to move the liquid at its current viscosity.

A selected pump must be able to meet the required suction head, discharge head, and flow volume. These design factors include how much suction head is provided to the pump, how much discharge head the pump is required to supply, and how much flow the pump is required to provide.

Other factors that determine pump selection include the required output of the pump, the speed (which affects capacity), the altitude (which affects head pressure), and the temperature (which determines the construction material of the pump). If a pump is required for slurry service, it must also be able to handle suspended solids.

Another factor affecting pump selection is specific gravity (density). Specific gravity affects the horsepower required by the pump.

Finally, the vapor pressure of the liquid affects the **net positive suction head (NPSH)** or the liquid pressure that exists at the suction end of a pump. NPSH is the minimum pressure needed at the pump suction for the pump to perform correctly. This can affect the pump being used.

Types of Pumps

The two main categories of pumps are dynamic and positive displacement. Within each of these categories are subcategories. Figure 6-1 shows a diagram of the different types of pumps and their subcategories.

Dynamic pumps, which convert the spinning motion of a blade or impeller into dynamic pressure in order to move liquids, are classified as either centrifugal or axial.

Positive displacement pumps, which use pistons, diaphragms, gears, or screws to deliver a constant volume with each stroke, are classified as either reciprocating or rotary. Unlike dynamic pumps, positive displacement pumps deliver the same amount of liquid, regardless of the discharge pressure.

FIGURE 6-1 Pump Family Tree

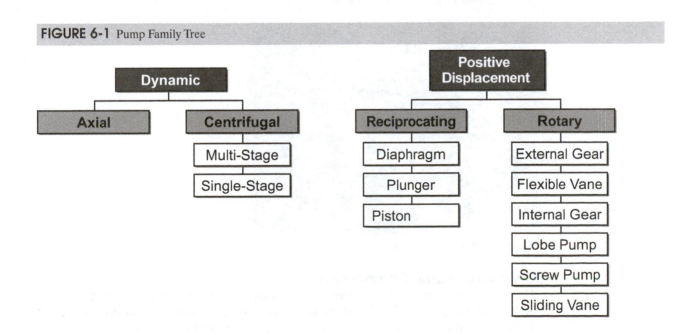

Dynamic Pumps

Dynamic pumps are nonpositive displacement pumps that convert centrifugal force to dynamic pressure to move liquids (as opposed to positive displacement pumps, which use a piston to push liquids). They are classified as either centrifugal or axial.

CENTRIFUGAL PUMPS

Centrifugal pumps use an impeller on a rotating shaft to generate pressure and move liquids. In a centrifugal pump, an impeller spins, creating **centrifugal force** (the force that causes something to move outward from the center of rotation). This force moves the liquid to the outer casing within the pump and then out through the discharge shaft. Pressure is created in the liquid as it passes through a widening in the spiral casing in the discharge section known as a **volute**. Figure 6-2 and Figure 6-3 show the outside and cutaway views of a centrifugal pump.

Centrifugal pumps are used to move large volumes of liquid with a low velocity. Centrifugal pumps differ from positive displacement pumps because the amount of liquid they deliver depends on the discharge pressure, not the size of the chamber.

FIGURE 6-2 Picture of the Outside of a Centrifugal Pump Showing Complete System

Courtesy of Brazosport College

FIGURE 6-3 Cutaway of a Centrifugal Pump

Courtesy of Brazosport College

Centrifugal Pump Components

The main components of a centrifugal pump include the suction, inlet (suction eye), outlet (discharge), impeller, bearings and seals, shaft, and casing or housing. Figure 6-4 illustrates and identifies the various components of a centrifugal pump.

FIGURE 6-4 Centrifugal Pump Components

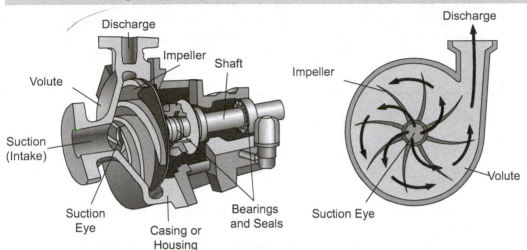

In a centrifugal pump, liquid enters the suction and flows through the suction eye to a spinning impeller driven by a rotating shaft. An **impeller** is a vaned device that rapidly spins a liquid in order to generate the centrifugal force necessary to move the liquid. The **vanes** on an impeller are raised ribs designed to accelerate a liquid during impeller rotation. This rotation results in centrifugal force that causes the liquid to spin outward toward the outer edge of the impeller. Figure 6-5 shows an example of an impeller. Figure 6-6 shows an actual cutaway of a pump impeller.

As the liquid is forced through the pump, pressure is created as the liquid passes through a widening of the casing known as a volute (a spiral casing in a centrifugal pump that is designed so that speed will be converted to pressure without shock). The pressure that is developed is called **liquid head**. The amount of liquid head developed is a function of the tip speed of the impeller (impeller revolutions per minute and diameter). The higher the tip speed, the higher the liquid head.

The amount of head developed by a centrifugal pump is directly related to the impeller diameter and revolutions per minute (rpm). This maximum head occurs at

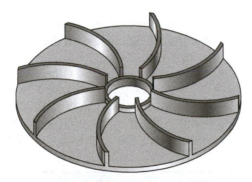

FIGURE 6-5 Impeller

FIGURE 6-6 Pump Impeller and Seals Cutaway

Courtesy of Brazosport College

IMPORTANT NOTE: Process technicians should never dead head a pump! Dead heading a pump causes friction in the liquid that is converted to heat. This heat can cause vapor lock or cavitations that can ruin the pump, so always follow proper procedures and make sure the discharge valve is open before starting a pump.

zero flow and is called the **dead head**. The head developed by a centrifugal pump does not increase above the dead head even if the discharge valve is closed.

Figure 6-7 shows the path of the liquid as it enters a pump, leaves the impeller, and approaches the vane tip velocity. The following explains what is happening at each step in Figure 6-7:

- Step 1: The liquid enters into the suction and is picked up by the motion of the impeller.
- Step 2: As the liquid moves at high speed from the close clearance area (impeller to pump case) to the wider clearance area of the volute, the velocity energy in the liquid converts to pressure energy as the liquid slows.
- Step 3: The high-pressure liquid approaches the discharge nozzle. In the discharge nozzle, the vortex (the cone formed by a swirling liquid or gas) is broken.

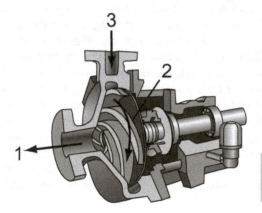

FIGURE 6-7 Vertical Centrifugal Pump Showing Flow Through the Pump

Some centrifugal pumps, including most multistage centrifugal pumps, utilize stationary diffuser vanes installed within the casing (instead of the volute) to slow the liquid's velocity and convert the velocity to pressure.

When working with pumps, it is important to know that pumps cannot remove gas (air) from the suction line, so a pump must be primed before it is started. **Priming** a pump involves filling the suction of the pump with liquid to remove any vapors that might be present. This reduces the likelihood of the pump cavitating, becoming vapor-bound, or losing suction. Pump priming is discussed in more detail later in this chapter.

Did You Know?

When a pump cavitates, it sounds as if rocks are being poured into the suction line.

These bursts of liquid and gas create a "liquid hammer" that can seriously damage the inside of the pump and the impeller.

Pumps that have been damaged by cavitation frequently appear as though the internal workings have been sandblasted or pitted, particularly on the outer edges of the vanes.

Single-Stage Versus Multistage Pumps

Centrifugal pumps may be single-stage or multistage (a stage is one pressure level increase). In a single-stage pump, the liquid enters the pump, the pressure is increased one time, and then the liquid exits the pump. In a multistage pump, the liquid enters the pump and the pressure is increased multiple times (e.g., two, three, or four times) before it exits the pump.

Single-stage centrifugal pumps (shown in Figure 6-8) consist of a disk-shaped impeller mounted on a shaft and fitted into a case. The impeller's role is to impart kinetic energy (the energy associated with mass in motion) and velocity (speed) to the pumped liquid. An everyday example of a single-stage centrifugal pump is the cooling water pump in an automobile engine.

FIGURE 6-8 Dissectible Centrifugal Pump

Courtesy of Design Assistance Corporation (DAC)

A **multistage centrifugal pump** (shown in Figure 6-9) is used in high-volume, high-pressure applications, such as boiler-feed water pumps. Multistage pumps contain two or more impellers on a single shaft. The discharge of one impeller enters into the suction of the next to increase the pressure of the liquid in sequential steps, so a multistage pump is like having several pumps in one.

FIGURE 6-9 Multistage Centrifugal Pump

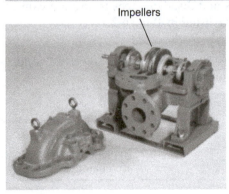

Courtesy of Design Assistance Corporation (DAC)

There are advantages and disadvantages to using a multistage centrifugal pump. Those advantages and disadvantages are listed in Table 6-1.

Specialty Centrifugal Pumps

In addition to the centrifugal pumps already mentioned are other types of specialty pumps, including canned, magnetic drive, high-speed centrifugal, and jet pumps.

TABLE 6-1 Advantages and Disadvantages of Using a Multistage Centrifugal Pump

Advantages	*Disadvantages*
• Effective	• High initial cost
• Requires a small amount of space	• Parts can be expensive
• Minimal maintenance	• Proper installation and operation are critical
	• More complex bearing coolers, seal flush systems, and operating guidelines
	• Increase in operating problems due to the complexity

Canned Pump   **Canned pumps** are seal-less pumps that ensure zero emissions. These pumps are used with liquids that are difficult to seal, such as liquefied gases and liquids regulated by the Environmental Protection Agency (EPA). Some states also require the use of seal-less pumps for certain processes. Canned pumps can provide zero emissions because there are no seals to leak to the atmosphere. Figure 6-10 shows an example of a canned pump.

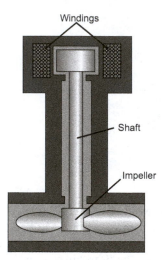

Windings

Shaft

Impeller

FIGURE 6-10 Canned Pump

In a canned pump, the rotating part of the electric motor attaches to the pump's shaft in an enclosure similar to a can. Windings surround the can and induce a magnetic field that causes the entire sealed assembly to rotate when electrical current is applied.

There are advantages and disadvantages to using a canned pump. Those advantages and disadvantages are listed in Table 6-2.

TABLE 6-2 Advantages and Disadvantages of Using a Canned Pump

Advantages	*Disadvantages*
• Seals are not required because everything is contained within the pump	• High cost
	• Pumped liquid must be compatible with motor components
	• Typically cheaper to replace the pump than to repair it

Magnetic Drive Pumps   **Magnetic (mag) drive pumps** use magnetic fields to transmit torque to an impeller. Like canned pumps, magnetic drive pumps are used with liquids that are difficult to seal, such as liquefied gases and liquids regulated by the EPA. Figure 6-11 shows an example of a magnetic (mag) drive pump.

In a magnetic drive pump, a magnet magnetically couples the pump to the driver to eliminate shaft leakage and the need for mechanical seals. A magnetic force passing through a stainless-steel canister drives the inner coupling (the inner coupling includes the magnets on the drive shaft and the magnets on the pump shaft). These magnets are

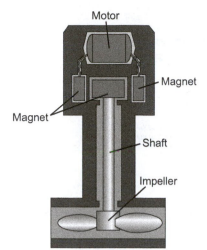

FIGURE 6-11 Magnetic (Mag) Drive Pump

FIGURE 6-12 Pump Shaft Coupling Used to Connect Shafts for Power Transmission

Courtesy of Brazosport College

separated by a shroud that keeps the process contained inside the pump. Figure 6-12 is an example of a pump shaft coupling.

There are advantages and disadvantages to using a magnetic drive pump. Those advantages and disadvantages are listed in Table 6-3.

TABLE 6-3   Advantages and Disadvantages of Using a Magnetic Drive Pump

Advantages	*Disadvantages*
• Elimination of costs associated with mechanical seal repair • Elimination of expensive monitoring equipment for EPA-regulated liquids	• Expensive to purchase • Subject to slippage (magnets failing to rotate at the same speed) • Must be "bump-started," i.e., started and stopped quickly to ensure the magnets on the drive shaft line up with the magnets on the pump shaft

High-Speed Centrifugal Pumps   High-speed centrifugal pumps are used to generate high pressures. Because the pressure generated by a centrifugal pump is proportional to the square of the impeller rotations per minute (rpm), a tenfold increase in impeller rpm results in a 100-fold increase in discharge pressure. Impeller speeds of 20,000 to 30,000 rpm can be achieved using a gearbox coupled to the drive shaft. This results in relatively high discharge pressures with a relatively small pump.

Jet Pumps   Jet pumps are used to lift liquids from wells. A jet pump consists of a centrifugal pump that recycles up to three-fourths of its discharge into a **Venturi** (a device consisting of a converging section, a throat, and a diverging section) installed in its suction line. Figure 6-13 shows an example of a Venturi pump.

A Venturi can be inserted deep into a well. By introducing a restriction in the line, the Venturi increases the velocity of the recycled liquid, causing a drop in its pressure

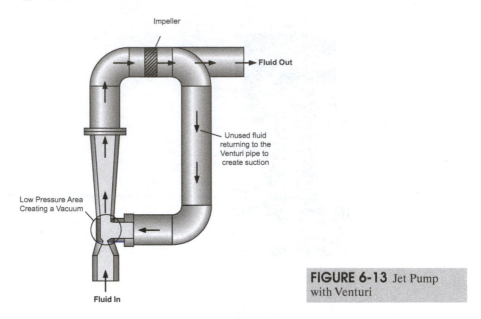

FIGURE 6-13 Jet Pump with Venturi

(Bernoulli's Principle). This adds motive power to lift liquid from the well into the pump's suction.

AXIAL PUMPS

Axial pumps are dynamic pumps that use a propeller or row of blades to propel liquids along a shaft (as opposed to centrifugal pumps, which use an impeller to force liquids to the outer wall of a chamber). Figure 6-14 shows the main components of an axial pump.

FIGURE 6-14 Axial Pump Showing Labels

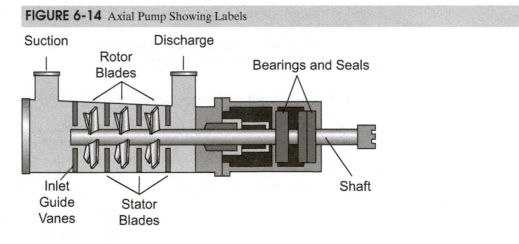

Positive Displacement Pumps

Positive displacement pumps (shown in Figure 6-15) use pistons, diaphragms, gears, or screws to deliver a constant volume with each stroke. Unlike dynamic pumps, positive displacement pumps deliver the same amount of liquid, regardless of the discharge pressure.

Positive displacement pumps move liquid by trapping it in the space between the pumping elements. As the pump's motion reduces the size of the space, the liquid is forced out. Positive displacement pumps develop pressure by displacement of volume. The pressure that is developed is not a function of pump speed. There is essentially no theoretical limit to the amount of pressure developed by a positive displacement pump. For this reason, a positive displacement pump must never be dead headed or blocked in while running. (*Note:* Dead heading and blocking in are discussed in more detail later in this chapter.)

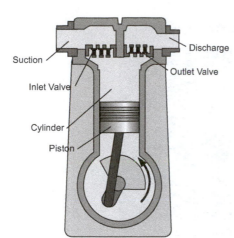

FIGURE 6-15 Positive Displacement Reciprocating Pump Showing Flow in and Out of the Cylinder

The flow rate of a reciprocating positive displacement pump is equal to the piston size, times the stroke length, times the stroke rate. By varying the stroke rate, the flow rate can be precisely controlled. Because of this, these pumps are frequently used as metering pumps (precision pumps that deliver an exact amount of liquid with each stroke or revolution of the handle).

The two types of positive displacement pumps are rotary and reciprocating. Rotary pumps displace the liquid at a constant rate by using two elements that rotate in contact. Reciprocating pumps use a piston or plunger that moves back and forth to displace the liquid.

Because positive displacement pumps displace volume to create pressure, they are self-priming (i.e., they can remove gas or air from the suction line). This makes them very useful in many applications where a centrifugal pump would not work (e.g., removing waste from a drum).

ROTARY PUMPS (POSITIVE DISPLACEMENT)

Rotary pumps are positive displacement pumps that move liquids by rotating a screw or a set of lobes, gears, or vanes. As these screws, lobes, gears, or vanes rotate, the liquid is drawn into the pump (intake) by lower pressure on one side and forced out of the pump (discharged) through higher pressure on the other side.

Gear Pumps

Gear pumps contain two rotating gears with teeth in opposing directions, allowing the liquid to enter the space between the teeth of each gear. As the teeth rotate, the liquid is moved around the casing to the outlet. The advantages and disadvantages to using a gear pump are listed in Table 6-4. Figure 6-16 and Figure 6-17 show cutaways of gear pumps.

TABLE 6-4   Advantages and Disadvantages of Using a Gear Pump

Advantages	*Disadvantages*
• Suitable for medium pressures	• Not suitable for solids
• Quiet operation	
• Accommodates a variety of materials	
• Used at high speeds	

Lobe Pumps

Lobe pumps consist of two or three lobes that are larger than, but operate similarly to, the gear teeth on a gear pump. The lobes are fitted with vanes to increase the seal of the pump. As the lobes rotate, liquid is trapped between the lobes and the wall of the pump.

FIGURE 6-16 Cutaway Picture of a Gear Pump

Courtesy of Bayport Training & Technical Center

FIGURE 6-17 Picture of a Gear Pump Showing Gear Wheels

Courtesy of Bayport Training & Technical Center

This carries the liquid from one side of the pump to the other. Figure 6-18 provides an internal view of a lobe pump, and Figure 6-19 shows a drawing of a lobe pump with labels.

There are advantages and disadvantages to using a lobe pump. Those advantages and disadvantages are listed in Table 6-5.

FIGURE 6-18 Internal Picture of a Lobe Pump

Courtesy of Baton Rouge Community College

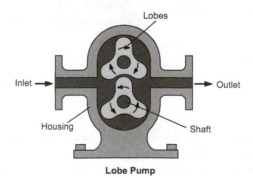

FIGURE 6-19 Drawing of Lobe Pump with Components Labeled

TABLE 6-5 Advantages and Disadvantages of Using a Lobe Pump

Advantages	*Disadvantages*
• Can handle slurries with large particle size	• Not suitable for low-viscosity liquids
	• Inferior loading characteristics
	• Low suction ability
	• May require factory service for repair

Vane Pumps

Vane pumps are rotary pumps that have flexible or rigid vanes that are designed to displace liquids. Vane pumps are used for low-viscosity liquids, and they can be run dry for short periods of time without damage. Vane pumps can develop a good vacuum, and the vanes can be replaced or reversed when they become worn. Everyday examples of vane pumps are the power-steering and automatic-transmission pumps of many automobiles. Figure 6-20 shows a cutaway of a sliding vane pump, and Figure 6-21 displays a drawing of a sliding vane pump with the vanes and the direction of flow labeled.

FIGURE 6-20 Rotary Vane Pump Cutaway

Courtesy of Bayport Training & Technical Center

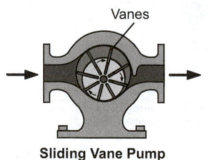

Vanes

Sliding Vane Pump

FIGURE 6-21 Sliding Vane Pump with Vane Labeled and Direction of Flow Identified

Did You Know?

The screw pump is one of the oldest pump designs. It was created by Archimedes around 200 B.C., and it is still in use today.

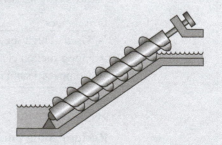

Many coastal cities still use large screw-type pumps and spillways to prevent flooding during extreme weather.

Vane pumps consist of a cylindrical rotor that is eccentrically mounted (i.e., it has an axis or point of support that is not centrally placed) in a cylindrical case. Because the rotor and case are eccentric, there is a crescent-shaped void between the rotor and the case. Small plates, or vanes, are set into slots in the rotor. Centrifugal force and/or springs, hydraulic pressure, or other forces cause the vanes to slide in and out of the rotor as it turns, thereby maintaining a tight seal. As the rotor turns, liquid enters the case at one narrow end of the crescent-shaped void and is swept along by the vanes and out the discharge at the other narrow end of the crescent.

Screw Pumps

Screw pumps are rotary pumps that displace liquid with a screw. This screw design accommodates a wide range of liquids with varying viscosities, pressures, and flows. Screw pumps, such as those illustrated in Figures 6-22 and 6-23, consist of a casing with a rotating screw impeller containing helical threads. The liquid enters the pump, is trapped between the case and the helical threads, and is then pushed out the discharge.

FIGURE 6-22 Screw Pump Cutaway

Courtesy of Bayport Training & Technical Center

FIGURE 6-23 Screw Pump

Courtesy of Baton Rouge Community College

There are advantages and disadvantages to using a screw pump. Those advantages and disadvantages are listed in Table 6-6.

TABLE 6-6 Advantages and Disadvantages of Using a Screw Pump

Advantages	*Disadvantages*
• Used at high speeds	• Expensive
• Good suction characteristics	• Low level of efficiency
• Compact design that creates low vibration	

RECIPROCATING PUMPS (POSITIVE DISPLACEMENT)

Reciprocating pumps are positive displacement pumps that use the inward stroke of a piston or diaphragm to draw liquid into a chamber, followed by a subsequent outward stroke to positively displace that liquid. Figure 6-24 shows how a reciprocating pump operates and how liquid is drawn in and discharged.

Most reciprocating pumps consist of a liquid end and a drive end. The liquid end contains a device that displaces a fixed liquid volume (volume to the pump suction) for each stroke from the drive end. The suction and discharge flows typically are directed by the positioning of check valves.

FIGURE 6-24 Piston-Type Reciprocating Pump Showing Rotation in the Cylinder

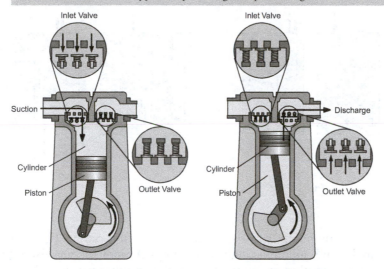

Check valves open and close to prevent a backflow of liquid. As the stroke of the pump draws liquid from the suction (suction stroke), the check valve allows the suction to fill while the **discharge check valve** closes. This prevents backflow. As the stroke transfers liquid to the discharge (discharge stroke), the discharge check valve opens to allow the discharge to flow through while the **suction check valve** closes.

To better illustrate the actions of a reciprocating pump, think of a syringe. As a syringe plunger is pulled out of its housing, liquid is drawn in (intake). As the plunger is pushed back into the housing, liquid is forced out (discharged).

Piston Pumps

Piston pumps, like the one shown in Figure 6-24, are reciprocating pumps that are driven by either a motor or direct-acting steam. Piston pumps use cylinders mounted on bearings within a casing to move liquids. Motor-driven pumps drive the piston by a crankshaft to convert the rotary motion of the driver to the back-and-forth motion of the piston. Steam-driven pumps drive the piston directly (i.e., the steam driver's piston moves the pump's piston). Discharge volume is equal to the volume displaced by the piston, less slippage past the piston packing rings. Simplex and multiplex designs are available. Direct-acting steam pumps allow flexible pressure and flow by throttling the steam inlet (the point at which the steam enters the driver).

A simplex design contains a single piston for the suction and discharge. A multiplex design has multiple pistons for suction and discharge (e.g., duplex pumps have two pistons, triplex have three, and quadraplex have four). Figure 6-25 shows examples of simplex and multiplex pumps.

FIGURE 6-25 Simplex and Multiplex Pump Designs

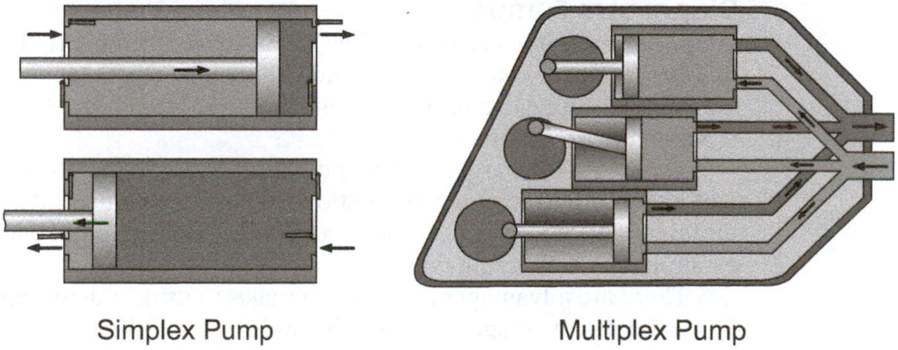

Simplex Pump

Multiplex Pump

FIGURE 6-26 Comparison of Single- and Double-Acting Piston Pumps

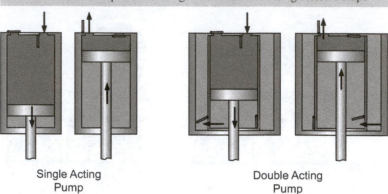

Single Acting
Pump

Double Acting
Pump

Piston pumps can also be single- or double-acting. Figure 6-26 shows examples of single- and double-acting piston pumps. A **single-acting piston pump** pumps by reciprocating motion on every other stroke. In other words, it fills the cylinder only when the piston moves in one direction (called the suction stroke), and forces the liquid out of the cylinder as the piston moves in the other direction (called the discharge stroke).

A **double-acting piston pump** pumps by reciprocating motion on every stroke (one on one side of the piston, and one on the other side). In other words, as one end is filling, the other end is discharging.

Plunger Pumps

Plunger pumps (shown in Figure 6-27) displace liquid using a piston and are generally used for moving water and pulp. Plunger pumps maintain a constant speed and torque. There is no change in the capacity of the pump when it is maintained at the prescribed setting. The plunger is driven through a crankshaft either by a motor or by steam. The discharge per stroke is equal to the volume displaced by the plunger.

FIGURE 6-27 Positive Displacement Plunger Pump

Courtesy of Design Assistance Corporation (DAC)

Diaphragm Pumps

Diaphragm pumps are mechanical or air-driven pumps that consist of two flexible diaphragms connected by a common shaft. The diaphragms, which are forced by a mechanical linkage, compressed air, or some other liquid, move in a reciprocating (back-and-forth) motion.

The pump discharges with one diaphragm, while the other diaphragm exhausts air and allows liquid to enter the suction side of the chamber. Adjusting the compressed air inlet valve and pressure allows a variable capacity and head. Figure 6-28 shows an example of a diaphragm pump.

There are advantages and disadvantages to using a diaphragm pump. Those advantages and disadvantages are listed in Table 6-7.

CHAPTER 6 Pumps 117

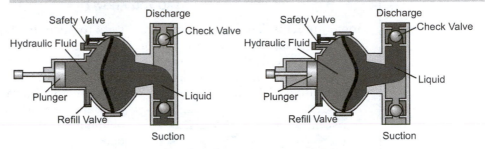

FIGURE 6-28 Diaphragm Pump Intake and Discharge

TABLE 6-7 Advantages and Disadvantages of Using a Diaphragm Pump

Advantages	Disadvantages
• No contact between liquid and the reciprocating piston, which eliminates the possibility of leaks	• Limited in head and capacity range
	• Require check valves in the suction and discharge nozzles

Suction Pumps

Suction pumps are pumps that pull liquid using a partial vacuum created by a piston on the upstroke. Suction pumps use atmospheric pressure to push the liquid under a retreating valve piston. Figure 6-29 shows an example of a suction pump.

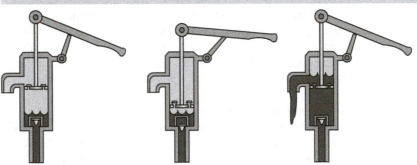

FIGURE 6-29 Suction Pump

Operating Principles

It is important to understand how pumps operate. Specifically, process technicians need to be familiar with inlet flows, outlet flows, head pressure, and what is happening inside the pump.

HEAD PRESSURE

Head pressure is the pressure of a liquid exerted upon a system; the amount of head pressure is determined by the height of the liquid. **Suction head** is the pressure required to force liquids into a pump. **Static head** is the amount of pressure the liquid is placing on the pump. (*Note:* Static head pressure is created by the weight of the liquid). **Dynamic head** is the amount of "push" a pump must have in order to overcome the pressure of the liquid. Figure 6-30 shows a comparison of static and dynamic head.

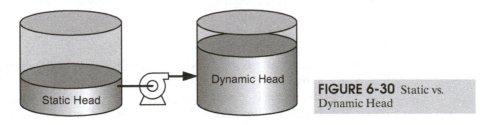

FIGURE 6-30 Static vs. Dynamic Head

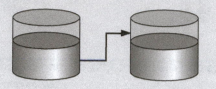
INLET (SUCTION)

The inlet (suction) is the point at which the liquid enters the pump. Generally, the installed pump is for a particular service. If the pump is not designed for the liquid it is pumping, damage to the pump can occur and the pump may not move the desired quantity of liquids. The types of liquids that are pumped may include water, process chemicals, slurry, and chemical waste water.

Some pumps require suction screens (screens that filter out loose particles) to prevent damage to the pump. In the case of waste streams or slurries, pumps and suction screens are monitored for plugging and the screens are cleaned as needed. Suction screens are also used when starting up a new facility to capture undesirable items that may have been left in the piping systems by construction or maintenance activities (e.g., bolts, tools, or rags).

Sometimes it is necessary to pump a liquid from above (e.g., from a sump). This is referred to as suction lift, the process of using vacuum to lift a liquid. If you pump from the same level or below the liquid being pumped (e.g., from a tank) and/or a pressurized suction line (i.e., from a pressurized vessel or the discharge from another pump), it is referred to as suction head. Remember, it is important to provide enough NPSH in order to prevent cavitation. When designing a pump, suction conditions such as liquid height, liquid temperature, and other factors must be considered.

WHAT HAPPENS INSIDE THE PUMP

Centrifugal Pumps

In centrifugal pumps, pressure is added to the liquid by increasing its velocity through centrifugal force. Liquid enters the suction pipe and flows into the impeller eye. The impeller eye is located at the center of the pump and picks up the liquid by the impeller vanes (raised ribs on the impeller used to catch and sling the liquid) and is accelerated in the direction of rotation. As the liquid leaves the impeller, its velocity approaches the vane tip velocity (the amount of velocity depends on the pump). Next, as the liquid moves at high speed from the impeller eye to the volute, the velocity energy in the liquid is converted to pressure energy. As the high-pressure liquid approaches the discharge nozzle, it is directed through the nozzle by a cutwater (a thick plate in the discharge nozzle of the pump that breaks the vortex).

Reciprocating Pumps

Reciprocating pumps use the inward stroke of a piston or diaphragm to draw (intake) liquid into a chamber, followed by a subsequent outward stroke to positively displace (discharge) the liquid. Reciprocating pumps consist of a liquid end and a drive end. The liquid end contains a device that displaces a fixed liquid volume for each stroke from

the drive end. The suction and discharge flows are typically directed by the positioning of check valves. As the stroke of the pumps draws liquid from the suction, the check valve allows the suction to fill while the discharge check valve closes to prevent back-flow. As the stroke transfers the liquid to the discharge, the discharge check valve opens to allow flow through the discharge, while the suction check valve closes.

OUTLET (DISCHARGE)

Several variables affect the discharge flow rate, including piping size, downstream vessel size, discharge head, pressure differential, and wear rings. Causes of no or low discharge flow include reduced head, closed valves, line blockage, system redesigns, and blocked suction screens.

Associated Utilities/Auxiliary Equipment

Pumps have associated utilities and auxiliary equipment that are important to their operation. These utilities and pieces of equipment include stuffing box seals, mechanical seals, seal flushes, seal pots, and wear rings.

STUFFING BOX SEALS

Seals are devices that hold lubricants and process fluids in place while keeping out foreign materials. Seals are especially critical in preventing environmental hazards because they help contain the liquid in the pump and prevent leakage. **Stuffing box seals** are seals around a moving shaft or stem that contains packing material designed to prevent the escape of process liquids. Figure 6-31 displays an example of stuffing box seals.

Stuffing box seals are used in nontoxic, nonflammable, and nonpolluting applications, such as process water and cooling water. Stuffing box seals are expensive, but the packing associated with them is cheaper than mechanical seals. Because of this, they are often used in nontoxic service.

Packing is used inside the packing gland of a pump to prevent leakage. Another feature of packing is that it must allow some liquid to exit the pump in order to provide cooling to the packing and pump shaft. If the packing is tightened too much and liquid cannot exit, the shaft can overheat, causing the packing to become brittle or the shaft to break. The shaft can also rub against overly tightened packing and become damaged by wear.

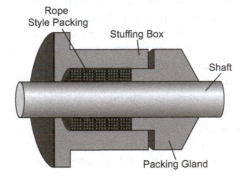

FIGURE 6-31 Stuffing Box Seals

MECHANICAL SEALS

Mechanical seals are two flat surfaces (faces) that are in contact with each other to prevent leaks. One rotates with the shaft while the other remains stationary. The faces are held together mechanically using springs that provide tension between a soft surface and a hard surface. The seals are flexible and can be moved to make up for static (pump is down) and dynamic (pump is running) misalignments and wear. Figures 6-32, 6-33, and 6-34 show examples of mechanical seals used to prevent leaks.

FIGURE 6-32 Mechanical Seal Used to Prevent Leaks
Courtesy of Brazosport College

FIGURE 6-33 Mechanical Seals

FIGURE 6-34 Mechanical Seals Used to Prevent Leaks
Courtesy of Brazosport College

SEAL FLUSHES AND SEAL POTS

Many pumps use some type of seal flush to keep the seal lubricated and free of debris. A **seal flush** is a small flow (slip stream) of pump discharge or externally supplied liquid that is routed to the pump's mechanical seal. This acts as a barrier liquid between the two faces of the seal to reduce friction and remove heat. A flush can use the liquid being pumped through the process (self-contained) or external liquids. Typically, the self-contained flush is used when the liquid being pumped is clean, noncorrosive, and nontoxic. Figure 6-35 shows an example of a seal flush.

Many pumps use a reservoir called a seal pot to contain a secondary seal liquid. The seal pot pressure is slightly higher than the pump pressure. A pressure differential or a seal pot that empties too quickly can indicate a pump seal leak. If a seal leak occurs, the leaking material is the material from the seal pot (which is normally nonflammable and nontoxic). Figure 6-36 shows an example of a condensate seal pot.

There are basically two types of seal systems that employ a seal pot: tandem seal and double seal. A tandem seal has a primary seal and a backup seal. If the primary seal leaks, the pressure in the seal pot increases dramatically. If the backup seal fails, the seal pot is drained of its contents. In both instances, maintenance needs to be performed.

A double seal uses pressurized liquid from the seal pot to pressurize the seal in two directions: internally to the process, and externally to the outside. If the primary seal fails, the neutral liquid inside the seal pot will leak into the process, indicating there is a

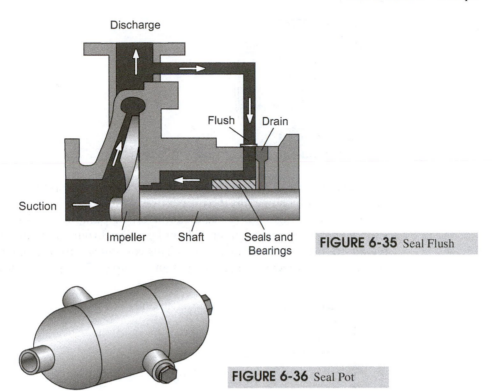

FIGURE 6-35 Seal Flush

Discharge

Flush Drain

Suction

Impeller Shaft Seals and Bearings

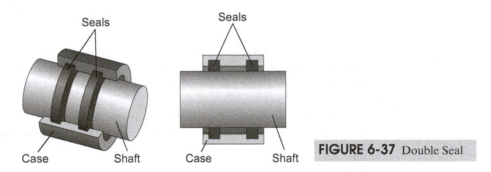

FIGURE 6-36 Seal Pot

seal failure. Because of this, a double seal is highly effective on flammable, toxic, and corrosive materials and those with solids and abrasive material. Figure 6-37 shows an example of a double seal.

Seals Seals

Case Shaft Case Shaft

FIGURE 6-37 Double Seal

WEAR RINGS

Wear rings (shown in Figure 6-38) are rings that allow the impeller and casing suction head to seal tightly together without wearing each other out. Wear rings are close-running, noncontacting, replaceable, pressure breakdown devices located between the impeller and casing of a centrifugal pump. One ring, called the impeller ring, rotates with the impeller. The other, called the casing ring, is stationary. Wear rings increase efficiency by minimizing discharge-to-suction recirculation. They control axial thrust (back-and-forth motion of the shaft) by reducing the discharge pressure acting on the impeller. They also minimize seal and chamber pressure and can be replaced to restore clearances between the moving and stationary rings.

Purpose of a Pump Curve

It is important for a process technician to understand the basic concept of a pump curve. A **pump curve** is a specification that describes the capacity (horsepower) and head needed for correct operation of the pump. The manufacturer determines the pump curve based on the type of pump and the related process. With the system curve

FIGURE 6-38 Wear Ring

and head criteria, a manufacturer can determine which pump is the best one for the process (i.e., which one has the proper operating capacity).

Each centrifugal pump has a characteristic pump curve. Liquid head is the pressure developed from the pumped liquid passing through the volute. As the flow rate is increased, the liquid head decreases. This continues until the maximum pumping rate of the pump is reached. Figure 6-39 is an example of a typical centrifugal pump curve.

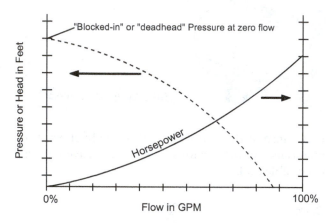

FIGURE 6-39 Typical Centrifugal Pump Curve

The maximum liquid head occurs at zero flow (called the dead head pressure). Some centrifugal pumps (especially very high speed and multistage pumps) may have excess vibration or undesired overheating if they are dead headed, so dead heading should be avoided.

Centrifugal pump manufacturers supply capacity pump curves for their pumps. These curves are normally based on water and contain the following information:

* Head
* Flow rate
* Efficiency
* Horsepower
* NPSH required

Figure 6-40 provides a pump curve for a pump operating at 3,500 rpm with various impeller diameters. If you look at the pump curve data in Figure 6-40, you will notice that the dead head pressure (no flow through the pump) on the 7¼″ impeller is 230 feet. At a flow rate of 400 gpm, the head is 215 feet, the horsepower is 33, the required NPSH is 13 feet, and the efficiency is 68 percent.

HEAD PRESSURE

Pump head pressure is an expression of a pressure in terms of the height of the liquid. In other words, it is the amount of pressure a liquid is exerting based on the height of the liquid. The head of the pump is normally on the Y axis and is normally stated in feet. The head developed by the pump does not depend on the material being pumped (e.g., water or organic).

To convert the head to pressure (psi), you must multiply by the density of the liquid. For water, the conversion factor is .433 psi/ft (62.4/144). If the liquid has a specific

FIGURE 6-40 Sample Pump Curve Graph

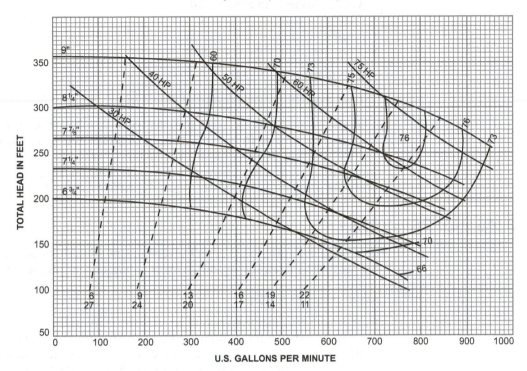

Sample Pump Curve Graph

gravity that is less than 1.0, then the pump develops less pressure than when pumping water and consumes less horsepower.

FLOW RATE

The flow rate is normally on the X axis and is usually stated in gallons per minute (gpm). The maximum flow rate that can be achieved increases as the impeller diameter is increased. The maximum head occurs at zero flow. As the flow rate is increased, the head decreases until the maximum flow rate of the pump is achieved.

EFFICIENCY

The efficiency is the ratio of the liquid horsepower to the brake horsepower. Pumps are designed so the best efficiency point (BEP) is at design operating condition (the condition at which the pump was designed to run). Operations at other points on the curve result in a reduced efficiency (i.e., more brake horsepower for the same liquid horsepower).

HORSEPOWER

The horsepower required is normally based on water. The horsepower increases as the impeller diameter increases or as the flow rate is increased. If the specific gravity of the material is less than 1.0, then the horsepower required decreases. If the specific gravity of the material is greater than 1.0, the horsepower required increases.

The viscosity of the liquid also affects the horsepower required. Water has a viscosity of one centipoise (a common measure of the viscosity of a liquid) at room temperature. If the liquid is more viscous than water, the required horsepower increases. If the liquid is less viscous than water, the required horsepower decreases.

NPSH REQUIRED

NPSH is the net positive suction head. The NPSH given on the pump curve is the required NPSH for the pump. The available NPSH from the process must be calculated and must be greater than the NPSH required or the pump will cavitate. NPSH is usually stated in feet and is calculated using the following factors:

- Gas pressure above the liquid
- Vapor pressure of the liquid
- Liquid height above pump suction
- Pressure drop in the suction line

The formula for NPSH is:

NPSH available = gas pressure + liquid height − vapor pressure − pressure drop

If the NPSH available is too low, the technician has several options to increase it. One option would be to increase the gas pressure above the liquid. Another option would be to raise the liquid level. The technician can also reduce liquid temperature, which lowers the vapor pressure. Finally, the technician can reduce the flow rate, which lowers the pressure drop and decreases the NPSH required.

Potential Problems

When working with pumps, process technicians should always be aware of potential hazards such as overpressurization, overheating, cavitation, and leakage.

OVERPRESSURIZATION

Pump overpressurization can occur if the valves beyond the pump are incorrectly closed or blocked. Consider the example in Figure 6-41. In the figure, valves A and B are both closed. Because both valves are closed at the same time, the liquid has no place to go. The result is excessive back pressure or dead heading. Both can cause serious personal injury, pump seal failures, or pump damage.

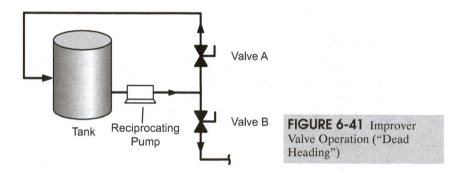

FIGURE 6-41 Improver Valve Operation ("Dead Heading")

OVERHEATING

Pump overheating is often caused by improper lubrication. Without lubrication, bearings fail and equipment surfaces rub together. As these surfaces rub against one another, friction is produced and heat is generated. This can cause mechanical failures, swelling, leakage, and decomposition of the process liquid.

If the pump is dead headed, the liquid in the pump is heated by the mechanical energy of the motor. This can cause vapor lock and cavitation. To prevent either condition, recycle loops (minimum flow lines) may be added to allow flow through the pump, even if the valves downstream are closed to prevent overheating of the liquid.

Process technicians should always monitor rotating equipment for excessive heat because operating equipment under these conditions can lead to permanent equipment damage or personal injury (e.g., burns).

LEAKAGE

Process technicians should always check pumps for leaks because leaks can introduce slipping hazards, exposure to harmful or hazardous substances, and process problems (e.g., inferior product produced as a result of improper feed supply). Leaks most frequently occur where the pump shaft exits the pump housing. Packing or mechanical seals are normally used to prevent this leakage from occurring. Leaks occurring at this location are normally corrected by tightening or replacing the packing or replacing the mechanical seal.

CAVITATION

Cavitation is a condition inside a pump wherein the liquid being pumped partly vaporizes due to variables such as temperature and pressure drop. These variable changes cause vapor pockets (bubbles) to form and collapse (implode) inside a pump.

Cavitation occurs when the pressure on the eye of a pump impeller falls below the boiling pressure of the liquid being pumped. This is a very serious problem in dynamic pumps, especially centrifugal pumps. Cavitation is also a problem in vacuum operations because low pressure liquids boil at lower temperatures.

Key characteristics of cavitation include large pressure fluctuations, inconsistent flow rate, and severe vibration. Cavitation has been described as sounding like the pump is pumping rocks. Technicians who identify cavitation should always try to eliminate it as quickly as possible because it can cause excessive wear on the pump seal, impeller, bearings, and casing.

To prevent cavitation, a pump should always be primed before it is started. Priming is the process of filling the suction of a pump with liquid to remove any vapors that might be present. This reduces the likelihood that the pump will become vapor-bound or lose suction.

Cavitation can still occur, however, even if a pump was properly primed. For example, if the liquid becomes too hot, vapor bubbles can be created. The way to correct this is by cooling the liquid. Another way to stop cavitation is to raise the level of the liquid in the suction line to increase the suction pressure on the pump. Table 6-8 displays causes of and solutions for cavitation.

TABLE 6-8   Causes of and Solutions for Cavitation

Cause	*Solution*
Suction pressure reduction	Increase suction pressure, slow pump down, check the level of the suction vessel, and check the design. Check for restrictions
Liquid temperature increase	Lower liquid temperature
Flow rate too high	Reduce flow rate
Separation and contraction of flow due to a change in viscosity	Check design, and find cause of viscosity change
Undesirable flow conditions due to obstructions or sharp turns	Locate obstructions or sharp turns and correct
Pump is not suitable for the system curve	Check design
Low liquid level	Increase liquid level

FLASHING SEAL LEAKING INTO A PUMP

In abnormal operating conditions, deviation can occur in the form of a flashing seal leak. A flashing seal leak can cause pump cavitation. Table 6-9 shows the causes and solutions for flashing seals leaking into the pump.

TABLE 6-9   Causes and Solutions for Flashing Seal Leaking into a Pump

Cause	*Solution*
Overheated seal	Check operation of seal flush or seal pot system; install a seal cooling system
Change in liquid vapor pressure	Determine reason for change in liquid resulting in vapor pressure change. Check pumped liquid composition and temperature

EXCESSIVE VIBRATION

Excessive vibration can cause mechanical damage and pump shutdown, especially on a pump equipped with vibration sensors. Because of this, pumps with worn bearings or couplings should be given a vibration analysis. When misalignment occurs due to vibration, alignment and realignment must be implemented.

PUMP SHUTDOWN

Pumps can shut down if a deviation from normal conditions occurs. This shutdown can cause a loss of process flow. Table 6-10 shows the causes and solutions for the most common pump shutdown.

TABLE 6-10   Causes and Solutions for Pump Shutdown

Cause	*Solution*
Overspeed trip (steam-driven)	Check turbine overspeed trip for proper operation; check suction for proper flow; check discharge lines for large failure
Power failure	Determine cause of power failure (e.g., breaker, fuse, loss of commercial power). Repair and restore power as appropriate
Flow, pressure, or vibration (on a system with sensors)	Determine cause (e.g., leak, valve closed, and instrumentation elsewhere in the process); correct problem to return pump operation
Motor trip out	Pump is using too much horsepower, flow rate is too high, or mechanical problems exist. Correct problem to return pump operation

VAPOR LOCK

When a pump suffers from vapor lock, the pump loses its liquid prime (the liquid being pumped). This causes vibrations and abnormal noises. The solution (especially if the pump is improperly vented) is to shut down the pump, bleed off the vapor, and restart the pump.

Safety and Environmental Hazards

There are a number of safety and environmental concerns associated with pumps, including chemical hazards and hazards caused by the equipment itself. Hazards with normal and abnormal pump operations can affect equipment, personal safety, plant operations, and the environment.

PERSONAL SAFETY HAZARDS

Pumps can cause personal injury if attention is not paid to safety regulations. Possible hazards involving pumps include exposure to hazardous liquids and materials, slipping or tripping due to leaks, overpressurization, and the physical hazard of rotating equipment. When hazardous chemicals such as acids, caustics, reactives, and hot petroleum liquids are pumped at high rates and pressures, a small problem can quickly escalate to a serious hazard.

During normal and abnormal pump operations, process technicians should always wear personal protective equipment (PPE), including safety glasses or goggles, a hard hat, and hearing protection. Additional PPE may also be required based on the liquid being pumped and other factors.

During routine and preventive maintenance, additional safety measures should be taken. These measures may include housekeeping duties such as removing spilled liquids to prevent fires or exposure to others in the facility, cleaning up seal leaks, and removing trash around pumps.

Lockout/tagout and isolation procedures must also be followed and the pump should be placed in a zero energy state before maintenance is performed. Zero energy state means that the pump is secured of all energy (e.g., pressure, rotation, chemical hazards). This protects maintenance personnel from hazards while working on the pump.

EQUIPMENT OPERATION HAZARDS

There are several equipment operation hazards associated with pumps during normal and abnormal operations. Table 6-11 lists some of these hazards and their consequences.

TABLE 6-11 Equipment Operation Hazards During Normal and Abnormal Operations

Situation	*Consequences*
Head/capacity	Causes excessive wear on pump parts, causes seal leaks, and affects process operations and product quality.
Insufficient NPSH (cavitation)	Causes seal damage, bearing damage, and impeller damage. Cavitation is caused by inadequate suction pressure.
Speed	Affects pump capacity if abnormal. An extreme change in direction causes mechanical damage. The pump is designed to operate at a determined speed. Excess speed results in overheating and increasing the temperature of the product, thereby increasing the vapor pressure. This leads to cavitation and eventual damage to the pump internals. Low speed affects the pump output. (*Note:* Most pumps are motor-driven, and are constant speed devices.)
Multistage	These pumps are more sensitive to operating variables than a single impeller design. Most multistage pumps, for instance, require a minimum flow through the pump to prevent damage caused by overheating and therefore must never be run against a closed discharge.
Vibration	Causes mechanical failure of pump components if operation continues. Vibration is caused by cavitation, worn bearings, or misalignment of the pump.
Temperature	Very high and very low temperatures can cause severe burns if personnel are exposed to piping or liquid. Abnormal operating conditions could result in damage to the pump and hazards from the chemical being too hot. Heat can cause an explosion or decomposition of the chemicals. Exposure to the process technician (high or low temperature) can also occur, creating a personal safety hazard.
Overpressure	Caused by starting with a positive displacement or multistage pump with the discharge closed. Actual catastrophic failure in such a case would be an unlikely, yet possible, outcome; more probably, motor overload and/or internal equipment damage would result.
Lack of lubrication	Causes overheating or scoring of rotating parts.

FACILITY OPERATION HAZARDS

Reliable pump operation is vital to the operation of the facility. In continuous operations (e.g., refineries), process equipment depends on continuous liquid flow for proper operation. Loss of liquid flow can cause process equipment failure, which can result in fires, explosions, or fouling. Loss of production can cause facility shutdowns, pump replacement costs due to damage, system downtime, or loss of product.

Standby pumps are used in many cases to maintain flow if the primary pumps fail. Standby pumps may be steam driven in the event the pump failure is due to power loss. Many of these standby pumps start automatically when a given process variable (e.g., flow rate or pressure) reaches a set low point.

ENVIRONMENTAL HAZARDS

Environmental impact depends on the liquid used in the pump. Impact can range from excess water usage (due to leaks) to facility or community evacuations if the liquid is extremely hazardous (e.g., toxic or flammable). The Occupational Safety and Health Administration (OSHA) and the EPA have set limits for the discharge of specific toxic and flammable liquids. Violations of these limits result in heavy fines and penalties.

Potential problems associated with the environment include discharge onto the ground or into the sewer, discharge into the air, or fines and legal issues from not following regulations. Additional problems are seal leaks where environmental regulations limit the amount of fugitive emissions from chemicals into the environment. If these limits are exceeded, the leak is reported to the regulatory agencies involved. If there is a massive leak, it could affect the area and community outside the facility.

Process Technician's Role in Operation and Maintenance

The process technician plays a vital role in the operation and maintenance of the pumps within a process industry. Process technicians must have a basic understanding of the pumps within their process area and be aware of the potential problems that can occur with pumps at any given time.

Process technicians should be trained on the operational and emergency procedures associated with pumps, including startup and shutdown. In addition, monitoring must be conducted on a regular basis to eliminate potential problems or hazards before they occur.

When working with pumps, process technicians should look, listen, and feel for the items listed in Table 6-12. Failure to perform proper maintenance and monitoring could affect the process and result in equipment damage.

TABLE 6-12   Process Technician's Role in Operation and Maintenance

Look	*Listen*	*Feel*
• Check oil levels to make sure they are satisfactory.	• Listen for abnormal noises.	• Feel for excessive vibration.
• Check to make sure water is not collecting under the oil (water is not a lubricant, so it can cause bearing failure).		• Feel for excessive heat.
• Check seals and flanges to make sure there are no leaks.		

Typical Procedures

Typical procedures are required when working with pumps. For example, operations and maintenance required by the process technician may include monitoring, lockout/tagout, routine and preventive maintenance, startup, and shutdown.

MONITORING

When monitoring pump operation, the process technician typically is directed to perform the following steps:

1. Check and record suction/discharge pressures.
2. Check lubricant levels.
3. Check seal flush system (if applicable).
4. Check for abnormal noise and vibration, using senses (sight, hearing, touch, etc.) to ensure proper operation.
5. Check for proper temperatures.
6. Check pump rates and instrument outputs to the flow controllers of the pump.
7. Check for missing coupling guards.

STARTUP

Steps can vary from equipment to equipment. The process technician should always follow the proper startup procedure for a given pump. A general startup procedure may include the following steps:

1. Ensure all downstream equipment is lined up and ready.
2. Ensure the pump is ready for startup.
3. Ensure all auxiliary systems (seal flush, cooling water, etc.) are lined up.
4. Warm up all equipment as needed.
5. Start pump.
6. Ensure the flow rate on the pump discharge is as expected.
7. Conduct routine inspections to check suction/discharge pressures, to check the machine operation, and to check for leaks.

LOCKOUT/TAGOUT

Lockout/tagout procedures are performed to isolate the equipment from any energy sources. The basic process includes the following steps:

1. Close the pump suction and discharge so liquid does not enter the pump.
2. De-energize the driver by opening the breaker to the motor or blocking the inlet and exhaust steam valves on the turbine.
3. Drain the pump to the designated collection system.
4. Shut down any auxiliary systems, such as lube oil and seal flush systems.
5. Lock and/or tag the valves and breaker for the motor or valves on the turbine (or the driver used).
6. Flush/purge the pump to clear the pump of the chemical contained in the pump.

SHUTDOWN

Steps can vary from equipment to equipment. The process technician should always follow the proper shutdown procedure. A general shutdown procedure may include the following steps:

1. Ensure the pump is ready to be shut down (the standby pump is running or the process is being shut down).
2. Shut off the pump.
3. If the pump is being shut down following the startup of a standby pump, immediately check the rate. Loss of rate from the standby pump can occur if the discharge check valve on the pump shutdown sticks open. If this occurs, quickly closing the discharge block valve on the shutdown pump restores the rate.
4. Close the suction and discharge valves if the pump is being removed from service for maintenance. The valves should be left open if the pump is to remain in standby service.
5. Prepare the pump for maintenance as needed.

EMERGENCY PROCEDURES

Emergency procedures include additional actions by a technician beyond those normally required during a shutdown. Process technicians should always follow the specific emergency operating procedures for their unit and company.

Summary

Pumps play a major role in process operations. They move products through the piping systems, and they are figuratively the "heart" of a given process because they transfer liquids from one place to another and provide flow through process equipment. In continuous operations, such as refineries, process equipment depends on continuous liquid flow for proper operation. Because of this, pumps are essential to operations. In process critical flows, standby pumps are commonly used and may automatically start up in the event of a pump loss.

Two main categories of pumps include dynamic and positive displacement pumps. Dynamic pumps are nonpositive displacement pumps that convert centrifugal force to dynamic pressure to move liquids. They are classified as either centrifugal or axial. Positive displacement pumps are piston, diaphragm, gear, or screw pumps that deliver a constant volume with each stroke. Unlike dynamic pumps, positive displacement pumps deliver the same amount of liquid, regardless of the discharge pressure.

The internal operation of a pump depends on the pump type. Dynamic pumps operate differently than do positive displacement pumps. The selection of pumps for use within a process is based on a number of factors, including liquid viscosity, line pressures and flows, equipment speed, altitude and temperature, and whether the material is a slurry or liquid. Variables that affect pump discharge are calculated into the performance of the pump. Pump manufacturers supply optimum pump curves that are used in system design to size the pump or driver.

Process technicians play an important role in ensuring that pump operations remain at normal conditions. Close monitoring of conditions is required to ensure problems can be resolved before they escalate to a hazardous situation. Pumps can cause personal injury if attention is not paid to the required safety regulations. Environmental impact depends on the liquid used in the pump. Impact can range from excess water usage (due to leaks) to facility or community evacuations if the liquid is extremely hazardous (toxic or flammable).

Problems typically associated with pumps on a process include cavitation, flashing seal leaking into a pump, an unexpected pump shutdown, and excessive vibration.

Checking Your Knowledge

1. Define the following terms:
   a. Axial pump
   b. Centrifugal pump
   c. Diaphragm pump
   d. Dynamic pump
   e. Gear pump
   f. Positive displacement pump
   g. Rotary pump
   h. Screw pump
   i. Vane pump
   j. Cavitation
   k. Impeller
   l. Lobe pump
   m. Piston
   n. Piston pump
   o. Priming
   p. Pump curve
   q. Seal
   r. Vanes
   s. Viscosity
   t. Volute

2. Pump selection is based on (select all that apply):
   a. Viscosity
   b. Needed discharge pressure
   c. Suction pressure
   d. Flow rate

3. *(True or False)* A gear pump is a type of centrifugal pump.

4. Which of the following are rotary-type positive displacement pumps? (Select all that apply.)
   a. Gear
   b. Lobe
   c. Vane
   d. Screw

5. A _____ pump is a pump that moves liquids by rotating a screw or a set of lobes, gears, or vanes.
   a. plunger
   b. dynamic
   c. axial
   d. rotary

6. *(True or False)* A piston pump uses a cylinder mounted on bearings within a casing.

7. *(True or False)* Suction head refers to pumping from above the liquid (e.g., from a sump).

Match the potential cause of pump shutdown with the potential troubleshooting solution:

8. Power failure

9. Flow, pressure, or vibration

10. Overspeed trip

11. Motor trip out

a. Check turbine overspeed trip for proper operation; check suction for proper flow; check discharge lines for large failure.

b. Determine cause (e.g., breaker, fuse, loss of commercial power). Repair and restore power as appropriate.

c. Pump is using too much horsepower, flow rate is too high, or mechanical problems exist. Correct problem to return pump operation.

d. Determine cause (e.g., leak, valve closed, and instrumentation elsewhere in the process); correct problem to return pump operation.

12. When a pump suffers from _____ lock, the pump loses liquid prime, vibrates, and makes abnormal noises.
    a. vibration
    b. vapor
    c. cavitation
    d. bearing

13. Is the diagram below an example of a centrifugal pump or a positive displacement pump?

On the diagram below, identify the following parts of a centrifugal pump.

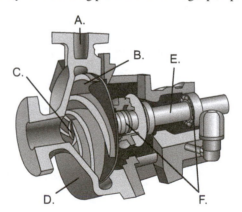

14. Bearings and seals
15. Casing
16. Discharge
17. Impeller
18. Shaft
19. Suction eye

Student Activities

1. Using a cutaway model of a reciprocating pump (or a cutaway drawing), locate the following:
    a. Inlet check valve
    b. Outlet check valve
    c. Piston/plunger
    d. Packing
    e. Crankshaft
    f. Connecting rod
    g. Cylinder
    h. Seal

2. Using the flowing pump curve in Figure 6-40, determine the head, horsepower, and NPSH required for a 9″ impeller at 600 gpm flow rate.

3. Using a pump model, demonstrate a task associated with typical pump operations, such as lockout/tagout, startup, or shutdown.
4. Demonstrate how you would perform the following maintenance and monitoring activities on a typical pump:
    a. Inspect for abnormal noise
    b. Inspect for excessive heat
    c. Check oil levels
    d. Check for leaks around seals and flanges
    e. Check for excessive vibration
5. Examine the concept of centrifugal force by doing the following:
    a. Obtain a small sand bucket or pail.
    b. Fill the pail with water until it is half full.
    c. Locate an open area away from other individuals or obstructions.
    d. In the open area, grasp the pail by its handle and swing it in a circular motion, arms fully extended, repeatedly rising over your head and back down to your knees.
    e. Examine what happens. Did the water stay in the pail or did it spill out?
6. Examine the actions of a reciprocating pump by doing the following:
    a. Obtain a syringe (without a needle) and a cup of water.
    b. Depress the syringe plunger all the way into the housing.
    c. Place the syringe in the cup of water and pull the plunger back until the syringe housing is full of water.
    d. Lift the syringe out of the water and then depress the plunger so the water is forced out of the syringe, into the cup.
    e. Repeat steps b to d several times. Each time, vary the amount of pressure applied to the plunger when forcing the water out of the syringe.
    f. Examine what happens. Did the amount of liquid discharged change as the pressure changed, or did it remain the same?

Compressors

Objectives

After completing this chapter, you will be able to:

- Explain the purpose of compressors in the process industries.
- Identify common compressor types and describe the operating principle of each.
- Identify compressor components and explain the purpose of each.
- Describe the theory of operation for a compressor.
- Identify potential problems associated with compressors.
- Describe safety and environmental concerns associated with compressors.
- Identify typical procedures associated with compressors.
- Describe the process technician's role in compressor operation and maintenance.
- Explain the purpose of a compressor performance curve and demonstrate its use.

Key Terms

Antisurge—automatic control instrumentation designed to prevent compressors from operating at or near pressure and flow conditions that result in surge.

Antisurge protection—protection that prevents damage to the compressor. Antisurge protection is calculated and designed by engineers to ensure proper and safe compressor operation.

Axial compressor—a dynamic compressor that contains a rotor with contoured blades followed by a stationary set of blades (stator). In this type of compressor, the flow of gas is axial (in a straight line along the shaft).

Centrifugal compressor—a dynamic compressor in which the gas flows from the inlet located near the suction eye to the outer tip of the impeller blade.

Cylinder—a cylindrical chamber in a positive displacement compressor in which a piston compresses gas and then expels the gas.

Demister—a device that promotes separation of liquids from gases.

Dry carbon ring—an easy-to-replace, low-leakage type of seal consisting of a series of carbon rings that can be arranged with a buffer gas to prevent the process gas from escaping.

Dynamic compressor—nonpositive displacement compressor that uses centrifugal or axial force to accelerate and convert the velocity of the gas to pressure. Dynamic compressors are classified as either centrifugal or axial.

Interlock—a type of hardware or software that does not allow an action to occur unless certain conditions are met.

Labyrinth seal—a shaft seal designed to restrict flow by requiring the fluid to pass through a series of ridges and intricate paths.

Liquid buffered seal—a close fitting bushing in which oil and water are injected in order to seal the process from the atmosphere.

Liquid ring compressor—a rotary compressor that uses an impeller with vanes to transmit centrifugal force into a sealing fluid, such as water, driving it against the wall of a cylindrical casing.

Lubrication system—a system that circulates and cools sealing and lubricating oils.

Multistage compressor—device designed to compress the gas multiple times by delivering the discharge from one stage to the suction inlet of another stage.

Positive displacement compressor—device that may use screws, sliding vanes, lobes, gears, or a piston to deliver a set volume of gas with each stroke.

Reciprocating compressor—a positive displacement compressor that uses the inward stroke of a piston to draw (intake) gas into a chamber and then uses an outward stroke to positively displace (discharge) the gas.

Rotary compressor—a positive displacement compressor that uses a rotating motion to pressurize and move the gas through the device.

Seal system—devices designed to prevent the process gas from leaking from the compressor shaft.

Separator—a device used to physically separate two or more components from a mixture.

Single-stage compressor—device designed to compress the gas a single time before discharging the gas.

Surging—the intermittent flow of pressure through a compressor that occurs when the discharge pressure is too high, resulting in flow reversal within a compressor.

Introduction

Compressors are an important part of the process industries. Compressors increase the pressure of gases and vapors so they can be used in applications that require higher pressures. For example, they can be used in a wide variety of applications like compressing gases such as carbon dioxide, nitrogen, and light hydrocarbons, or providing the compressed air required to operate instruments or equipment.

The two most common compressor types are positive displacement and dynamic. Positive displacement compressors use pistons, lobes, screws, or vanes to reduce a fixed volume of gas through compression and deliver a constant volume. Dynamic compressors use impellers or blades to accelerate a gas and then convert that velocity into pressure. Dynamic compressors are more commonly used than positive displacement compressors because they are less expensive, more efficient, have a larger capacity, and require less maintenance.

All compressors require a drive mechanism such as an electric motor or turbine to operate, and all are rated according to their discharge capacity and flow rate. Most compressors require auxiliary components for cooling, lubrication, filtering, instrumentation, and control. Some compressors require a gearbox between the driver and compressor to increase the speed of the compressor.

Selection of Compressors

A compressor is a mechanical device used to increase the pressure of a gas or vapor. The type of compressor that is used for a particular application depends on several factors. The factors include the type of gas being compressed, flow rates which are expressed as cubic feet per minute (cfm) or meters cubed/second (m^3/s), and discharge pressure which is expressed as pounds per square inch (psi) or Kilopascals (kPa).

Types of Compressors

Several types of compressors are used in the process industries (see Figure 7-1). The most common, however, are positive displacement and dynamic. Both of these types of compressors can be single- or multistage.

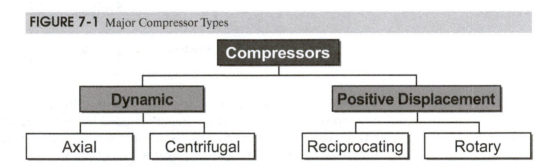

FIGURE 7-1 Major Compressor Types

DYNAMIC COMPRESSORS

Dynamic compressors are nonpositive displacement compressors that use centrifugal or axial force to accelerate and convert the velocity of the gas to pressure (as opposed to positive displacement compressors, which use a piston, lobe, or screw to compress gas).

CENTRIFUGAL COMPRESSORS

Centrifugal compressors are a dynamic type compressor in which the gas flows from the inlet located near the suction eye to the outer tip of the impeller blade. In a centrifugal compressor, the gas enters at the low pressure end and is forced through the impellers by the rotation of the shaft. As the gas moves from the center of the impeller toward the outer tip, the velocity is greatly increased. When the gas leaves the impeller and enters the volute, the velocity is converted to pressure due to the slowing down of the molecules.

Centrifugal compressors are used throughout industry because they have few moving parts, are very energy efficient, and provide higher flows than similarly sized reciprocating compressors. Centrifugal compressors are also popular because their seals allow them to operate nearly oil-free, and they have a very high reliability. They are also effective in toxic gas service when the proper seals are used, and they can compress high

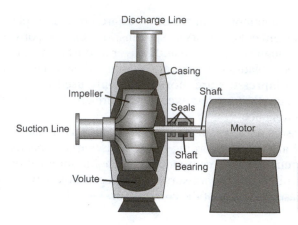

FIGURE 7-2 Centrifugal Compressor and Volute with Driver

volumes at low pressures. The primary drawback is that centrifugal compressors cannot achieve the high compression ratio of reciprocating compressors without multiple stages. Figure 7-2 shows an example of a centrifugal compressor.

Centrifugal compressors are more suited for continuous-duty applications such as ventilation fans, air movers, cooling units, and other uses that require high volume but relatively low pressures.

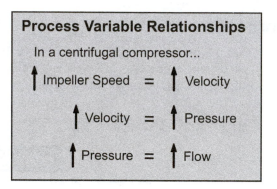

Process Variable Relationships

In a centrifugal compressor...

↑ Impeller Speed = ↑ Velocity

↑ Velocity = ↑ Pressure

↑ Pressure = ↑ Flow

In a centrifugal compressor, there is a direct relationship between impeller speed, velocity, pressure, and flow. As the impeller speed increases, velocity increases. As velocity increases, pressure increases. As pressure increases, flow increases.

Centrifugal compressors may be single-stage or multistage, and the stages may be contained in one casing or several different casings. (*Note:* Multistage compressors are discussed in more detail later on in this chapter.) Figure 7-3 shows a cutaway of a single-stage centrifugal compressor.

FIGURE 7-3 Cutaway of a Single-Stage Centrifugal Compressor

Courtesy of Bayport Training & Technical Center

The main components of a centrifugal compressor include bearings, a housing (casing), an impeller, an inlet and outlet, a shaft, shaft couplings, and shaft seals. Figure 7-4 shows another example of a centrifugal compressor and its components.

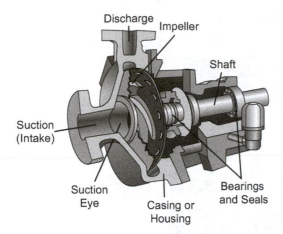

Discharge Impeller

Shaft

Suction (Intake)

Suction Eye

Casing or Housing

Bearings and Seals

FIGURE 7-4 Centrifugal Compressor with Parts Labeled

Did You Know?

As the temperature of a gas increases, so does the pressure.

Heat causes gases to expand. Thus, if you want to compress more gas into the same space, you must cool it first.

AXIAL COMPRESSORS

Axial compressors are dynamic-type compressors in which the flow of gas is axial (in a straight line along the shaft). A typical axial compressor has a rotor that looks like a fan with contoured blades followed by a stationary set of blades, called a stator.

Rotor blades attached to the shaft spin and send the gas over stator blades, which are attached to the internal walls of the compressor casing. These blades decrease in size as the casing size decreases. Rotation of the shaft and its attached rotor blades causes flow to be directed axially along the shaft, building higher pressure toward the discharge of the unit.

Each pair of rotors and stators is referred to as a stage. Most axial compressors have a number of such stages placed in a row along a common power shaft in the center. Figure 7-5 shows an example of an axial compressor with rotor and stator blades.

The stator blades are required to ensure efficiency. Without these stator blades, the gas would rotate with the rotor blades, resulting in a large drop in efficiency. Each stage is smaller than the last because the volume of air is reduced by the compression of the preceding stage. Axial compressors therefore generally have a conical shape, widest at the inlet. Compressors typically have between nine and fifteen stages. The main components of an axial compressor include the housing (casing), inlet and outlet, rotor and stator blades, shaft, and inlet guide vanes.

Axial compressors are very efficient compressors; however, they are not as frequently used in industry as reciprocating and centrifugal compressors because of the high initial and maintenance costs. Regardless of the type, all compressors are rated by dividing the discharge pressure by the suction pressure. This is called the compression ratio.

Calculating Compression Ratio

Discharge Pressure ÷ Suction Pressure = Compression Ratio

Example: A compressor with a suction pressure of 400 PSIG and a discharge pressure of 2000 PSIG has a compression ratio of 5.0.

2000 PSIG ÷ 400 PSIG = 5.0 Compression Ratio

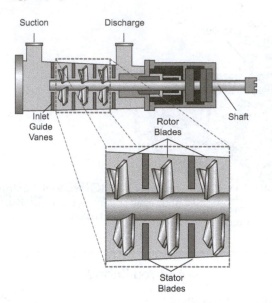

Suction   Discharge

Inlet
Guide
Vanes

Rotor
Blades

Shaft

Stator
Blades

FIGURE 7-5 Axial
Compressor Showing Rotating
and Stator Blades

Positive Displacement Compressors

Positive displacement compressors (see Figure 7-6) are devices that may use screws, sliding vanes, lobes, gears, or a piston to deliver a set volume of gas with each stroke. Positive displacement compressors work by trapping a set amount of gas and forcing it into a smaller volume. The two main types of displacement compressors are reciprocating and rotary, with reciprocating being the most commonly used.

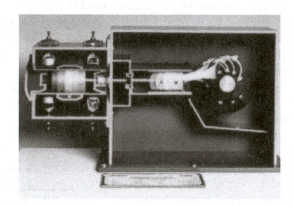

FIGURE 7-6 Positive
Displacement Compressor
Showing Motion

Courtesy of Design Assistance
Corporation (DAC)

RECIPROCATING COMPRESSORS

The term *reciprocating* refers to the back-and-forth movement of the compression device (a piston or other device is positioned in a cylinder). **Reciprocating compressors** use the inward stroke of a piston to draw (intake) gas into a chamber and then use an outward stroke to positively displace (discharge) the gas. A common application for the reciprocating compressor is in an instrument air system.

In a reciprocating compressor, a piston receives force from a power medium (e.g., a drive shaft) and then transfers that power to the gas being compressed. In a piston-type reciprocating compressor, the gas is trapped between the piston and the cylinder head and then compressed. The **cylinder** is the cylindrical chamber in which a piston

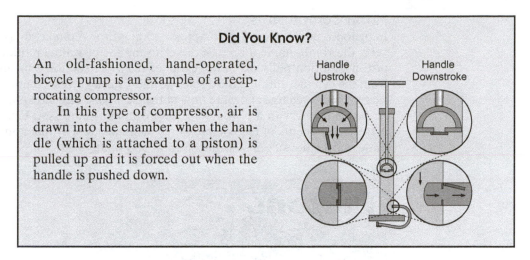
compresses gas and from which gas is expelled. Figure 7-7 displays a cutaway picture of a reciprocating compressor.

FIGURE 7-7 Cutaway of a Reciprocating Compressor

Courtesy of Bayport Training & Technical Center

In theory, reciprocating compressors are more efficient than centrifugal compressors. They are also cheaper to purchase and install than a centrifugal compressor. However, problems with pulsation and mechanical reliability cause these compressors to be less desirable than centrifugal compressors for most industrial applications. More potential problems associated with compressors are discussed later in this chapter.

Another type of reciprocating compressor is the diaphragm compressor. Diaphragm compressors can be used for a wide range of pressures and flows (very low to very high). In a diaphragm compressor, a fluid is forced against one side of the diaphragm, which flexes the diaphragm into the free space above it, thereby compressing and pressurizing the gas on the other side of the diaphragm. Figure 7-8 displays the basic design of a diaphragm compressor with labels.

Because the process gas in a diaphragm compressor does not come in contact with the fluid, process purity is assured. This is useful in laboratory or medical applications.

FIGURE 7-8 Diaphragm Compressor with Parts Labeled

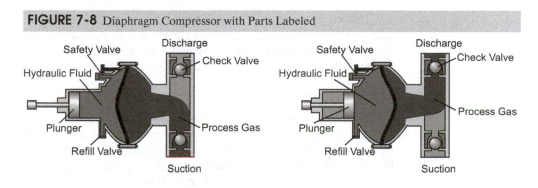

Piston Compressors

In piston-type reciprocating compressors, the pistons connect to a crankshaft that converts the rotational motion of a driver to the reciprocating motion of the piston. The piston's motion pulls gas into a cylinder from the suction line, and then displaces it from the cylinder through the discharge line. Check valves (compression valves) on the suction and discharge allow the flow of the gas in one direction only. Figure 7-9 displays an actual cutaway of a piston-type reciprocating compressor. Figure 7-10 displays the major components of the compressor, which include a cylinder, inlet, outlet, inlet and outlet valves, housing, piston, piston rings, and shaft.

FIGURE 7-9 Reciprocating Compressor

Courtesy of Baton Rouge Community College

FIGURE 7-10 Gas Flow in a Piston-Type Reciprocating Compressor

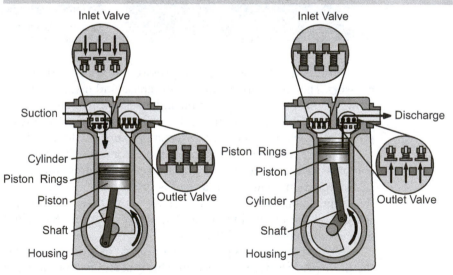

Piston-type compressors can be single- or double-acting. Double-acting compressors trap the gas during the suction stroke on one side of the piston, while compressing the gas on the discharge side of the piston at the same time. Figure 7-11 shows an example of both double- and single- acting compressors.

ROTARY COMPRESSORS

Rotary compressors (shown in Figure 7-12) move gases by rotating a set of screws, lobes, or vanes. As these screws, lobes, or vanes rotate, gas is drawn into the compressor by negative pressure on one side and forced out of the compressor (discharged) through positive pressure on the other. Rotary compressors do not require a constant suction pressure to produce discharge pressure.

FIGURE 7-11 Examples of Single- and Double-Acting Compressors

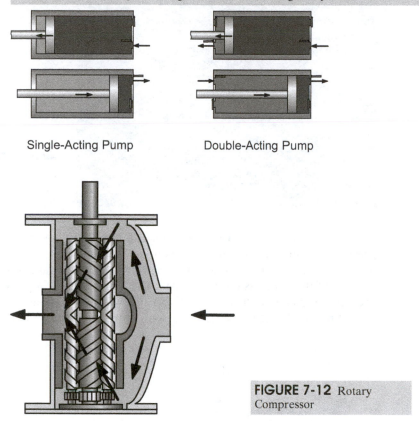

Single-Acting Pump   Double-Acting Pump

FIGURE 7-12 Rotary Compressor

In other positive displacement compressors, lobes or gears displace the gas from a cavity created between the rotors and the compressor body. If the suction pressure is lower than the original compressor design capacity, the compressor will still work, but with lower-than-design capacity results. Because of this, these compressors are appropriate for processes in which the inlet pressures change over a wider range than centrifugal compressors can operate. Figure 7-13 shows three diagrams of rotary compressors. Figure 7-14 is an example of a screw compressor, and Figure 7-15 shows a lobe compressor.

FIGURE 7-13 Examples of Rotary Compressors

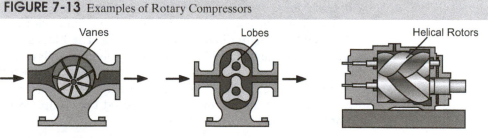

Sliding Vane Compressor   **Lobe Compressor**   **Rotary Screw Compressor**

FIGURE 7-14 Screw Compressor

Courtesy of Baton Rouge Community College

FIGURE 7-15 Lobe Compressor

Courtesy of Baton Rouge Community College

Another kind of rotary compressor is a liquid ring compressor. A **liquid ring compressor** uses an eccentric impeller with vanes to transmit centrifugal force to a sealing fluid (e.g., water), driving it against the wall of a cylindrical casing. The liquid moves in and out of the vanes as the rotor turns. The liquid is used in place of a piston that compresses the gas without friction. Gas is drawn into the vane cavities and is expelled against the discharge pressure. Figure 7-16 shows an example of a liquid ring compressor.

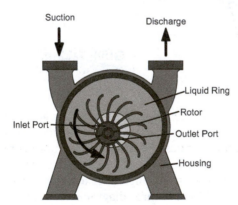

FIGURE 7-16 Liquid Ring Compressor

The sealing fluid in a recycling system is replenished and cooled in an external reservoir. In a once-through cooling system, the sealing fluid is removed and replaced with fresh fluid.

A sliding vane compressor employs a rotor filled with blades that move freely in and out of the longitudinal slots in the rotor. The blades are forced out against the housing wall by centrifugal force, creating individual cells of gas that are compressed as the eccentrically mounted rotor turns. As the vanes approach the discharge port, they have reduced the chamber volume and compressed the gas, which is discharged at a pressure much higher than when it entered the compressor. Figure 7-17 shows an example of a sliding vane compressor.

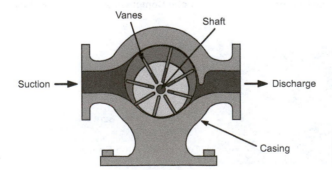

FIGURE 7-17 Sliding Vane Compressor Components

Operating Principles

Compressors are used in a variety of industries and have many uses. They are designed and selected based on inlet pressure, gas flow, and final discharge pressure. Inside a compressor, mechanical components compress gases and vapors for use in a process system.

When selecting compressors, unit designers must take into consideration the capability of the compressor and the process conditions. The difference between the operations of compressors and the operation of pumps is the physical properties of gases and liquids. While they operate similarly, compressors cannot move liquids, and pumps cannot move gases. Positive displacement compressors are called displacement compressors because they trap a volume of gas into a chamber, compress it with a device such as a piston or rotor, and then force (displace) it through the discharge valve and into the discharge piping.

SINGLE-STAGE VERSUS MULTISTAGE COMPRESSORS

Single-stage compressors, which compress the entering gas one time, are generally designed for high gas flow rates and low discharge pressures.

Multistage compressors compress the gas multiple times by delivering the discharge from one stage to the suction inlet of another stage. Figure 7-18 shows an example of a multistage compressor. Figure 7-19 is an example of a multistage compressor with parts labeled.

Multistage compressors raise the gas to the desired pressure in steps or stages (chambers), cooling the gas between each stage. Between stages, the gas from a multistage compressor is cooled. During this cooling phase, liquids are frequently condensed.

FIGURE 7-18 Functional Diagram of a Multi-Stage Compressor

First Stage   Second Stage

Pre-Cooler Heat Exchanger   Inter-Cooler Heat Exchanger   After-Cooler Heat Exchanger

Eccentric

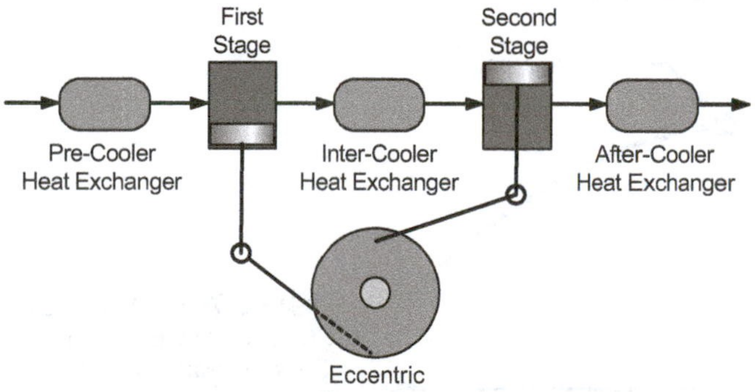

FIGURE 7-19 Multi-Stage Compressor with Labels

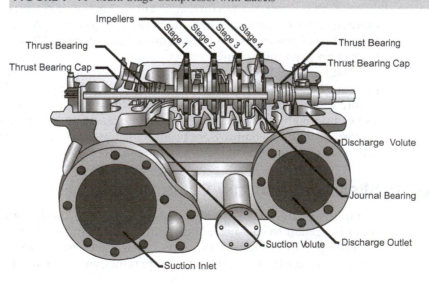

Impellers

Stage 1   Stage 2   Stage 3   Stage 4

Thrust Bearing

Thrust Bearing Cap

Thrust Bearing

Thrust Bearing Cap

Discharge Volute

Journal Bearing

Suction Volute   Discharge Outlet

Suction Inlet

These liquids must be removed and not allowed to enter the compressor because liquids are noncompressible and could cause severe damage to the compressor.

Frequently in the process industries, the desired discharge pressure is more than ten times that of the inlet pressure. In cases such as these, a single-stage compressor cannot be used because of the high temperature generated, so a multistage reciprocating compressor with cooling after each stage is required.

In all compressors, the temperature of a gas increases as it is compressed. The amount of temperature increase is a function of the gas and the compression ratio (the ratio of the discharge pressure to inlet pressure in absolute pressure units). To avoid extremely high discharge temperatures, the compression ratio in compressors is usually limited to around 3:1 or 5:1.

Multistage compressors can be centrifugal, axial, or reciprocating piston compressors. Large, multistage compressors can be extremely complex, with many subsystems including bearing oil systems, seal oil systems, and extensive vibration detection systems.

Centrifugal Compressor Performance Curve

The design and selection of compressors involve several factors, including physical property data and a description of the process gas. When a compressor is selected, a performance curve is provided with the compressor so engineers can track the operational performance of the equipment.

Centrifugal compressors must operate on a performance curve just like centrifugal pumps. A typical curve is shown in Figure 7-20. The vertical axis is feet of head (H_p), which means feet of head for a gas being compressed. The horizontal axis is actual cubic feet per minute (cfm) of gas compressed. The compressor operates on its specified curve unless a problem occurs (e.g., it malfunctions, is mechanically defective, or is compressing dirty gas).

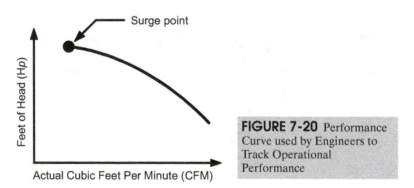

FIGURE 7-20 Performance Curve used by Engineers to Track Operational Performance

For all compressor performance curves, as the compressor discharge pressure increases, the H_p increases and the volume of gas compressed (ACFM) decreases. When the volume of gas compressed drops below a critical flow, then the compressor is backed up to its surge point.

Data such as flow rate, temperature, and pressure conditions are collected on compressors. These data are provided to engineers, who analyze the data on a daily basis to ensure that the compressor is operating at the designed output. If the data begins to show abnormalities, the compressor may be assigned to maintenance for repair. It is important to keep compressors properly maintained to prevent compressor failure, reduce maintenance costs, increase equipment life, and prevent process upset.

Associated Utilities and Auxiliary Equipment

Compressors are part of process systems. As a result, a wide variety of auxiliary equipment is needed to enhance equipment productivity. Auxiliary equipment associated with compressors includes lubrication systems, seal systems, antisurge protection, intercooler, after-cooler, heat exchangers, separators, and surge bottles.

LUBRICATION SYSTEM

Normally, when gases are compressed, the bearings and seals become hotter. Some of the heat is transmitted to the seals and bearings. This heat is removed by cooling the lubricants.

Lubrication systems circulate and cool sealing and lubricating oils. In some applications, the sealing and lubricating oils are the same. Figure 7-21 shows an example of a lubricant cooling system.

FIGURE 7-21 Diagram of Lubrication System Components

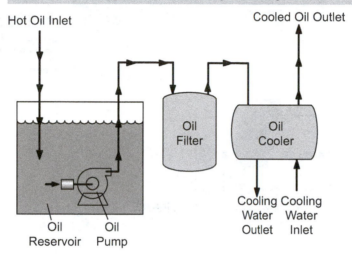

SEAL SYSTEM

Seal systems (shown in Figure 7-22) prevent process gases from leaking to the atmosphere where the compressor shaft exits the casing. Compressor shaft seals can include labyrinth seals, liquid buffered seals, and dry carbon rings.

A **labyrinth seal** (shown in Figure 7-23) is designed to restrict flow by requiring the fluid to pass through a series of ridges and intricate paths. A purge of an inert gas (barrier gas) is provided at an intermediate injection point on the seal to prevent external leakage of process gas.

A **liquid buffered seal** is a close-fitting bushing in which a liquid is injected in order to seal the process from the atmosphere.

Dry carbon rings (see Figure 7-24) are a low-leakage type of seal that can be arranged with a buffer gas that separates the process gas from the atmosphere.

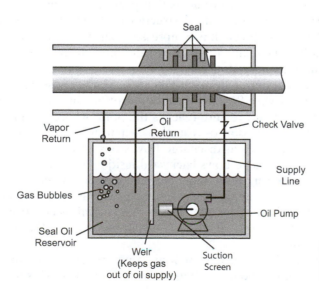

FIGURE 7-22 Typical Seal Oil System

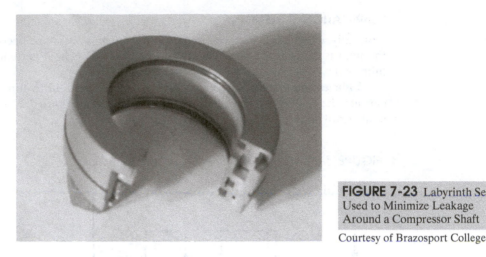

FIGURE 7-23 Labyrinth Seal Used to Minimize Leakage Around a Compressor Shaft

Courtesy of Brazosport College

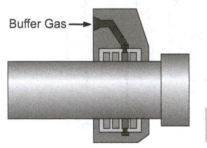

Buffer Gas →

FIGURE 7-24 Example of Dry Carbon Rings

Packing is used in reciprocating compressors to prevent leakage where the piston rod passes through the crank-end of the cylinder. Typical packing consists of rings of fibrous material fitted around the piston rod (contained in a stuffing box) and compressed by an adjustable gland. Friction between the packing and piston rod is reduced by the injection of lubricating oil into the stuffing box.

ANTISURGE DEVICES

When the flow rate on a compressor is low enough to match the minimum critical flow rate of the maximum flow stage, a recycle valve located downstream of the discharge opens and spills flow back into the suction. This avoids reverse flow through the compressor.

Some compressor installations use variable speed motors to control the discharge pressure. Steam turbine drives use steam flow to control the compressor speed. Larger machines have antisurge instruments that prevent surging because surging can cause bearing and shaft failures and the destruction of the machinery. **Surging** is the intermittent flow of pressure through a compressor that occurs when a stage fails to pump. Surging results when the compressor falls into an unstable condition. Antisurge devices are used to prevent surging.

In **antisurge** devices (shown in Figure 7-25), the pressure differential across the machine increases when flow is reduced by throttling in either the suction or discharge. When this differential pressure reaches the point where the machine can no longer maintain it, the compressor loses suction. When this happens, the check valve on the discharge quickly closes as discharge flow stops and the discharge pressure before the discharge check valve flows backward inside the compressor toward the suction. This causes the pressure differential between the suction and discharge to equalize, so the compressor once again begins moving the gas. The discharge pressure again reaches the critical value, and the process repeats. This is called surging. Surging is discussed later in this chapter. If allowed to continue surging, the compressor can experience severe damage.

Because surging is so detrimental to equipment, centrifugal compressors are equipped with antisurge protection to prevent damage. This antisurge protection is

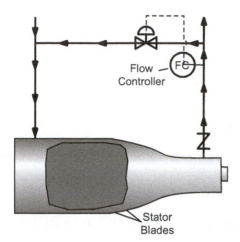

FIGURE 7-25 Diagram of a Antisurge Control System

Flow Controller

Stator Blades

designed by engineers to ensure proper and safe compressor operation. Compressors are also designed with automatic shutdown initiators (emergency shutdown switches) to prevent damage to the compressor during abnormal conditions.

One form of antisurge device is a minimum-flow recycle. In this type of device, the compressor discharge flow is measured. When a flow rate is low enough to match the minimum critical flow rate of the maximum flow stage, a recycle valve located downstream of the discharge opens and spills flow back to the suction. This avoids surging of the compressor.

Another form of antisurge protection is a variable speed driver, which controls the discharge pressure. When steam turbine drives are used, steam flow to the turbine is used to control the compressor speed.

COOLERS

Coolers are used to cool gases at various stages in the process. Figure 7-26 shows a functional diagram of a multistage compressor with precooler, intercooler, and after-cooler.

FIGURE 7-26 Functional Diagram of a Multi-Stage Compressor

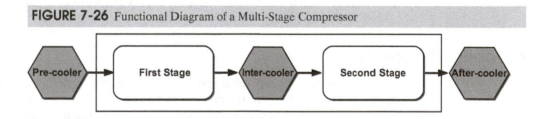

Pre-cooler → First Stage → Inter-cooler → Second Stage → After-cooler

Precoolers are used to cool the gas before it enters a compressor. Intercoolers are used in multistage compressors to cool gas between each stage of compression. In other words, intercoolers cool the discharge of the first stage before it enters the suction to the second stage. Interstage coolers prevent the overheating of the process gas and equipment. After-coolers cool the gas downstream from the last stage once the compression cycle is complete.

SEPARATORS

Separators physically separate two or more components from a mixture and collect excess moisture vapor from the compressor application. Examples of separators include demisters and desiccant dryers.

A **demister** (shown in Figure 7-27) is a device that promotes separation of liquids from gases. Demisters contain a storage area designed to collect liquid from gas and separate it from the gas in the system to prevent damage.

Desiccant dryers (shown in Figure 7-28) use chemicals (desiccants) to remove or absorb moisture. Desiccant dryers are usually used in tandem (one in service and one

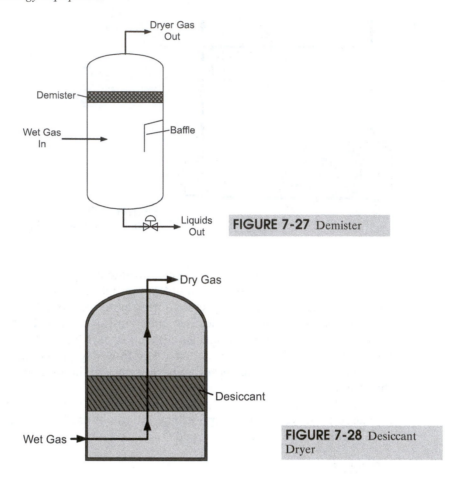

FIGURE 7-27 Demister

FIGURE 7-28 Desiccant Dryer

out of service). When the desiccant in one dryer becomes saturated, the dryer is taken out of service and hot gas is sent through the dryer bed to dry it out. Because they are more efficient than demisters, desiccant dryers are often used for sensitive applications like instrument air.

Potential Problems

The role of the process technician is vital to the production and safety of any process unit. The responsibility of the process technician is to have a clear understanding of the process in which he or she is working and always maintain a keen awareness of the surroundings. Process technicians must have a basic understanding of troubleshooting techniques to prevent damage to a compressor in a time of malfunction.

To prevent potential problems, compressors are equipped with various safety protection devices. Problems that can arise include overpressurization, overheating, surging, leaks, loss of lubrication, vibration, interlock systems, loss of capacity, motor overload, and governor malfunction. Process technicians must be able to recognize these conditions and respond appropriately.

OVERPRESSURIZATION

Compressor overpressurization can occur if the valves associated with the compressor are incorrectly closed or blocked. Consider the example in Figure 7-29. In this example, valve A is open and valve B is closed. This means the gas is flowing through valve A, and then back into the tank. If a technician were to redirect the flow of the gas through valve B, he or she must open valve B before closing valve A. If both valves are closed at the same time, shown in Figure 7-30, the gas has no place to go. This is referred to as dead heading. The result is backflow pressure that can damage the compressor or cause serious personal injury.

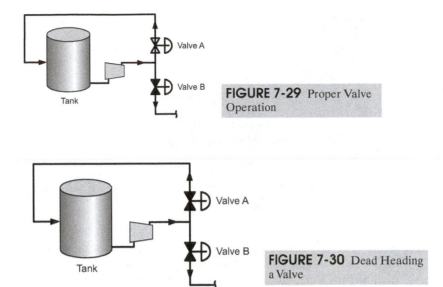

FIGURE 7-29 Proper Valve Operation

FIGURE 7-30 Dead Heading a Valve

OVERHEATING

Overheating is when excess heat is generated. Excessive amounts of overheating can be very detrimental to processes and equipment. Compressor overheating can be caused by a loss of cooling, improper lubrication, or valve malfunction. Compressor overheating is often caused by improper lubrication. Without lubrication, bearings fail and equipment surfaces rub together. As these surfaces rub against one another, friction is produced and heat is generated. Process technicians should always monitor compressors for unusual sounds and excessive heat because operating compressors under these conditions can lead to permanent equipment damage and/or personal injury (e.g., burns).

SURGING

Another operational hazard is surging (explained previously in the antisurge device section). Surging, which is typically associated with centrifugal compressors, is a temporary loss of flow to one or more impellers or stages of the compressor. This loss of flow causes the compressor speed to fluctuate wildly and vibration to increase dramatically. As discharge flow drops below acceptable levels, antisurge protection activates to ensure there is no reverse flow through the compressor.

Surging can result when the compressor throughput and head fall into an unstable region of the head/capacity curve. This can cause many problems, including bearing failures, shaft failures, and destruction of the machinery.

A compressor's surge point is a variable function of gas gravity, system pressure, machine speed, compressor differential pressure, and compressor rate. Surging can occur at a low gas gravity, system pressure, and compressor rate or at a high speed and high differential pressure.

Surge prevention is critical to prevent damage, so process technicians should adjust the variables listed above (to the extent they have control of them) in a direction away from that which leads to surging.

SEAL OIL PROBLEMS

Seal oil problems can cause loss of process gases. Examples of seal oil problems include dirty hydraulic oil (which causes plugging of passages), vibration or cracking of internal oil tubing, and internal seal failure.

LEAKS

Process technicians should always check compressor cylinders or housing for leaks because leaks can introduce harmful or hazardous substances into the atmosphere and

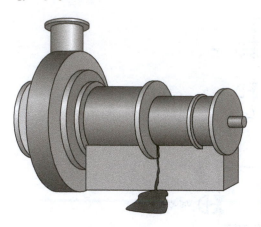

FIGURE 7-31 Compressor Showing Oil Leak

create process problems. Excessive flows of barrier gas purges indicate seal leakage into or out of the process. Monitoring process gas leakage to the atmosphere allows technicians to determine the condition of the atmospheric seal. Figure 7-31 shows a compressor with an oil leak.

Compressors may be damaged if liquid, which is noncompressible, is inadvertently introduced into compression chambers or between compressor components. Because of this, compressors employ a liquid removal step for gases that may condense inside the compressor, and interstage liquid removal points are provided to prevent the liquid from causing compressor damage.

When equipment is operating under vacuum conditions, monitoring process fluid for oxygen is necessary to determine whether air is leaking into the compressor.

LUBRICATION FLUID CONTAMINATION

Lubrication fluids can be contaminated with materials, such as water, process fluids, and metal particulates. As the contamination increases, the cooling and lubricating capacity of the fluid decreases. Because of this, frequent monitoring of the seal and bearing lubricants is necessary to avoid problems. Monitoring may include frequent inspection of bearing areas with temperature probes to prevent serious heat damage. It may also include a check for low flow or pressure of lubricant to the bearings.

VIBRATION

It is important to recognize excessive vibration because of the potential problems vibration can cause. Common causes of vibration include impeller imbalance or damage, worn bearings, piping pulsation, a misaligned shaft, damaged couplings, and bearing or seal damage. In most cases, vibration monitoring devices are attached to equipment to give an alarm when vibration becomes excessive. However, some installations depend on process technicians to detect vibration.

To prevent major damage, larger compressors are typically equipped with interlock systems that monitor and record the vibration of each compressor. These interlocks are designed to shut the compressor down if high vibration or excessive shaft movement is detected.

INTERLOCK SYSTEM

An **interlock** is a type of hardware or software that does not allow an action to occur if certain conditions are not met. For example, modern automobiles contain interlocks that prevent you from taking the key out of the ignition unless the transmission is in park. Interlocks are used to ensure that a proper sequence is followed or, if it is important enough, to shut down a process. Larger compressors typically contain interlock systems to prevent major damage.

Examples of interlock protection in compressors include indicators and shutdown systems for low lubrication, low oil pressure or flow, excessive vibration, high liquid level in a suction vessel, high discharge temperature, high differential pressure, or high power consumption. Many compressors are designed with emergency shutdown (ESD) instrumentation to shut them down if an abnormal condition exists. These types of interlock systems protect against a variety of failure modes. Most of these systems are designed to fail safe (i.e., shutdown the system in a manner that causes little or no equipment damage).

Voting sensor systems are used to reduce the frequency of false shutdowns. In a voting system, multiple temperature transmitters monitor the discharge temperature. The voting sensor system uses two-out-of-three logic to determine an action. In other words, if two out of three transmitters indicate that the temperature is too high, the compressor will shut down. This logic prevents the failure of one transmitter from shutting down the compressor unnecessarily. It also allows faulty transmitters to be repaired without shutting down the compressor.

LOSS OF CAPACITY

Most machines are designed with safety factors that allow continuous operation up to the driver's rated capacity. Exceeding the design rate may cause damage to the compressor due to the possibility of liquid carryover from the suction drum into the machine. Unless the liquid carried into the suction is in the form of a slug (a large amount given all at once), the protective instrumentation should be able to protect against significant damage. However, liquid droplets in small quantities carried into suction lines can damage the machine's components over time. Rotary compressors are frequently designed to handle the liquid phase in the suction.

Loss of capacity in a reciprocating compressor can be caused by leaking compression valves and/or leaking piston rings. A leaking compression valve can be indicated by higher-than-normal temperature on the valve's cover plate located on the exterior of the cylinder. Leaking piston rings may also be identified by a loss of rate with normal or lower-than-normal valve cover temperatures.

MOTOR OVERLOAD

Compressor motor overload and motor interlock shutdown can occur for a variety of reasons, including higher than designed discharge pressure, high motor amperage, or bearing failure. In some cases, the motor overload may occur if the compressor is used to compress a suction gas with a molecular weight above what the compressor is designed to handle. Compressors with variable speed electric drivers can suffer motor overload situations if the torque demand is high when the revolutions per minute (rpm) are low. This situation is most likely to occur with rotary compressor types.

HIGH/LOW FLOW

Lubricant flow that is too high or too low may indicate eminent bearing or seal failure. Because of this, lubricant flow monitoring systems are designed to sound an alarm should flow that is too high or too low be detected.

Gas flow that is too high through the compressor may be caused by a compressor operating at excessive rpm, which could result in compressor damage. Gas flow that is too low through the compressor may indicate suction side restrictions or the onset of compressor surge conditions.

Safety and Environmental Hazards

Hazards associated with normal and abnormal compressor operations can affect personal safety, equipment, and the environment. Table 7-1 lists some of these hazards and their effects.

Process Technician's Role in Operations and Maintenance

When working with compressors, process technicians should always conduct monitoring and control activities to ensure the compressor is not overpressurizing, overheating, or leaking. On a periodic basis, process technicians should check vibration, oil flow, oil

TABLE 7-1 Safety and Environmental Concerns Associated with Compressors		
Personal Safety Hazards	*Equipment*	*Environment*
• Hazardous fluids/ materials	• Excessive noise	• Discharge on ground or in sewer sump (seal leaks)
• Excessive noise levels	• Rotating equipment	• Discharge of harmful gas or fluid into the air
• Slipping/tripping due to leaks	• Electric shock	
• Chemical exposure	• Flange gasket or seal leakage or failure	
• Skin burns	• Fires resulting from seal blowout	
• Rotating equipment hazards	• Hot surfaces or cryogenic material releases	
• Hazardous fluids/ materials	• Excessive vibration	
• Leaks	• Inadequate lubrication	
• High or very low temperatures	• Overpressure causing equipment failure	
• High-voltage electricity	• Blown cylinder heads	
• High pressure		

level, temperature, and pressure. They should also make sure that connectors, hoses, and pipes are in proper condition; they should examine overspeed trips and oil levels, and check for leaks at seals, packing, and flanges.

Failure to perform proper monitoring and control could affect the process and result in equipment damage. All process technicians should monitor equipment and guard against hazards by following the appropriate procedures, which may include the following:

- Wear the appropriate personal protective equipment and observe all safety rules.
- Observe the compressor operation for signs of compressor failure or maintenance needs during normal operation.
- Check and record pressures, temperatures, and flow rates periodically to identify a problem trend.
- Check rotating equipment frequently for mechanical problems, such as lack of lubrication, seal leakage, overheating of compressors and drivers, and excessive vibration.
- Check lubrication systems to ensure proper temperatures and lubricant levels because seals become worn and weak and because motor, turbine, and compressor bearings become worn and overheat.

When working with compressors, process technicians should look, listen, and feel for the items listed in Table 7-2. In addition to operational maintenance, process technicians may perform additional tasks for routine and preventive maintenance.

Typical Procedures

Process technicians should be aware of the various compressor procedures, including startup, shutdown, lockout/tagout, and emergency procedures.

STARTUP AND SHUTDOWN

Compressors should be started under no load or low load conditions by opening the recycle valves and establishing minimum flow. Compressor drivers and the type of compressor being used determine automatic control schemes. Automatic control systems control compressor driver speed by either turbine or variable speed motors. These control systems make it easier to establish the operating conditions of the compressor.

TABLE 7-2 Process Technician's Role in Operation and Maintenance

Look	*Listen*	*Feel*
• Check oil levels to make sure they are satisfactory. • Check seals and flanges to make sure there are no leaks. • Check vibration monitors to ensure they are within operating range. • Verify that the liquid level in the suction drum is within limits.	• Listen for abnormal noises.	• Feel for excessive vibration. • Feel for excessive heat.

Startup procedures vary considerably from machine to machine. However, the following steps apply to most compressors:

1. Assure that the machine is completely assembled and ready for startup, and that the process is ready for the compressor to come on line.
2. Place ancillary lubricating and seal oil systems, housing vent lines, and other components into operation.
3. Use an inert gas, such as carbon dioxide or nitrogen, to purge the compressor case or cylinders of air, if required.
4. With pressure on the machine, drain the compressor's case or cylinders to remove any liquid, and then check for leaks.
5. Adjust the controls for minimum driver load at startup (e.g., unload reciprocating compressor cylinders).
6. Put the driver into condition to be started (e.g., prepare steam systems and superheaters and/or remove tagout devices from electrical switches).
7. Start the machine at low load. As applicable for variable speed drivers, bring the machine up to operating speed. Avoid running the equipment in its critical speed ranges.
8. For reciprocating compressors, adjust the cylinder loadings for efficient operation once you are satisfied that the machine is running smoothly.
9. Closely monitor operation as the machine runs in and comes to normal operating conditions.

LOCKOUT/TAGOUT

The following is a generic lockout/tagout procedure similar to the ones followed by process technicians preparing a compressor for maintenance:

1. De-energize the driver to all equipment per an energy isolation plan. If possible, physically disconnect the driver from the compressor.
2. Close all block valves to isolate the compressor following lockout/tagout (LOTO) and standard operating procedure (SOP).
3. De-pressurize the compressor.

Again, the procedures may differ from facility to facility, so process technicians should always follow the procedures specific to their unit. Figure 7-32 shows an example of a lockout/tagout sign.

EMERGENCY PROCEDURES

Emergency procedures include additional actions by a technician beyond those normally required during a shutdown. Process technicians should always follow the specific emergency operating procedures for their unit and company.

FIGURE 7-32 Lockout/
Tagout Sign

Summary

Compressors are used in the process industries to compress gases and vapors so that they may be used in a system that requires a higher pressure. Compressors are similar to pumps in the way that they operate. The primary difference is the service in which they are used. Pumps are used to move liquids and slurries, while compressors are used in gas or vapor service.

The two most common types of compressors are positive displacement and dynamic. Positive displacement compressors use screws, sliding vanes, lobes, gears, or pistons for compression. Positive displacement compressors include reciprocating and rotary compressors.

Reciprocating compressors use an inward stroke to draw gas into a chamber to positively displace the gas. Piston type reciprocating compressors use pistons connected to a crankshaft that converts the rotational motion of a driver to the reciprocating motion of the piston.

Rotary compressors move gas by rotating a screw, a set of lobes, or a set of vanes. Liquid ring compressors use an eccentric impeller with vanes to transmit centrifugal force into a sealing fluid and drive it against the wall of a cylindrical casing.

Dynamic compressors are nonpositive displacement compressors that use centrifugal or rotational force to move gases. Types of dynamic compressors include centrifugal and axial compressors. In a centrifugal compressor, gas enters through the suction inlet and goes into the casing where a rotating impeller spins, creating centrifugal force. Axial compressors use a series of rotor and stator blades to move gas along the shaft.

Compressors can be single- or multistage. In a single-stage compressor, the gas enters the compressor and is compressed one time. In a multistage compressor the gas is compressed multiple times, thereby delivering higher pressures.

Engineers use a compressor curve supplied by the manufacturer to design a compressor system. Process technicians use the curve to monitor the performance of compressors.

Auxiliary equipment associated with compressors includes lubrication systems, seal systems, antisurge devices, intercooler, after-cooler, and separators.

Potential problems associated with compressors include overpressurization, overheating, surging, seal oil problems, leaks, lubrication fluid contamination, vibration, interlock systems, loss of capacity, motor overload, high/low flow, and governor malfunction.

Process technicians are required to be aware of safety and environmental hazards associated with compressors and to follow the standard operating procedures for conditions of the unit operation.

Checking Your Knowledge

1. Define the following terms:
   a. Axial compressor
   b. Centrifugal compressor
   c. Dynamic compressor
   d. Liquid ring compressor
   e. Multistage compressor
   f. Positive displacement compressor
   g. Reciprocating compressor
   h. Rotary compressor
   i. Single-stage compressor
   j. Cylinder
   k. Interlock
   l. Packing
   m. Piston
   n. Shaft
   o. Surging

2. Which type of compressor uses pistons, lobes, screws, or vanes to compress gases?
   a. Dynamic
   b. Positive displacement

3. Which type of compressor uses impellers to generate centrifugal force?
   a. Dynamic
   b. Positive displacement

4. Which type of compressor uses sliding vanes to force gases out of a chamber?
   a. Axial
   b. Centrifugal
   c. Positive displacement

5. *(True or False)* In an axial compressor, the flow of gas moves parallel to the shaft.

6. On the diagram below, identify the following parts of a centrifugal compressor.
   a. Casing
   b. Discharge
   c. Impeller
   d. Shaft
   e. Suction
   f. Bearings

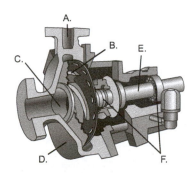

7. Which type of rotary compressor uses an impeller with vanes to transmit centrifugal force into a sealing fluid (e.g., water), driving it against the wall of a cylindrical casing?
   a. Screw
   b. Lobe
   c. Sliding vane
   d. Liquid ring

8. *(True or False)* In a centrifugal compressor, a rotating impeller spins, creating centrifugal force that decreases the velocity of the gas.

9. A lubrication system in a compressor:
   a. provides circulation and cooling of sealing and lubrication oils
   b. minimizes total leakage
   c. increases the pressure on process gases
   d. transfers heat to the seals

10. Surging can cause (select all that apply):
    a. seal leakage
    b. vibration
    c. destruction of the machinery
    d. dead heading conditions

11. *(True or False)* Interlock systems are used to bypass emergency shutdown systems in order to keep production running so a particular unit can meet its production goals.

Student Activities

1. Given a cutaway or a diagram of a centrifugal compressor, identify the following components:
   a. Casing
   b. Discharge
   c. Impeller
   d. Shaft
   e. Suction
2. Given a cutaway or diagram of a reciprocating compressor, identify the following components:
   a. Casing
   b. Connecting rods
   c. Suction valve
   d. Discharge valve
   e. Piston
3. Using a P&ID provided by your instructor, locate and identify the type of compressor. Prepare a written report discussing potential problems that can arise from a compressor malfunction and the systems it can affect.
4. Using a P&ID provided by your instructor, locate the compressors. Prepare a presentation about the auxiliary equipment associated with each compressor and how it can enhance production.
5. Write a report about surging and the importance of antisurge prevention in compressors. Draw a diagram of an antisurge device for a compressor system or, using a P&ID provided by your instructor, design an antisurge device to improve the existing system shown on the P&ID.

CHAPTER

8

Turbines

Objectives

After completing this chapter, you will be able to:

■ Explain the purpose of turbines in the process industry.

■ Identify the common types and applications of turbines.

■ Identify and explain the purpose of turbine components.

■ Describe the operating principles of turbines.

■ Describe safety and environmental hazards associated with turbines.

■ Identify typical procedures associated with turbines.

■ Describe the process technician's role in turbine operation and maintenance.

■ Identify potential problems associated with turbines.

Key Terms

Carbon ring—a seal component located around the shaft of the turbine that controls the leakage of motive fluid (typically steam) along the shaft or the entrance of air into the exhaust.

Casing—a housing component of a turbine that holds all moving parts, (including the rotor, bearings, seals) and is stationary.

Chevrons—V-shaped blades found on a turbine rotor.

Combustion chamber—a chamber between the compressor and the turbine where the compressed air is mixed with fuel. The fuel is burned, increasing the temperature and pressure of the combustion gases.

Condensing steam turbine—a device in which exhaust steam is condensed in a surface condenser. The condensate is then recycled to the boiler.

Fixed blade—a blade inside a turbine, fixed to the casing, that directs steam or combustion gases.

Gas turbine—a device that uses the combustion of natural gas to spin the turbine rotor.

Governor—a device used to control the speed of a piece of equipment such as a turbine.

Hydraulic turbine—a device that is operated or affected by a liquid (e.g., a water wheel).

Impulse movement—movement that occurs when the steam first hits a rotor and the rotor begins to move.

Impulse turbine—a device that uses high-pressure steam to move a rotor.

Mechanical energy—energy of motion that is used to perform work.

Moving blade—rotor blades that are connected to the shaft and that move or rotate when a gas is applied.

Multistage turbine—a device that contains two or more stages used as a driver for high-differential pressure, high-horsepower applications, and extreme rotational requirements.

Noncondensing steam turbine—a device that acts similar to a pressure-reducing valve by converting the energy released into mechanical energy.

Nozzle—restrictive component used to drive the blades or rotor of the turbine. The nozzle converts the steam or motive fluid from pressure to velocity.

Overspeed trip mechanism—an automatic safety device designed to remove power from a rotating machine if it reaches a preset trip speed.

Radial/thrust bearing—designed to support and hold the rotor in place while offering minimum resistance to free rotation. The bearing prevents and offsets axial and radial movement.

Reactive turbine—a steam device with fixed nozzles and an internal steam source and that uses Newton's third law of motion.

Rotor—a rotating member of a motor or turbine that is connected to the shaft.

Sentinel valve—a spring-loaded, high-whistling safety valve that opens when the turbine reaches near maximum conditions.

Shaft—a metal rotating component (spindle) that holds the rotor and all rotating equipment in place.

Single-stage turbine—a device that contains one set of blades (two that turn and one that is fixed) called a stage.

Steam chest—area where steam enters the turbine.

Steam strainer—a mechanical device that removes impurities from steam.

Steam turbine—a device that is driven by the pressure of high-velocity steam, which is discharged against the turbine's rotor.

Trip throttle valve—a component designed to shut down the turbine in the event of excess rotational speed or vibration.

Wind turbine—a device that converts wind energy into mechanical or electrical energy.

Introduction

Turbines are used in the process industries to convert the kinetic and potential energy of a motive fluid into the mechanical energy necessary to drive a piece of equipment or generate electricity. Gas and steam turbines are activated by the expansion of a fluid on a rotor attached to a central shaft. These motive fluids provide the input power necessary to rotate shaft-driven equipment. Hydraulic turbines use pressurized liquids as their motive energy source.

While each of the different turbine types varies in how they work, all of them use the same basic principles. That is, they utilize a fluid force to turn a rotor. The rotor, in turn, powers process devices that are connected to the turbine shaft.

Common Types and Applications of Turbines

A turbine is a machine used to produce power and rotate shaft-driven equipment such as pumps, compressors, and generators. The four main types of turbines are steam, gas, hydraulic, and wind. Turbines are classified according to how they operate and the fluid that turns them (i.e., steam, gas, liquid, or air). The two most common turbine types used in industry are steam turbines and gas turbines.

STEAM TURBINE

A **steam turbine** is a device that is driven by the pressure of high-velocity steam discharged against the turbine's rotor. **Mechanical energy** is the energy of motion that is used to perform work. Steam turbines use the temperature and pressure energy of steam to turn a rotor and produce mechanical energy, so there is a direct relationship between the boiling point of the water and the pressure of the steam.

Steam turbines are the drivers for a variety of equipment such as pumps, generators, and compressors. Steam turbines have many advantages over electrical equipment. For example, they are free from spark hazards, so they are useful in areas where volatile substances are produced. In addition, they do not require electricity to run, so they can perform during power outages. They are also suitable for damp environments that might cause electrical equipment to fail. Figures 8-1 and 8-2 show cutaways of steam turbines.

As steam enters a turbine, the steam passes through a nozzle. The nozzle restricts the flow and increases the pressure of the steam. This pressurized steam passes through the turbine against the blades, causing rotation.

FIGURE 8-1 Cutaway of a Steam Turbine

Courtesy of Bayport Training & Technical Center

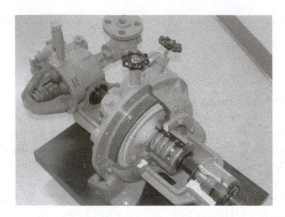

FIGURE 8-2 Cutaway of a Steam Turbine with Wheels Shown and Governor on Left

Courtesy of Baton Rouge Community College

The blades in a turbine are attached to a wheel. The combined blades, wheel, and shaft is called the rotor assembly. As the steam progresses through the rotor assembly, the kinetic and potential energy of the steam is transformed to mechanical energy.

Steam turbines can be classified as either impulse or reactive, condensing or non-condensing, and single-stage or multistage.

Simple Reactive Turbines

A simple **reactive turbine** (shown in Figure 8-3) is a steam device with fixed nozzles and an internal steam source, and uses Newton's third law of motion. Newton's third law of motion states that for every action there is an equal but opposite reaction.

In a simple reactive turbine, water is placed in a globe that contains two opposing nozzles. As the water is heated, steam, temperature, and pressure are produced, and the pressure forces the steam out of the nozzles. As the steam exits, propulsive (rotational) force is created. This propulsive force causes the globe to spin.

FIGURE 8-3 Reactive Turbines are Acted Upon by An Internal Steam Source

Impulse Turbines

Impulse turbines are another type of turbine. Like a reactive turbine, an **impulse turbine** uses high-pressure steam to move a rotor. However, instead of generating their own steam the way reactive turbines do, impulse turbines are acted upon by an external steam source. Figure 8-4 shows an example of an impulse turbine.

In an impulse turbine, steam is channeled through a steam nozzle onto the turbine blades. As the steam passes through the nozzle, its pressure and temperature energy is converted to velocity energy. This force causes the rotor to turn, thereby rotating the shaft and any equipment coupled to the shaft.

Condensing Versus Noncondensing Turbines

A method for classifying steam turbines is based on whether they are condensing or noncondensing. A **condensing steam turbine** is a device in which exhaust steam is condensed in a surface condenser, and the condensate is then recycled to the boiler.

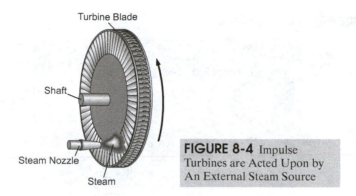

Turbine Blade

Shaft

Steam Nozzle

Steam

FIGURE 8-4 Impulse Turbines are Acted Upon by An External Steam Source

A **noncondensing steam turbine** is a device that acts similar to a pressure-reducing valve by converting the energy released into mechanical energy. The lower pressure exhaust steam still retains 90 percent of its heat energy and can be used for other industrial applications requiring heat.

Single-Stage Versus Multistage Turbines

A third way to classify turbines is single-stage or multistage. A **single-stage turbine** is a device that contains a set of blades (two that turn, and one that is fixed) called a stage. Figure 8-5 shows an example of a single-stage turbine.

In a single-stage turbine, the **chevrons** (V-shaped blades) on the first turning blade direct the steam toward the fixed blade. The chevrons on the fixed blade then direct the steam to the chevrons on the second turning blade. This causes **impulse movement** (movement that occurs when the steam first hits a rotor and the rotor begins to move).

A **multistage turbine** is a device that works like a single-stage turbine but contains multiple stages. This type of turbine is typically used as a driver for high-differential pressure, high-horsepower applications, and extreme rotational requirements. Each stage in a multistage turbine provides additional energy to the rotor. Figure 8-6 shows an example of a multistage turbine.

In a multistage turbine, when one turbine stage turns, they all turn because they are all fixed to the same rotor. As the steam moves down the line in a multistage turbine, the amount of steam energy decreases. This pressure differential is what helps provide motive force. (*Note:* The greater the pressure differential, the more pressure or force you have.)

Steam Turbine Components

Turbine components vary depending on the type of turbine and its application. However, the most common components are turbine wheels, seals, bearings, steam

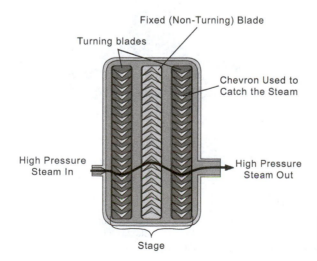

Fixed (Non-Turning) Blade

Turning blades

Chevron Used to Catch the Steam

High Pressure Steam In

High Pressure Steam Out

Stage

FIGURE 8-5 Single-Stage Turbine

FIGURE 8-6 Multistage Turbine

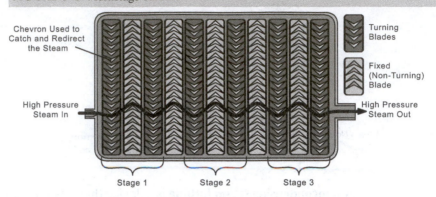

inlets and outlets, a throttle valve, and a governor. Figure 8-7 shows a cutaway photograph of steam turbine components. Figure 8-8 shows a cutaway illustration of steam turbine components.

The **casing** is the housing around the internal components of a turbine. The casing holds all moving parts, including the rotor, bearings, and seals, and is stationary. Some

FIGURE 8-7 Cutaway of a Turbine

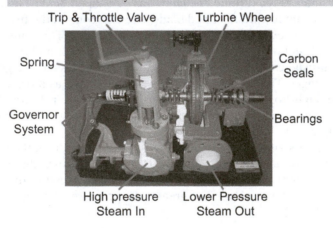

FIGURE 8-8 Steam Turbine Components

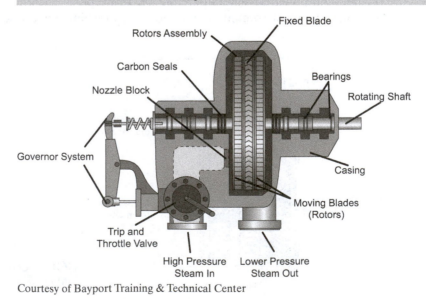

Courtesy of Bayport Training & Technical Center

FIGURE 8-9 Turbine
Housing Covering the Rotors

Courtesy of College of the
Mainland

turbine designs use split casings for ease of rotor, seal, and bearing maintenance. Figure 8-9 shows a casing on a turbine.

The **steam chest** is the area in a turbine where steam enters, and the **steam strainer** removes impurities from steam. The **shaft** is a metal rotating component (spindle) that holds the rotor and all rotating equipment in place. It is suspended by bearings. The **moving blades** are rotor blades that are connected to the shaft and that move or rotate when steam is applied. The rotor is the moving part that extracts work from the motive fluid and transfers it to the shaft. The **fixed blades** (stators) are the blades inside a turbine that are fixed to the casing, and they direct steam or combustion gases. Because they are fixed, these vanes redirect the motive fluid so it enters more efficiently in the direction of the moving blades. Fixed blades also convert velocity back to pressure.

A **rotor** is the rotating member of a motor or turbine that is connected to the shaft. The **governor** is a device used to control the speed of the turbine. It may also be used to restrict steam flow as directed by the overspeed trip mechanism. A **nozzle** is a small spout or extension on a hose or pipe that directs the flow of steam. This restrictive component is used to drive the blades or rotor of a turbine. The nozzle converts the steam or motive fluid from pressure to velocity. Steam enters the nozzle from the steam inlet, and exits the turbine through the steam outlet.

The **trip throttle valve** is a component designed to shut down the turbine in the event of excess rotational speed or vibration. These types of valves work in conjunction with the governor to control the speed of the turbine. Figure 8-10 shows an example of a throttle valve used to control turbine speed.

Steam turbines are complex in nature and have additional components that are not listed in Figure 8-8. These additional components may include bearings, seals and overspeed trip devices.

Radial/thrust bearings are bearings designed to support and hold the rotor in place, while offering minimum resistance to free rotation. These bearings prevent and offset axial and radial movement.

FIGURE 8-10 Throttle Valve
That is Used to Control
Turbine Speed

Courtesy of Bayport Training &
Technical Center

FIGURE 8-11 Top View of the Carbons Seals

Courtesy of Bayport Training & Technical Center

Carbon rings are components of seals that are located around the shaft of the turbine and that control the leakage of motive fluid (typically steam) along the shaft or the entrance of air into the exhaust. Seals are mechanical devices that hold lubricants and process fluids in place, while keeping out foreign materials. In a turbine, seals keep fluids from leaking out past the rotor shaft at the point where it exits the case. Figure 8-11 shows the top view of the carbon seals that are used to prevent steam from leaking out of the turbine casings.

An **overspeed trip mechanism** is an automatic safety device designed to remove power from a rotating machine if it reaches a preset trip speed. The overspeed trip is the last-resort stop valve for inlet fluid to prevent damage to the turbine caused by excessive rotating speed. The overspeed trip protects the turbine when the governor fails to maintain the correct speed.

GAS TURBINE

A **gas turbine** is a device that uses the combustion of natural gas to spin the turbine rotor. Gas turbines are coupled to an auxiliary electric motor or small steam turbine and are used as starter motors to start the turbine. Figure 8-12 shows a drawing of a gas turbine with labels showing the shell and internal components of the turbine.

Gas turbines are more efficient, but typically more complex, than steam turbines. In a gas turbine, an electric motor is used to take the shaft and get the air compressor up to speed and start the machine. Once the turbine is started, it provides sufficient power to drive the integral air compressor as well as the connected process equipment

FIGURE 8-12 Gas Turbine with Internal Components Labeled

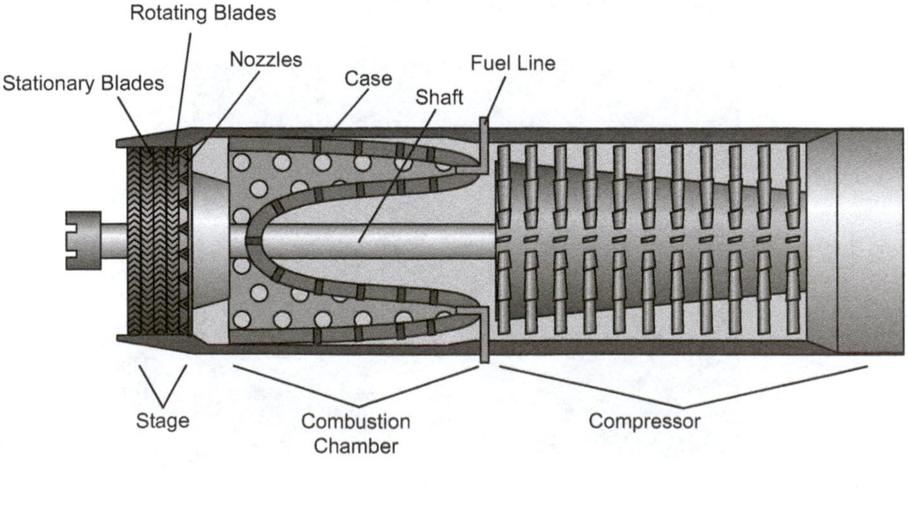

Stationary Blades

Rotating Blades

Nozzles

Case

Shaft

Fuel Line

Stage

Combustion Chamber

Compressor

(e.g., pumps, compressors, and generators). To conserve energy, the hot exhaust gases produced by the gas turbine may be used as a heat source for other parts of the process.

The compressor in a gas turbine pressurizes air for entry into the combustion chamber. All gas turbines have a compressor. In some cases, as the air is compressed through several stages, it is beneficial to cool the air to reduce the volume to be compressed in later stages. Heat exchangers called intercoolers are used between the stages to cool the gases before they reach the next stage.

Combustion chambers are chambers between the compressor and the turbine where the compressed air is mixed with fuel. The fuel is burned, increasing the temperature and pressure of the gases. These hot gases flow into the turbine, where energy is extracted. The greater portion of the energy drives the air compressor. The smaller portion of the turbine-generated energy is available as shaft work output.

HYDRAULIC TURBINE

A **hydraulic turbine** is a turbine that is moved, operated, or affected by a liquid. In a hydraulic turbine, a liquid flows across the rotor blades, forcing them to move. The faster the liquid flows, the faster the wheel turns. An example of a simple hydraulic turbine is a water wheel like the one shown in Figure 8-13.

FIGURE 8-13 Hydraulic Turbine

WIND TURBINE

A **wind turbine** is a mechanical device that converts wind energy into mechanical or electrical energy. A windmill, like the one in Figure 8-14, is an example of a wind turbine.

In a wind turbine, air currents move across fan-like blades, causing them to turn. As they turn, a shaft is rotated. The rotation of the shaft drives devices such as pumps or electrical generators. Figure 8-15 shows an example of a simple wind turbine and how it might be used to power a water pump.

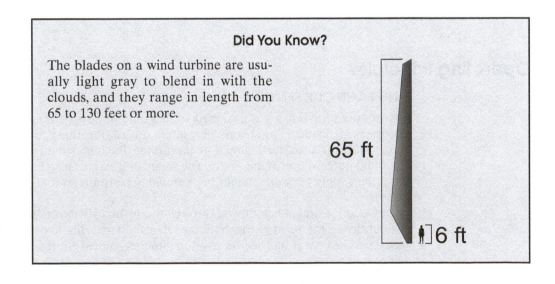

Did You Know?

The blades on a wind turbine are usually light gray to blend in with the clouds, and they range in length from 65 to 130 feet or more.

65 ft

6 ft

FIGURE 8-14 Wind Turbines Generating Electricity

FIGURE 8-15 Wind Turbine Actuating a Pump

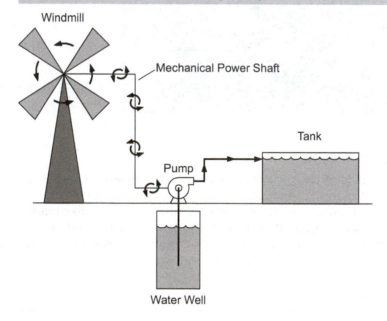

Operating Principles

INLET AND OUTLET FLOWS

Most steam turbines are designed to use only dry superheated steam. Any water droplets in the incoming steam can result in severe vibration problems and can damage the turbine. Before starting up a steam turbine, the inlet steam line and turbine casing must be hot and free of any water. The steam or gas must also be free of contaminants, and the supply pressure should be reasonably constant so that governor operation is stable.

Once the steam has flowed through the turbine, it moves toward the outlet. The outlet flow is the point at which the steam or fluid pressure leaves the turbine after it is used and its energy has been converted into mechanical energy.

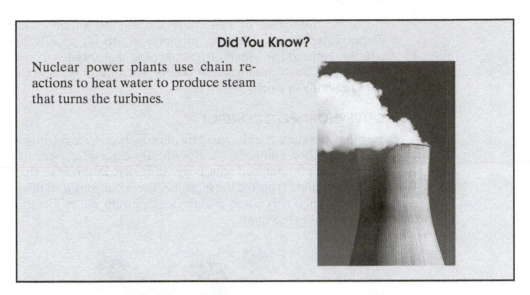

Did You Know?

Nuclear power plants use chain reactions to heat water to produce steam that turns the turbines.

WHAT HAPPENS INSIDE A TURBINE

A turbine operates because high-velocity steam, gas, or air turns blades that turn the shaft. The kinetic (velocity) energy and the potential (pressure) energy of the gas are converted into rotational mechanical energy that actuates (moves) process equipment.

Back Pressure versus Condensing

Inlet gas is at high pressure and low velocity. Once inside the turbine, the fluid pressure and flow are converted to kinetic energy (energy associated with mass in motion) through high-velocity passages. This kinetic energy is what turns the shaft. The power extracted from incoming fluid is maximized as the exhaust backpressure is minimized. However, as the pressure reduces, the volume of the gas expands. Because of this expansion, the passages and moving vanes in the lower pressure zones must be sufficiently large to minimize pressure drop.

In condensing steam turbines, the exhaust pressure is usually well below atmospheric pressure and is set mostly by the temperature of the cooling water in the condenser. The steam condenses to water under vacuum conditions.

Control Systems

The principal control systems inside a turbine include the governor, the extraction/induction system, the oil system, and the safety/trip system. The safety/trip system consists of all sensors (e.g., high vibration or low oil pressure) that cause turbine shutdown, and a logic controller that permits or dumps oil pressure off the trip/throttle valve. Dumping oil pressure off this valve closes off the steam supply to the machine.

Radial and Axial Shaft Movement

Radial movement is the side-to-side or up-and-down movement that occurs when the rotor starts to rotate. The use of radial bearings helps minimize the amount of vibration that radial movement can cause. Thrust is the axial, back-and-forth movement that is parallel to the shaft. The use of thrust bearings helps minimize the amount of vibration that thrust can cause. Figure 8-16 shows examples of radial and axial movements.

FIGURE 8-16 Illustration Showing Coupling Shaft Movement

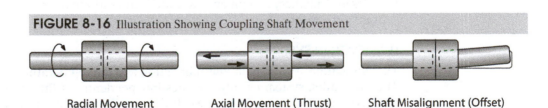

Radial Movement   Axial Movement (Thrust)   Shaft Misalignment (Offset)

Shaft misalignment can cause damage to shafts, bearings, turbine wheels, diaphragms, or casings. In large turbines, sensors are usually installed to continuously monitor vibration and shaft displacement so that shutdown occurs before the machine is damaged. Shaft misalignment occurs when the shaft is not in the proper position, either radially or axially.

GOVERNOR/SPEED CONTROL

Turbine operation is maintained by the governor system (shown in Figure 8-17), which is a closed-loop controller. A closed-loop controller measures variables, compares them to the setpoint, and adjusts performance. In this case, the setpoint is the revolutions per minute (rpm) of the machine. The input sensor of the governor is typically an electronic shaft revolution counter set to produce a millivolt or digital output signal proportional to the rpm.

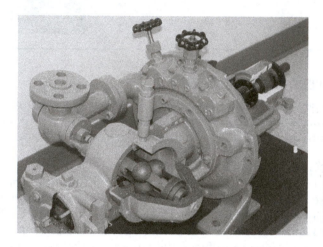

FIGURE 8-17 Cutaway of a Turbine Showing the Fly Ball Governor

Courtesy of Baton Rouge Community College

The governor system can also be a mechanical flywheel rpm device. The output of the governor system is important to the position of the control rod (the link between the governor and the throttle valve). When the governor is signaled to open or close the throttle valve, the governor adjusts the control rod to adjust the speed. For example, when you press the accelerator in your car, the accelerator is linked to the throttle on the carburetor or fuel injector. The more you apply the accelerator, the more fuel is fed to the carburetor or fuel injector and the faster the car runs.

The controller portion of the governor system is either a mechanical link or an electronic, analog, or digital controller. The mechanical link arrangement operates the shaft of the governor valve. Speed can be changed only by internal adjustment.

The **sentinel valve** is a spring-loaded, high-pitched, whistling safety valve that opens when the turbine reaches near maximum conditions. In other words, the sentinel valve acts as a warning device.

The speed setpoint can be selected either by the process technician or by a master controller. As the driven unit changes the amount of power demanded, the governor opens or closes the steam admission valve to maintain the required speed.

Auxiliary Equipment Associated with Turbines

Utilities and auxiliary equipment associated with turbine operation include lubrication systems and seals.

LUBRICATION SYSTEM

The lubrication system maintains lubrication and the temperature of turbine bearings. The lubrication system usually operates independently of the turbine, but at times it may use a shaft-driven oil pump. Shaft-driven oil pumps use power from the turbine shaft to drive a pump that forces oil into the bearings.

The lubrication system consists of a pump, oil filters, oil coolers, supply and return lines, and an oil reservoir. The lubrication system also contains an emergency pressure tank so that oil pressure can be temporarily maintained if the oil pump stops.

To prevent damage in larger machines, it is imperative that oil flow starts before shaft rotation begins and that the oil flow continues until after the shaft rotation stops. Because lubrication is so critical, many large machines are equipped with interlock devices to ensure that this lubrication is occurring.

SEALS

The shaft sealing system keeps the high-pressure motive fluid in the proper passages by preventing excessive leakage along the rotating and nonrotating boundaries. Seal designs depend on fluids, pressure, and temperatures. Typical seal designs are labyrinth and nonrotating carbon rings.

Labyrinth seals (shown in Figure 8-18) are close-fitting, rotating and nonrotating discs. Labyrinths consist of multiple rings through which oil cannot pass. The labyrinth seals are located along the turbine shaft to minimize leaks.

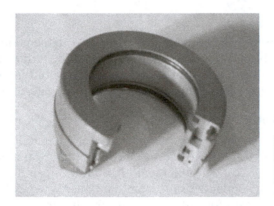

FIGURE 8-18 Example of a Labyrinth Seal Used to Keep Fluids Inside the Turbine

Courtesy of Brazosport College

In nonrotating carbon rings, sealing occurs when the rings are forced against the shaft, minimizing leakage. Nonrotating carbon rings are kept from rotating by a stop on each ring.

Sometimes buffer fluids, such as nitrogen gas, are used to regulate temperature at the seal and exclude motive fluids from oil systems. This allows some leakage of the buffer gas in order to prevent leakage of the motive fluid in the other direction.

Potential Problems

When working with turbines, process technicians should always be aware of potential problems such as turbine failure, loss of lubrication, bearing damage, impingement, excessive vibration, fouling, pitting, corrosion, assembly errors, and turbine overspeeding.

TURBINE FAILURE

Turbines are generally the prime movers of large and single-train machinery. Because of this, turbine failure usually shuts down the processing in a unit. This shutdown can result in flaring and a total loss of light gases in the process.

In the case of gas turbines and cogeneration facilities, the loss of a turbine and generator set does not cause widespread process unit outage. When electrical loss occurs, the backup utility power compensates for the loss of power. The boiler downstream of the turbine may or may not be able to run without the turbine operating. Because gas turbine installations are not inherently as reliable as steam turbines, they are typically in applications where an outage will not cause a widespread facility problem. However, the reliability of properly designed gas turbine installations is continually improving and their use as process equipment drivers continues to increase.

LOSS OF LUBRICATION

Loss of lubrication can result from oil pump failure or plugging of the oil filter. Turbine bearing failure can result from dirt in the oil or a block in the oil supply, so filtration must be adequate to retain particles whose size may exceed the oil film thickness.

When maintaining and lubricating turbines, process technicians should always utilize the correct oil as specified by the manufacturer or in the standard operating procedures (*Note:* Many operations label the correct lubricant on the machine to reduce the chance of an incorrect lubricant being used.) Using an incorrect lubricant (e.g., one that is too thick or too thin) can damage the equipment and create potentially hazardous situations such as reactivity with bearing and seal materials or process fluids.

BEARING DAMAGE

Bearing damage can result from excessive temperatures and pounding from vibration caused by shaft bowing or other shaft misalignment problems. Process technicians should pay close attention and report excessive vibrations so they can be eliminated before major damage occurs. Furthermore, process technicians should not depend entirely on vibration interlocks and detectors because these devices sometimes fail.

IMPINGEMENT

Impingement occurs when the steam quality decreases to the point that steam condensate exists in the turbine. Water droplets impinging on the turbine blades contribute to wear and erosion on the blades. The steam entering the turbine must be adequately superheated to avoid condensation and the resulting impingement.

FOULING, PITTING, AND CORROSION

Turbine fouling is caused by mineral deposits from the steam. The solution for fouling is a wheel wash or hot water wash to break down and carry off mineral deposits.

Pitting can be caused by a steam leak that allows steam to condense inside the turbine. As the steam condenses, salts from the boiler water collect on surfaces and cause pitting. The solution for pitting is to ensure that the steam supply has double block and bleed valves to prevent steam leaks to the turbine.

Most corrosion/erosion problems come from damage that takes place when the turbine is not running. Standby turbines should be kept free of moisture or other contaminants.

ASSEMBLY ERRORS

Excessive wear, turbine failure, or turbine loss can also be caused by assembly error. Process technicians should pay close attention to the assembly and maintenance of the turbine and its lubrication system.

TURBINE OVERSPEEDING

Overspeeding of a turbine can result in catastrophic failure of the machine. Destructive overspeeding results from loss of load. Deposits cause both the governor stem and the trip valve stem to stick when there is a loss of load. To prevent this, process technicians should always follow proper preventive maintenance procedures.

HUNTING

Hunting is a condition in which a control loop is improperly designed, installed, or calibrated, or when the controller itself is not properly aligned. Process technicians will observe this condition as a random behavior or cycling above and/or below the setpoint (the desired control point). Figure 8-19 shows an example of hunting.

OVERPRESSURIZATION

Sentinel valves are safety valves used to warn process technicians of equipment overpressurization. These types of valves alert the process technician, by making a very loud whistle, to take corrective actions and thus prevent turbine shutdown or relief valve activation.

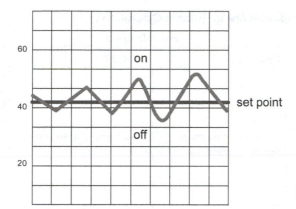

FIGURE 8-19 Hunting

Safety and Environmental Hazards

Hazards associated with normal and abnormal turbine operations can affect personal safety, equipment, production, and the environment. Table 8-1 lists several potential hazards and their effects.

Typical Procedures

Process technicians should be familiar with the appropriate procedures for monitoring, maintaining, and operating turbines.

STARTUP PROCEDURES

To ensure safety during turbine startup, process technicians must be familiar with proper startup procedures. The following is an example of a steam turbine startup procedure. Before attempting to start up a turbine, always refer to the startup procedure for the steam turbines in your unit.

1. Warm up steam lines and equipment to prevent water damage.
2. Place machines on slow roll (controlling turbine speed at low revolutions per minute) to bring the temperature up gradually and reduce the likelihood of thermal shock.
3. Properly commission the oil systems and, in the case of gas turbines, fuel systems.
4. Ensure that boiler equipment is ready to run in the case of gas turbines with cogeneration boilers.
5. Prepare low-pressure steam headers to receive letdown steam.
6. Make sure the boiler system is ready to increase load as the turbine is started.
7. Be familiar with the driven equipment as the turbine is started up so that the needs of each item are met.

While the turbine is down, the facility steam level letdown valves are open, holding the facility header pressure steady. As the turbine starts up, these letdown valves must close to maintain steady facility header pressures.

SHUTDOWN PROCEDURES

An example of a steam turbine shutdown procedure is as follows:

1. Switch to the spare.
2. Close the inlet valve to the steam turbine (*Note:* The inlet valve is normally closed first to prevent overpressurization of the turbine.)
3. Close the outlet steam valve from the turbine (if required) and open a bleed valve on the turbine casing (if required).

TABLE 8-1   Hazards Associated with Improper Steam Turbine Operation

Improper Operation	Possible Effects			
	Individual	*Equipment*	*Production*	*Environment*
Touching the external housing of a steam turbine that is full of steam	Burns caused by exposure to heat or steam			Hot surfaces should be insulated to avoid heat loss to the environment
Allowing hot steam to enter the turbine without following the proper warm-up procedure	Burn injury from steam	Equipment damage due to thermal shock	Downtime due to repair	
Running a turbine at a speed above or below the normal operating setpoint (the point or place where the control index of a controller is set)	If the overspeed trip does not activate, the turbine could fly apart, causing injury	Hunting may occur (when a turbine's speed fluctuates while the controller is searching for the correct operating speed) Turbine may shut down	Downtime due to repair Driven equipment (e.g., a pump or compressor) may not pump enough if the turbine is running too slowly Improper pump speed might cause product to go off-spec	Any abnormal operating condition may cause releases to the environment
Failure to lubricate the linkage between the governor and the governor valve		May cause hunting as a result of valve sticking or binding	Sub-optimum operations	
Failure to maintain sufficient inlet steam pressure		Hunting may occur due to insufficient steam pressure	Insufficient power to run the process	
Allowing "wet steam" or condensate into the turbine	Shrapnel from a turbine failure	Can cause total failure of the turbine	Lost production; cost of repair	Possible release to the environment
Failure to maintain critical speeds	Shrapnel from exploding machines	Turbine could be damaged as a result of harmonic vibrations if the turbine is not moved properly through critical speeds	Lost production; cost of repair	Releases to the environment
High-discharge steam pressure	Damage to hearing	Turbine slowdown and relief valve popping	Possible loss of production	Noise pollution
Steam leaks	Burn injuries; slippery surfaces	Loss of required output of turbine	Possible loss of production	Noise hazard
Improper lubrication	Possible injuries from turbine failure	Severe damage or violent machine failure	Possible loss of production	Possible releases or fire

EMERGENCY PROCEDURES

Process technicians should be knowledgeable about the emergency procedures relating to turbine operations for their units. Emergency procedures include steps to address loss of steam, loss of cooling water, loss of lube oil, or mechanical breakdown of the turbine.

LOCKOUT/TAGOUT

Depending on the type of maintenance required, lockout/tagout of the pump and turbine may be necessary. If lockout/tagout is required, make sure all energy-supplying devices are locked out or tagged out (this includes inlet and outlet steam, oil supply and return, cooling water supply and return, and any other energy sources).

Process Technician's Role in Operation and Maintenance

The process technician plays an important role in the safe operation and maintenance of turbines. Process technicians need to be alert to abnormal sounds that can signal steam leaks or bearing problems, and they must listen to the turbine to ensure that there is no rubbing or dragging inside the housing.

Process technicians should also monitor for thermal stresses, which can cause turbine parts to warp. Warped parts can wear against each other, leading to leaks that cause a loss of boiler feedwater. Leaks can also contaminate the lubrication system with condensate, ruining the bearings and forcing the turbine to shut down.

Additional responsibilities of the process technician include watching for excessive vibration and watching for and reporting surging or hunting, oil leaks, steam issuing from the oil reservoir, or any abnormal steam emissions.

It is important to check for inboard and outboard cooling of bearings and to test for changes in pressure because sudden changes could mean that steam nozzles have become plugged. Checking the lubrication system for corrosion is also important. Additional items that process technicians should monitor include the following:

- Inlet and exhaust steam temperatures (which could indicate dirty turbine blades) and pressures
- Oil levels, temperatures, flows, discoloration, foaming, filter pressure drops
- Speed governor inspection and lubrication
- Safety interlocks and devices

MONITORING

Process technicians routinely monitor turbines to ensure that they are operating properly. When monitoring and maintaining steam turbines, process technicians must always remember to look, listen, and feel for the items listed in Table 8-2.

Process technicians must also monitor oil pressure and quality. High-quality oil must be used for lubrication to keep high bearing temperatures from damaging turbine parts. High bearing temperatures and high turbine speeds, combined with the steam used in the process, can elevate temperatures even higher and cause the oil to break down. High-quality oil reduces the risk of parts fusing together, which can cause the turbine to fail.

Vibration must also be monitored on a regular basis. If a technician notices a vibration in the turbine, something is not working properly in the turbine or the turbine load

TABLE 8-2   Process Technician's Role in Operation and Maintenance

Look	*Listen*	*Feel*
• Check oil levels to make sure they are satisfactory and that oil color is normal	• Listen for abnormal noises	• Feel for excessive vibration
• Check cooling water or air flow to oil coolers and bearings and seals	• Wear ear protection around most turbines, but certainly around gas turbines	• Feel for excessive heat
• Check water flow, pressure and temperature to and from intercoolers		• Do not enter gas turbine enclosures or noise barriers without following established procedures and wearing personal protective equipment (PPE)
• Check seals and flanges to make sure there are no leaks		
• Check for water or foaming in the oil reservoir		• Feel lubricants for grit and smell for breakdown odor
• Check for proper steam pressure and temperature		
• Observe governor for proper operation		

shaft. Vibration sensors are installed on many large turbines to detect vibrations. However, process technicians should still visually inspect the sensor or the turbine to ensure that there is no vibration.

In addition to oil pressure and vibration, process technicians must also monitor speed, hunting, auxiliaries, and guards to ensure proper operation. Failure to perform proper maintenance and monitoring could affect the process and result in equipment damage or injury.

ROUTINE AND PREVENTIVE MAINTENANCE

Process technicians should be familiar with routine and preventive maintenance (procedures to ensure continued operation). For example, process technicians may need to check overspeed trip devices to ensure that the safety system is operating correctly. The overspeed trip is an automatic device that cuts power off to a rotating machine if it reaches a preset trip speed.

Startup and shutdown are required for routine and preventive maintenance as needed within the facility. Turbines should be started and shut down slowly to reduce physical stress (thermal shock and warping of the shafts and bearings). Most facilities have startup and shutdown procedures. If the turbines are driving complex or critical equipment, the procedures are specific for that piece of equipment. If the turbine is driving only a pump, the facility has standard operating procedures for the startup and shutdown.

Summary

The basic purpose of a turbine is to convert the kinetic or potential energy of steam, gas, or air into the mechanical energy required to drive equipment. Pumps, generators, and compressors are examples of equipment driven by turbines. High-velocity steam, gas, or air turns the blades that turn the shaft.

Steam turbines are driven by steam discharged at high pressures and temperatures against a set of turbine vanes. Reactive turbines use Newton's third law of motion, which states that for every action, there is an equal but opposite reaction. Impulse turbines use steam to move a rotor to generate their own steam. Condensing turbines use exhaust steam as it condenses in a surface condenser and the condensate returns to the boiler house.

Gas turbines produce hot gases in the combustion chamber and direct the gases toward the turbine blades, causing the rotor to move. Combustion chambers are used to boost the pressure and temperature of the mixed fuel and compressed air.

Hydraulic turbines allow a liquid to flow across the rotor blades, forcing them to move. The faster the liquid flows, the faster the wheel turns.

Wind turbines convert wind energy using air pressure and temperature to move a rotor. An example of a wind turbine is a windmill.

Turbine components consist of multiple parts that are similar in many of the turbines. A casing is the housing around the internal parts of the turbine. The shaft is the metal rotating component that holds the rotor and all the rotating equipment in place. Some turbines have moving blades and fixed blades. The governor is used to control speed, and the trip throttle valve is used to shut down the turbine at excess rotational speed or vibration.

Most turbines used in the process industries are either steam or gas. Steam turbines are used most often because of their convenient and efficient fit within the facility steam distribution system.

Utilities and auxiliary equipment associated with turbine operation include lubrication systems and seals. A turbine lubrication system requires high-quality oil to properly maintain turbine bearings, lubrication, and temperatures.

Improper operating conditions can result in severe bodily injury, total mechanical failure, lost product, lost revenue due to downtime, and environmental problems. Turbines are prime movers of typically large and single-train machinery, so turbine failure usually shuts down processing in a unit. Loss of lubrication, impingement, fouling, pitting, corrosion, assembly errors, and loss of overspeed shutdown systems are conditions that must be avoided and corrected.

Bearing damage can result from excessive temperatures, and pounding can result from vibration caused by shaft misalignment problems or improper lubrication. Vibration sensors are installed on many large turbines to alert the process technician of the excessive vibrations so he or she can make appropriate corrective actions.

Process technicians should monitor the unit frequently to search for unusual conditions. Personal protective equipment (PPE) should be worn at all times. Routine monitoring includes listening for unusual sounds; watching for excessive vibration; monitoring turbine speed; looking for oil or steam leaks; and maintaining proper oil levels, pressures, and temperatures. Process technicians must always watch for worn parts, and they must replace them or have them replaced when necessary.

Turbines should be started and shut down slowly using the manufacturer's recommended procedures to reduce stress, thermal shock, and the warping of shafts and bearings. During startup, the steam lines and equipment should be emptied of accumulated condensate and warmed up to prevent water damage, and the machines should be placed on slow roll to prevent thermal shock.

Process technicians must be familiar with the causes of and solutions for problems such as loss of lubrication, bearing damage, impingement, fouling, pitting or corrosion, assembly errors, and loss of overspeed shutdown systems. Performing proper preventive maintenance procedures can prevent or minimize most of these problems.

Checking Your Knowledge

1. Define the following terms:
   a. Gas turbine
   b. Hydraulic turbine
   c. Reactive turbine
   d. Steam turbine
   e. Wind turbine
   f. Impulse turbine
   g. Casing
   h. Governor
   i. Fixed blade
   j. Moving blade
   k. Nozzle
   l. Overspeed trip mechanism
   m. Rotor

2. *(True or False)* Gas and steam turbines are activated by the expansion of fluid on a series of curved rotor vanes attached to a central shaft.

3. Steam turbines are driven by the pressure of _____ discharged at a high pressure and temperature against the turbine vanes.
   a. air
   b. water
   c. gas
   d. steam

4. _____ energy is energy associated with mass in motion.
   a. Mechanical
   b. Kinetic

5. *(True or False)* Newton's third law of motion states that for every action, there is a less and similar reaction.

6. A(n) _____ turbine is used as a driver for high differential pressure and high horsepower, and can use different pressures at various inlet points to drive process equipment.
   a. condensing
   b. noncondensing
   c. multistage
   d. impulse

7. A(n) _____ is used to control the speed of the turbine as steam is channeled through the restrictive component called the nozzle.
   a. governor
   b. trip throttle valve
   c. thrust bearing
   d. overspeed trip

8. Inside a gas turbine, hot gases are produced in the _____ and are directed toward the turbine blades causing the rotor to move.
   a. fuel line
   b. combustion chamber
   c. stage
   d. compressor

9. Turbines can be powered by which of the following (select all that apply)?
   a. Steam
   b. Gas
   c. Liquid (hydraulic)
   d. Wind
10. *(True or False)* Shaft movement inside the gas turbine is generated as the combustion occurs, in turn generating the force.
11. *(True or False)* Labyrinth seals are single-layer seals used to allow small amounts of leakage inside the turbine.

Student Activities

1. Given a cutaway or a diagram of a steam turbine, identify the following components and explain the function of turbines.

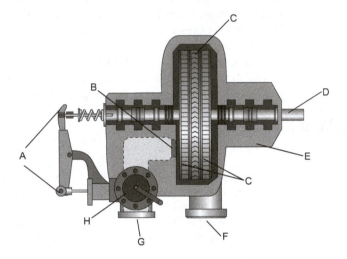

   a. Casing
   b. Shaft
   c. Moving and fixed blades
   d. Governor
   e. Nozzle block
   f. Inlet
   g. Outlet
   h. Trip and throttle valve
2. Write the theory of operation for a steam and a gas turbine.
3. Prepare a short presentation on how the governor inside a turbine helps control the speed and how it is related to the performance of the equipment.
4. Write a report on potential problems associated with turbines. In your report, list possible solutions and how the problems can affect other systems within a process facility.
5. Work with a classmate to write a startup, shutdown, emergency, or lockout/tagout procedure for a turbine.

CHAPTER 9

Electrical Distribution and Motors

Objectives

After completing this chapter, you will be able to:

- Identify and explain the components of an electrical distribution system.
- Explain the purpose of motors in the process industry.
- Identify the common types and applications of motors (AC/DC).
- Identify the components of a typical motor.
- Describe the operating principles of motors.
- Describe safety and environmental hazards associated with motors.
- Identify typical procedures associated with motors.
- Describe the process technician's role in operation and maintenance of motors.
- Describe potential problems associated with motors.

Key Terms

AC power source—a device that supplies alternating current.

Alternating current (AC)—electric current that reverses direction periodically, usually 60 times per second.

Ammeter—device used to measure the electrical current in a circuit.

Ampere (amp)—a unit of measure of the electrical current flow in an electrical circuit; similar to gallons of water flow in a pipe.

Bearing—a machine component that rotates, slides, or oscillates. Bearings reduce friction between the motor's rotating and stationary parts.

Circuit—a system of one or many electrical components that accomplish a specific purpose.

Circuit breaker—an electrical component that opens a circuit and stops the flow of electricity when the current reaches unsafe levels.

Conductor—materials that have electrons that can break free from the flow more easily than the electrons of other materials.

Direct current (DC)—electrical current that flows in a single direction through a conductor.

Disconnect—a large electrical switch used for isolation during system repairs and maintenance.

Electricity—the flow of electrons from one point to another along a pathway called a conductor.

Electromagnetism—magnetism produced by an electric current.

Electrons—negatively charged particles that orbit the nucleus of an atom.

Fan—a rotating blade inside a motor housing or casing that cools the motor by pulling air in through the shroud.

Frame—a structure that holds the internal components of a motor and motor mounts.

Fuse—a device used to protect equipment and electrical wiring from overcurrent.

Generator—a device that converts mechanical energy into electrical energy.

Induction motor—a motor that turns slightly slower than the supplied frequency and can vary in speed based on the amount of load.

Insulator—any substance that prevents the passage of heat, light, electricity, or sound from one medium to another. Insulators do not conduct electricity.

Load—the amount of torque necessary for a motor to overcome the resistance of the driven machine.

Motor control center (MCC)—an enclosure that houses the equipment for motor control, including isolation power switches, lockouts, fuses, overload protection devices, ground-fault protection, and sometimes meters for current (amperes) and voltage.

Ohm—a measurement of resistance in electrical circuits.

Rectifier—a device that converts AC voltage to DC voltage.

Semiconductor—a material that is neither a conductor nor an insulator.

Shroud—a casing over the motor that allows air to flow into and around the motor.

Static electricity—electricity that occurs when a number of electrons build up on the surface of a material but have no positive charge nearby to attract them and cause them to flow.

Stator—a stationary part of the motor where the alternating current supplied to the motor flows, creating a magnetic field using magnets and coiled wire.

Switch—an electrical device used to start, stop, or otherwise reconfigure the flow of electricity in a circuit.

Synchronous motor—a motor that runs at a fixed speed that is synchronized with the supply of electricity.

Transformer—an electrical device that takes electricity of one voltage and changes it into another voltage.

Universal motor—a motor that can be driven by either AC or DC power.

Volt—the electromotive force or a measure of current that establishes a current of one amp through a resistance of one ohm.

Voltage—a measurement of the potential energy required to push electrons from one point to another.

Voltmeter—a device that can be connected to a circuit to determine the amount of voltage present.

Watt—a unit of measure of electric power; the power consumed by a current of one amp using an electromotive force of one volt.

Introduction

In the process industries, motors provide power for rotating equipment such as pumps, compressors, fans, blowers, and conveyor drivers. Motors vary in size and strength. Because motors and other equipment require electricity in order to operate, electrical power distribution is an important concept in the process industries.

The electricity used by motors and other equipment can be either alternating current (AC) or direct current (DC). Alternating current (AC) oscillates back and forth like a sine wave, while direct current (DC) flows in a single direction. The power from a typical electrical receptacle is AC, while batteries supply DC power. AC power is the most common type of power used in the process industries.

While electricity is very beneficial, serious hazards are associated with it. For example, electricity can be present without being visible, and serious injury or death by electrocution can occur even at household voltages. When working around electrically powered equipment, process technicians should always follow safety procedures, be aware of potential hazards, and never perform any duties involving exposed live wires or connections.

What is Electricity?

Electricity is the flow of electrons from one point to another along a pathway called a conductor. Process technicians should have a firm understanding of electricity because it is integral to the functioning of a plant and its systems.

In chemistry, we learn that an atom is the smallest particle of an element that can combine with other elements. Atoms contain positive, negative, and neutrally charged particles. The negatively charged particles that orbit the nucleus of an atom are called **electrons**.

Electrons flow from one atom to another; this is called electrical current. Free electrons flow along a path like a river. This pathway is called a conductor. **Conductors** are materials that have electrons that can break free from the flow more easily than other materials. Metals, as well as liquids and some types of hot gases (e.g. plasmas), are good conductors.

Materials that do not conduct electricity or give up their electrons easily are called **insulators**. Insulators are poor conductors. Air, rubber, and glass are examples of insulators.

Materials that are neither conductors nor insulators are called **semiconductors**. Semiconductors are most commonly used in the electronics industry and are being used increasingly in power applications.

Another type of electricity is static electricity. **Static electricity** occurs when a number of electrons build up on the surface of a material but have no positive charge nearby to attract them and cause them to flow. When the negatively charged surface comes near or into contact with a positively charged surface, current flows until the charges on each surface become equalized, sometimes creating a spark. Lightning and the shock that occurs from touching a doorknob after shuffling across the carpet are both good examples of static electricity.

To better understand electricity, process technicians must be familiar with its principles and know that it is measured in watts, volts, amps, and ohms.

WATTS

A **watt** is the power consumed by a current of one amp using an electromotive force of one volt. The box below contains the formulas for determining wattage.

Wattage

The formula for determining how many watts a circuit uses, if you know the volts and amps, is:

$$P = VI$$

P = power in watts
V = potential difference in volts
I = current flow in amps

The formula for determining how many watts a circuit uses, if you know the amps and resistance, is:

$$P = I^2R$$

P = power in watts
I = current flow in amps
R = resistance to current flow in ohms

If you know the volts and amps of a circuit, you can figure how many watts a circuit uses by applying the formula **P = IR** (power in watts = volts times amps). For example, a 100-watt light bulb would consume .833 amps (120V × .833 amps = 100 watts).

VOLTS

Voltage is a measurement of the potential energy required to push electrons from one point to another. A **volt** is the electromotive force or a measure of current that establishes a current of one amp through a resistance of one ohm (measurement of resistance in electrical circuits).

Using the water analogy again, electric current flows like a river down a slope (the path of least resistance). The greater the angle of a slope, the faster the water flows. Electric current behaves in a similar manner. If the difference between positive and negative charges is low, electrons flow with little force. Increase the difference, and the electrons flow with greater force. The force that makes electrons flow is called voltage or electromotive force (EMF), and it is measured in units called volts (V). Voltage can be measured with a voltmeter (shown in Figure 9-1 and Figure 9-2).

A **voltmeter** is a device that can be connected to a circuit in order to determine the amount of voltage present. The symbol V is used to denote voltage. Sometimes the letter V is followed by AC or DC (e.g., VAC or VDC) to denote whether the voltage is from alternating current (AC) or direct current (DC).

AMPERES

Amperes, or amps, are a unit of measure of the electrical current flow in an electrical circuit, similar to gallons of water flow in a pipe. Amps describe how many electrons are flowing at a given time. The symbol I is used to denote amperes in equations.

FIGURE 9-1 Digital Voltmeter/Ammeter Combination

FIGURE 9-2 Analog Faceplate from a Voltage Meter

FIGURE 9-3 Faceplate of an Ammeter Used to Measure Current

An **ammeter** is a device used to measure the electrical current in a circuit. Ammeters must be connected in series to an electrical circuit in order to display actual amps.

Figure 9-3 shows an example of an ammeter faceplate.

Current capacity (in amperes) indicates how much work a circuit can do for a given voltage. In other words, amps show the capacity of a battery or other source of electricity to produce electrons.

OHMS

Ohms are a measure of resistance in electrical circuits. Ohm's law describes how volts, ohms, and amps act on each other. This law states that the amount of steady current through a conductor is proportional to the voltage across that conductor. This means a conductor with one ohm of resistance has a current of one amp under the potential of one volt. Simply put, volts equals amps times ohms ($V = IR$).

> ### Ohm's Law
>
> The formula for Ohm's law is:
>
> $$V = IR$$
>
> *V = volts*
> *I = current flow in amps*
> *R = resistance to flow in ohms*
>
> When working with Ohm's law, if you know the value of two units, you can always figure out the third using one of the following calculations:
>
> **I = V/R** (current = volts ÷ resistance)
> **R = V/I** (resistance = volts ÷ current)

CIRCUITS

A **circuit** is a system of one or many electrical components that accomplish a specific purpose. Circuits combine conductors with a power supply and usually some kind of electrical component (such as a switch or light) in a continuously conducting path. In a circuit, electrons flow along the path, uninterrupted, and return to the power supply to complete the circuit. Circuits fall into one of two types: series and parallel.

In a series circuit, the components are connected in a loop so the electrical current flows in a single path. Any break in the circuit stops the flow of current. An example of a series circuit is a string of Christmas lights. In some strings, if one light bulb burns out, the circuit (path) is interrupted so none of the bulbs will light. Figure 9-4 shows an example of a series circuit.

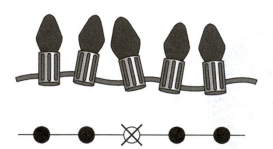

FIGURE 9-4 Series Circuit Example

In a parallel circuit, components are split into branches so the electrical current follows more than one path. A break in one branch does not stop current in the other branches. Some types of Christmas lights use this type of circuit. In a parallel circuit, if one bulb burns out, the others will continue to light. Houses are wired using parallel circuits.

Figure 9-5 shows an example of a parallel circuit.

FIGURE 9-5 Parallel Circuit Example

GROUNDING

Did You Know?

- Watt is named after James Watt, who invented the steam engine.

- Ampere is named after André Ampere, the first person to explain the electro-dynamic theory.

- Ohm is named after Georg Ohm, a German mathematician and physicist.

James Watt

André Ampere

George Ohm

All energized conductors supplying current to equipment are kept insulated from each other, from the ground (earth), and from the equipment user. Many types of equipment have exposed conductive parts, such as metal covers, that are routinely touched during normal operations. If these surfaces become energized, the process technician could complete the circuit and be shocked. For this reason, non-current-carrying conductive materials should be used to enclose electrical conductors, and the equipment should be properly grounded.

Grounding is the process of connecting an object to the earth using copper wire and a grounding rod to provide a path for the electricity to dissipate harmlessly into the ground. Grounding is typically accomplished by a separate conductor specifically designed for this purpose. With this conductor in place, if the equipment case becomes energized, a low resistance path for the flow of ground current back to the source is already in place, and it reduces or eliminates the shock hazard to the operator.

Figure 9-6 shows an example of a piece of electrical equipment that is attached to the earth (grounded) through a grounding wire.

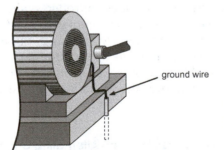
ground wire

FIGURE 9-6 Grounding Wire Attached to a Piece of Electrical Equipment

Types of Current

ALTERNATING CURRENT (AC)

Alternating current (AC) is electric current that reverses direction periodically, usually 60 times per second. The movement of alternating current is similar to water sloshing backward and forward in a pipe. When a negative charge is at one end of a conductor and a positive charge is at the other end, the electrons move away from the negative charge.

But if the charge (polarity) at the end of the conductors is reversed, the electrons switch directions (alternate). In the United States, the AC power supply changes direction 60 times per second. This cycling is called frequency or cycles per second.

Electricity is almost always generated as three-phase electricity because it is more efficient. Three-phase electricity is delivered by a circuit consisting of two energized wires, each carrying one phase and a neutral or ground wire. The electricity between any one energized wire and the ground, or the electricity between any two power wires, is single-phase.

DIRECT CURRENT (DC)

Direct current (DC) is electrical current that flows in a single direction through a conductor. Direct current flows like water moving in one direction through a pipe. Batteries are DC electrical power sources.

Did You Know?

Thomas Edison was an inventor who invented the phonograph and improved the printing telegraph.

Edison also built the first practical DC (direct current) generator.

DC power is used in limited applications in the process industries. Most frequently, DC power is used where critical equipment must remain functional even during a power outage. This type of backup power is usually provided by batteries. Backup batteries supply enough power to allow personnel to place the facility in a safe condition. DC power is also used to power emergency lighting and for special applications such as high-torque motors.

Motors that use DC have large battery banks to provide backup power if the regular power flow fails. These battery banks are kept fully charged by battery chargers that use **rectifiers** (devices that convert AC voltage to DC voltage) to convert AC to DC.

In the late 1800s, Edison power stations (created by Thomas Edison) supplied DC power to customers scattered across the United States. However, DC had a limiting factor. It could be sent economically only a short distance (about a mile) before the electricity began to lose power due to the resistance in the power cable. To remedy these distance limitations, George Westinghouse introduced alternating current (AC) power systems (designed by Nikola Tesla) as an alternative to DC power.

AC has an advantage over DC because AC voltages can easily be stepped up (increased) or stepped down (decreased) using transformers. By employing transformers to raise voltage levels, AC systems can economically distribute electricity for hundreds of miles. Table 9-1 summarizes the similarities and differences between AC and DC power.

TABLE 9-1 Similarities and Differences Between Alternating Current (AC) and Direct Current (DC)

Alternating Current (AC)	*Direct Current (DC)*
Polarity is switched repeatedly.	Polarity is fixed.
Voltage varies during cycles.	Voltage remains constant.
Voltage can be stepped up or stepped down by a transformer.	Voltage cannot be stepped up or stepped down by a transformer.
More difficult to measure than DC current.	Easier to measure than AC current.
Heating effect is the same as DC current.	Heating effect is the same as AC current.

Electrical Transmission

Electricity is a form of energy that must come from a power source such as batteries or generators. **Generators** are devices that convert mechanical energy into electrical energy. For a power station, the most common way to manufacture electricity is by burning fuels that power turbines, which rotate magnetic fields inside generators to create electric current. Other methods include hydroelectric and nuclear power generation.

Electricity flows in a continuous current from a point of high potential (the power source) to a point of lower potential (e.g., your home or plant) through a conductor such as a wire. High-voltage electricity is transmitted from the plant to the power grid through a system of wires. During transmission, the high-voltage electricity is routed to a substation that steps the electricity down to a lower, safer voltage. The substation then distributes the electricity through feeder wires to a step-down transformer, a device that steps down the voltage again so it can be used by residential or commercial customers.

Once inside the home or facility, the stepped down electricity is used to do work, such as lighting a bulb or operating a motor. To make the system as safe as possible, safety devices such as fuses, protective relays, and ground-fault detectors are used throughout the power transmission process.

Components of an Electrical Distribution System

When power enters a process facility, it is often at very high voltage. It is normally changed to a lower voltage by transformers. To make the power useable, the power is routed through a transformer, which adjusts the voltage. From there, the power is routed to a motor control center that contains fuses, circuit breakers, and switches that allow for the activation and control of motors and other electrical equipment.

TRANSFORMERS

A **transformer** is an electrical device that takes electricity of one voltage and changes it into another voltage. Transformers that reduce the voltage are referred to as step-down transformers. When voltage is stepped down, there is a corresponding increase in current. However, the power (the product of voltage and current) remains constant. Transformers that raise voltage are called step-up transformers.

MOTOR CONTROL CENTERS (MCCS)

Motor control centers (MCCs), like the one shown in Figure 9-7, are enclosures that house the equipment for motor control, including isolation power switches, lockouts, fuses, overload protection devices, ground-fault protection, and sometimes meters for current (amperes) and voltage. Most industrial motors (except small motors) receive their power from a central MCC.

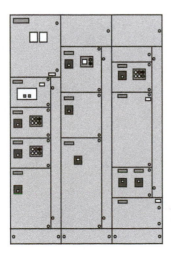

FIGURE 9-7 Motor Control Center

MCCs are maintained at cool temperatures and are usually isolated from the rest of the facility so they do not serve as an ignition source in the event of a spill or release of flammable material. An MCC houses the circuit breakers, disconnects, fuses, motor starters, meters, and other electrical accessories for all the large motors.

While a typical motor has a local start/stop switch located on the equipment in the field, a low-voltage control circuit powers the switch that, in turn, operates a control relay in the MCC. The control relay is what actually provides current to the equipment.

All motors should have controllers that protect the motor from instantaneous overload (e.g., a short circuit), as well as thermal overload from working beyond design limits. Motors should also have a way to isolate the circuitry during maintenance work. Typically, the motor's overload, starting and stopping devices, and disconnecting device are built together in one motor controller. A motor control center (MCC) is a grouping of these motor controllers.

FUSES

A **fuse** is a device used to protect equipment and electrical wiring from overcurrent. Fuses are designed to open or "blow" and stop current flow if the current becomes too great. Once the fuse is blown, the fuse must be replaced in order to re-energize the circuit. The primary parameter of a fuse is the current rating above which it will blow. Fuses are also rated for the maximum voltage of the circuit they can be used in and whether they are fast acting or slow acting (*Note:* For motors, slow-acting fuses must be used because there is a large current in-rush when a motor starts.)

Fuses are usually limited to DC and AC single-phase applications. In three-phase applications, or where ground-fault protection is needed, a circuit breaker is typically used instead of a fuse.

CIRCUIT BREAKERS

A **circuit breaker** (shown in Figure 9-8) is an electrical component that opens a circuit and stops the flow of electricity when the current reaches unsafe levels. Circuit breakers act as a reusable form of protection from overcurrent. When overcurrent occurs the circuit breaker trips and interrupts the flow through the circuit. The circuit breaker must then be reset to re-energize the circuit.

Many electronic components have a circuit breaker button that is used to reset the device for operations. An MCC may have one or more circuit breakers built in to protect each motor from current overload. These breakers have different response speeds and sensitivities.

FIGURE 9-8 Large Circuit Breaker Used in Process Industries

SWITCH

A **switch** is an electrical device used to start, stop, or otherwise reconfigure the flow of electricity in a circuit. The most common household application of switches is light switches.

In large motors, switches send a control signal to an MCC. Large electrical switches used for isolation during system repairs and maintenance are called **disconnects**. Disconnects cannot be operated under load. The power must first be interrupted using the start/stop switch and circuit breakers.

Purpose of Motors

A motor is a mechanical driver that converts electrical energy into useful mechanical work and provides power for rotating equipment. In the process industries, motors are used to provide power for pumps, compressors, fans, blowers, conveyor drivers, valves, and other equipment.

Types of Motors

Electric motors use electricity for motive energy. Electric motors may use either AC or DC electricity and may be single speed or variable speed. Table 9-2 lists examples of some of the various types of motors.

TABLE 9-2   Various Types of Motors

Motor Type	*Description*
Alternating current (AC)	The most common type of motor used in the process industries, primarily because of the simplicity of its construction.
Direct current (DC)	Most commonly used in situations where variable speed or high-torque applications are needed.
Single speed	Motors that run at a single rotator speed; generally AC motors.
Variable speed	Allows for the rotation speed of the motor to be adjusted; may be either an AC or a DC motor.

There are two basic types of AC motors: synchronous and induction. **Synchronous motors** run at a fixed speed that is synchronized with the supply electricity. Synchronous motors rotate at exactly the supply frequency or a fixed fraction of the supply frequency (e.g., 1/2 or 1/3). **Induction motors**, on the other hand, turn slightly slower than the supplied frequency and can vary in speed based on the amount of load. Most induction motors are squirrel cage design, though wound rotor design may be used where variable speed is necessary. Figure 9-9 shows an example of a squirrel cage rotor.

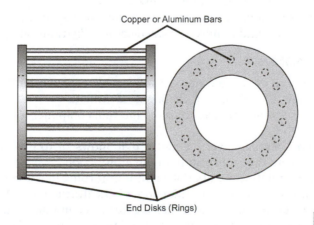

Copper or Aluminum Bars

End Disks (Rings)

Squirrel Cage Rotor

FIGURE 9-9 "Squirrel Cage" Induction Rotor

Wound rotor motors are larger, more expensive, and require significantly more maintenance than squirrel cage motors. Three-phase, single-speed squirrel cage AC induction motors are by far the most common type in the process industry.

DC motors are very useful for variable speed applications because their speed can be easily varied by adjusting the voltage. The amount of current DC motors draw is roughly proportional to the torque load they are driving. DC motors are also useful in applications that require very high torque at startup because series wound DC motors develop their greatest torque at slow speeds. This makes them particularly good as starting motors for small diesel engines and similar applications.

Another motor design is the universal motor. A **universal motor** can be driven by either AC or DC power. Universal motors are typically used in small, high-speed applications, such as angle grinders. However, they are suitable for only intermittent use because they require frequent maintenance.

SELECTION OF ELECTRIC MOTORS VERSUS OTHER DRIVERS

Various factors affect the selection of electric motors over other devices (e.g., internal combustion engines, steam turbines, hydraulic motors, and pneumatic motors). For example, internal combustion engines present toxic emission hazards and cannot be used in confined spaces. Likewise, pneumatic drivers that use nitrogen or other hazardous gases must not be used in confined spaces for fear of asphyxiation.

Some areas may prohibit the use of a gas or diesel engine due to the flammability of the fuel and the potential for fire. In these situations, an electric motor may be the motor of choice. An explosive environment may require a special classification of motor (e.g., intrinsically safe motors that do not produce sparks or thermal effects that will ignite flammable vapors) or an alternate type of driver.

In some applications, steam turbines may be more suitable than electric motors. In other applications, hydraulic motors may be the motor of choice because they are smaller than electric motors. The availability of the energy source is also considered when deciding which type of driver to use. Steam or hydraulic power may not be practical due to the piping required. Also, the physical location (e.g., elevation from the ground) and the available space to safely place a motor is a factor.

Motors must be placed in a location where they are safe from external factors yet remain accessible for repairs. Power is also a requirement for electric motors. Because of this, fire pumps and other critical devices are often diesel- or gasoline-driven so they will still function in the event of a power outage.

Long-term costs are often the determining factor when choosing a driver. Electric motors are often preferred because of their reliability, ease of use, low maintenance cost, and ease of repair.

In many processes, the operation of the drivers represents a significant portion of production costs, so it is important to understand how an operation affects process or final products. In these cases, long-term costs of operation versus reliability and convenience must be evaluated carefully. In many instances, the cost goes beyond the process unit boundary. For example, a motor that exceeds the capacity of the facility electrical distribution system would require a costly electrical distribution upgrade.

MOTOR COMPONENTS

AC motors are composed of a frame, shroud, stator, rotor, fan, and bearings. Figure 9-10 displays an AC motor with these components identified.

The **frame** is a structure that holds the internal components of a motor and motor mounts.

The **shroud** is a casing over the motor that allows air to flow into and around the motor. The air keeps the temperature of the motor cool.

The rotor is a rotating member of a motor or turbine that is connected to the shaft; it usually consists of an iron core with copper bars attached to it. When the stator creates an electric current in the rotor, this creates a second magnetic field in the rotor. The magnetic fields from the stator and rotor interact, causing the rotor to turn.

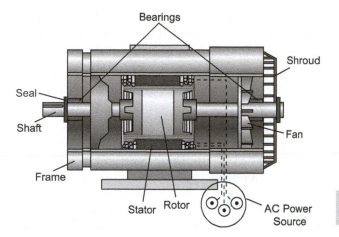

FIGURE 9-10 AC Motor Components

The shaft is a metal rotating component (spindle) that holds the rotor and all rotating equipment in place.

The **stator** is a stationary part of the motor where the alternating current supplied to the motor flows, creating a magnetic field using magnets and coiled wire.

The **fan** is a rotating blade inside a motor housing or casing that cools the motor by pulling air in through the shroud.

The **bearing** is a machine component that rotates, slides, or oscillates. Bearings reduce friction between the motor's rotating and stationary parts.

The **AC power source** is a device that supplies alternating current to the stator.

Figure 9-11 shows an example of a fan. Figure 9-12 contains a frame, and Figure 9-13 provides an example of a rotor and a motor.

FIGURE 9-11 Example of a Fan used Inside a Motor

FIGURE 9-12 Example of a Frame used in a Motor

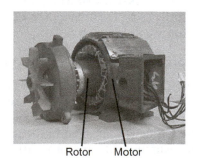

FIGURE 9-13 Example of a Rotor and Motor

Operating Principles of Electric Motors

Early pioneers in electricity discovered the principle of electromagnetism. When current is run through a coil of insulated wire that is wrapped around a soft iron bar, a strong magnetic field is created. When the current is removed, the magnetic field collapses. This process is called **electromagnetism** (magnetism produced by an electric current).

Electromagnetism plays an important role in electric motors. Whether AC or DC, electric motors operate on the same three electromagnetic principles:

1. Electric current generates a magnetic field.
2. Like magnetic poles repel each other (i.e., positive repels positive and negative repels negative), while opposite poles attract each other (i.e., positive attracts negative).
3. The direction of the electrical current determines the magnetic polarity.

A motor consists of two main parts: a stationary magnet, called a stator, and a rotating conductor, called a rotor. The stator is a stationary part of the motor where the alternating current supplied to the motor flows, creating a magnetic field using magnets and coiled wire. A rotor is the rotating member of a motor or turbine. A typical rotor usually consists of an iron core with copper bars attached to it. These bars are very conductive (i.e., they conduct electricity easily). The stator creates an electric current in the rotor, and this in turn creates a second magnetic field in the rotor. The magnetic fields from the stator and rotor interact, causing the rotor to turn. Figure 9-14 is an AC motor with a field magnet, stator, field coils, and rotor.

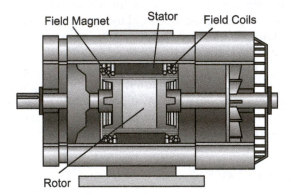

FIGURE 9-14 AC Motor Field Magnet, Stator, Field Coils and Rotor

Electrical meters are used to measure the performance of electrical equipment by showing the amount of electricity drawn (used) by the equipment. Electrical meters can be used to measure the different aspects of the electrical use such as voltage (volts), current (amperes), and power (watts).

LOAD

A **load** is the amount of torque necessary for a motor to overcome the resistance of the machine it drives. The two types of loads are full loads and no loads. A full load is when the motor is using the full amount of torque to which it is rated. No load occurs when the motor is not using torque.

THREE-PHASE VERSUS SINGLE-PHASE MOTORS

Most motors in process facilities use three-phase electricity instead of the single-phase electricity found in most homes. Three-phase electricity is more efficient than single phase. Three-phase motors can rotate clockwise or counterclockwise, depending on how the wiring is connected to the motor. Because of this, process technicians should always ensure that the motor rotation is correct before putting the motor in service, especially if any work was performed that involved the wiring to the motor or if it is a new installation.

Potential Problems

When working with electrical distribution and motors, process technicians should always be aware of potential problems such as high levels of vibration, high temperatures, and the motor not starting. Table 9-3 lists some potential problems associated with electrical distribution and motors.

TABLE 9-3   Potential Problems and Corresponding Causes Associated with Electrical Distribution and Motors

Conditions	*Problems/Causes*	
High vibration	Problem:	Severe misalignment should cause coupling failure before damage occurs. A coupling failure is often dangerous because of flying shrapnel. Coupling guards must always be in place as a safety measure. Larger motors have high temperature alarms or shutdowns.
	Cause:	Consequences of excessive vibration include damage to machinery. Vibration is caused by: • Rotor out of balance (e.g., loss of rotor counterweights) • Bearing failure • Misalignment
High temperatures	Problem:	Larger motors have high temperature alarms or shutdowns.
	Cause:	Typical causes of high temperatures include: • Bearings failure • Overload condition • Loss of cooling airflow
Motor will not start	Problem:	Motor will not start.
	Cause:	When a motor does not start, possible causes are: • Interlock protection • Frozen bearing due to high temperature • Locked rotor • Motor overload trip • Breaker open or racked out • A blown fuse • Faulty start/stop relay circuitry • No power to the MCC panel • Internal open circuit in the motor

Safety and Environmental Hazards

There are hazards associated with normal and abnormal motor operations. The hazards could affect the personal safety of the process technician, the equipment, facility operations, and the environment. The most common hazards associated with normal and abnormal operations include shock or electrocution, injury caused by moving parts, and burns.

The most common personal protective equipment (PPE) used during motor and engine operations include flash suits and Nomex coveralls (shown in Figure 9-15). Additional safety precautions that process technicians may be responsible for include lockout/tagout, isolation, and grounding procedures.

Along with electrical hazards, motors also present mechanical hazards to workers. Motors have moving parts that can pull, tear, or rip clothing or skin if they are not properly de-energized before process technicians work on them. Shroud coupling guards and other safety guards should always be in place during normal operation. Figure 9-16 is the equipment hazard symbol that warns of the hazards of moving parts. Motors can also cause a loss or complete interruption in production if they are performing inefficiently or if a complete failure occurs.

Motor control centers (MCCs) represent a substantial safety feature for large motors. First, they provide protection against short circuits and other electrical faults. Second, they protect the motor against overloading that could result in overcurrent, which could damage the motor. MCCs can also protect against motor overspeed and can

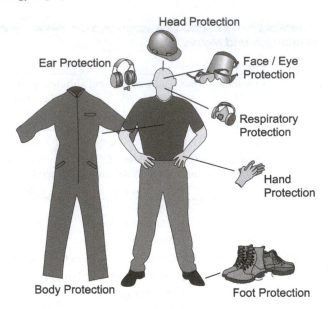

FIGURE 9-15 Personal Protective Equipment

FIGURE 9-16 Equipment Hazard Symbol

provide interlocks that prevent the motor from running in configurations (e.g., discharge valve closed) that would damage it or other process equipment. And finally, MCCs are used to electrically isolate motors for personnel protection during maintenance.

Studies have shown that the single most frequent cause of accidents is human error. Electrical-energy-related accidents are no different. If process technicians follow proper safety procedures when working on electrical circuits and electrical devices, the occurrence of these types of accidents can be significantly decreased.

ELECTRICAL CLASSIFICATION

Since electricity presents a possible ignition source, some areas have been classified by the National Electric Code (NEC) as possibly hazardous. These hazards are broken down by class, group, and division.

CLASS

The first level of electrical hazard classification is the class designation. Class designations are grouped by explosive/ignitable materials:

Class I—Gases, vapors, and liquids that can be present in explosive or ignitable mixtures.

Class II—Dust, combustible dust that can be present in amounts that could produce potentially explosive mixtures.

Class III—Fibers that are easily ignitable.

Group Designations

Another level of classification is the group designation. Group designations are used to group the material by relatively similar hazardous characteristics. Groups are assigned letters A through G. A thru D are for Class I, and E through G are for Class II.

Divisions

The third level of classification is division. Divisions are used to evaluate the possibility of the hazardous material being present in ignitable mixtures. Division I means it is normally present, and Division II means it is present during upset conditions. For example, a possible classification for an area with ignitable amounts of acetone present during normal operations would be Division I—Class I—Group D.

The classification of the area dictates the type of motor required for that area. The most intense motor category is explosion-proof, which means that the motor enclosure is capable of withstanding an explosion of a specified vapor or gas, thereby preventing the ignition of additional gas vapors in the surrounding area. Division I areas frequently require the use of explosion-proof motors. Process technicians should be aware of the electrical classification for their operating areas and should use only the proper electrical equipment in that operating area.

Typical Procedures

Process technicians are responsible for the safe and cost-effective operation of equipment. Typical procedures associated with motors include proper startup, monitoring, shutdown, lockout/tagout, and routine and preventive maintenance.

Before starting any motors or equipment, the process technician must be aware of how this equipment affects other process systems in the unit. Once the motor has been started, the process technician should ensure that the other process systems are responding as anticipated.

STARTUP

The typical procedure for starting an electric motor after maintenance work has been completed is as follows:

1. Determine how startup affects the rest of the facility and other facilities that interact with this facility.
2. Notify the control room personnel that the motor is about to be placed on line.
3. Confirm with the control room personnel that the motor has been returned to operation by maintenance if repairs were performed.
4. Inspect the circuit breaker and other electrical devices in the MCC associated with the motor, and ensure that all are in a proper state of repair.

Did You Know?

Approximately 15 percent of the Occupational Safety and Health Administration (OSHA) general industry citations relate to electrical hazards.

5. Ensure that the start/stop switch on the motor is in the "off" position.
6. Inspect to make sure that all wiring, insulation, motor connections, and switch covers are secure.
7. Inspect the driven equipment to ensure that the coupling guard is secure; the piping is connected; and all tools, scaffold boards, clothing, gloves, and toolboxes are out of the area.
8. Close all bleeds and vents. Perform the proper valve line up on the driven equipment per standard operating procedures.
9. Check all lubrication levels, central lubrication systems, oil filters, and coolers to ensure that all are ready for use.
10. Turn on the circuit breaker in ("rack in") the MCC.
11. "Bump" (quickly turn on and off) the motor to ensure that it is rotating in the proper direction.
12. If the motor is rotating in the proper direction, place the motor in normal operation. If the motor is not rotating in the proper direction, notify appropriate personnel.
13. Notify the control room that power is being turned on, and place the start/stop switch in the "on" (start) position.
14. Perform a walk-around inspection, looking for leaks, listening for unusual noises, and feeling for high temperatures of the equipment.
15. Inform the control room personnel that the motor is on line and in normal operation.

SHUTDOWN

Process technicians should follow the manufacturer-, site-, and unit-specific standard operating procedures for shutting down the equipment for the process facility.

EMERGENCY

Emergency procedures depend on the nature of the emergency. Process technicians should be aware of all the emergency procedures that are required for their job, and they should be able to follow them in an emergency situation.

LOCKOUT/TAGOUT

When performing maintenance on motors, it is necessary to perform electrical lockout/tagout procedures for the motors. This lockout/tagout should be performed at the MCC, the motor, and the driven equipment. If the driven equipment (e.g., a pump), can also present a hazard by rotating the motor, then it must also be isolated through lockout/tagout.

Lockout/tagout procedures vary according to the facility and the equipment. However, the following are some general lockout/tagout procedures:

1. Stop the equipment.
2. Isolate the equipment (i.e., block, tag, and lock all process lines).
3. Isolate and de-energize all sources of energy to the equipment.
4. Tag the start/stop station.
5. Centralize the isolation keys in an area with restricted access (e.g., a lockout/tagout lockbox).
6. Confirm isolation with maintenance and add maintenance tags.

Once the lockout/tagout is complete, perform verification procedures (verification techniques can be a combination of the following items, but not limited to only these items):

- A walk-through of the affected work area by all workers.
- Checking the de-energization of the electrical circuit by pushing the start button on the start/stop station.
- Checking depressurization through bleeder valves.
- Checking chained isolation block valves.

Process Technician's Role in Operation and Maintenance

Process technicians are a key link in a series of checks and balances that protect people and equipment during shutdown periods and normal operations. It is the process technician's responsibility to ensure that the equipment is operating properly. Proper training by the employer should cover all of the necessary information required to ensure this. Table 9-4 lists some routine observations that process technicians should make when working around electric motors.

TABLE 9-4 Process Technician's Role in Operation and Maintenance

Look	*Listen*	*Feel*
• Inspect wires to ensure the insulation is not cracked or cut. • Check for loose covers and shrouds. • Look for signs of corrosion. • Lubricate bearings when scheduled. • Visually observe that the cooling fan is running.	• Listen for abnormal noises. • Listen to the bearings; investigate when noisy.	• Feel for excessive heat. • Check for excessive vibration.

For example, scheduled maintenance on electric motors is an essential part of keeping equipment in a condition that minimizes or avoids equipment failure. Process technicians must know how to maintain communications and partnerships with maintenance personnel and coordinate preventive maintenance.

Process technicians have specific roles with regard to the operation and maintenance of motors. This includes the monitoring of lubrication, the condition of seals, and proper housekeeping. Process technicians may also be required to add or replace oil or other lubrication on moving parts and bearings as part of a routine maintenance schedule. Failure to do this can result in equipment damage or failure.

Housekeeping is another important factor in the maintenance of motors. Pump areas must be cleaned on a regular schedule and after repair work has been completed.

Instrumentation installed on motors should be regularly checked and recorded on a log sheet (if required by the employer). Recorded data should include vibration monitors, amp and power meters, and temperature indications. Abnormal readings should be promptly reported and investigated.

To prevent overcurrent, a process technician should monitor the amps that the electric motor is consuming with electrical meters, operate the motor at no more than its full load capacity, and use the right size motor for the work being performed.

Many motors cannot withstand frequent stops and starts, which can cause failure of the motor's starter. Because of this, process technicians must adhere to the operating procedures for each motor type.

Summary

Motors and electrical distribution are important components of industrial manufacturing facilities. Motors are mechanical drivers that convert electrical energy into useful mechanical energy and provide power for a variety of equipment.

Electricity either oscillates (alternating current) or flows in a single direction (direct current). Components of an electric distribution system include three-phase and single-phase systems, transformers, motor control centers (MCCs), circuit breakers, fuses, and switches.

Electric motors can be either alternating current (AC) or direct current (DC). AC motors can be either variable-speed or single-speed, whereas DC motors are usually variable-speed. Single-speed AC motors may include synchronous and induction motors. Induction motors, which are the most common design in the process industries, are usually of a squirrel cage design. AC wound rotor designs are variable speed. DC motors are typically variable speed.

The main components of an electric motor include a frame, shroud, rotor, stator, fan, and bearings. Both AC and DC motors work off the same principles of electromagnetism.

Process technicians should inspect motors and perform scheduled preventive maintenance in order to keep motors in good condition and prevent equipment failure. Process technicians should also be familiar with proper operating procedures before starting any piece of equipment, and should verify that the actual responses correspond to the anticipated responses.

Checking Your Knowledge

1. Define the following terms:
   a. Alternating current (AC)
   b. Direct current (DC)
   c. Ammeter
   d. Circuit breaker
   e. Fuse
   f. Load
   g. Motor control center (MCC)
   h. Stator
   i. Switch
   j. Transformer
   k. Ampere
   l. Ohm
   m. Volt
   n. Watt

2. *(True or False)* AC motors are the most common type of motor used in the process industries due to their simplicity.

3. *(True or False)* DC motors have the ability to adjust the rotor speed of the motor.

4. *(True or False)* Synchronous motors are the most common type of AC motor.

5. A(n) _____ is the amount of torque necessary for a motor to overcome the resistance of the machine it drives.
   a. watt
   b. amp
   c. volt
   d. load

6. Which of the following is *not* a factor that affects motors?
   a. Emission
   b. Location
   c. Rotation
   d. Size limits

7. A coupling failure is often dangerous because of _____.
   a. fire
   b. vibration
   c. a blown fuse
   d. flying shrapnel

8. Which of the following is *not* a potential hazard associated with motors?
   a. Corrosion
   b. Electrocution
   c. Burns
   d. Injury from moving parts

9. Where can a list of additional restrictions be found for hazards?
   a. National Electric Code information
   b. Employee handbook
   c. Motor control center
   d. All of the above

10. What three things must process technicians do as they perform routine tasks on electric motors?
    a. Anticipate, explore, and think
    b. Look, listen, and feel
    c. Prevent, understand, and consider
    d. Question, answer, and know

Components	Description
11. Shroud	a. A structure that holds the internal components of a motor and motor mounts
12. Frame	b. A rotating member of a motor or turbine that is connected to the shaft
13. Stator	c. A rotating blade inside a motor casing that cools the motor by pulling air in through the shroud
14. Rotor	d. A casing over the motor that allows air to flow into and around the motor
15. Fan	e. The stationary part of the motor where alternating current supplied to the motor flows, creating a magnetic field using magnets and coiled wire

Student Activities

1. Describe the basic principles of electricity, including the difference between AC and DC current, and identify which type is most commonly used in the process industry.
2. Write a one-page paper explaining how an electric motor works.
3. Given a picture or a cutaway of an AC motor, identify the following components:
    a. Frame
    b. Shroud
    c. Rotor
    d. Fan
    e. Bearings
    f. Power supply
4. Brainstorm a list of possible problems associated with motors. Be prepared to discuss these problems as a class.

CHAPTER 10

Engines

Objectives

After completing this chapter, you will be able to:

- Explain the purpose of engines in the process industry.
- Identify the common types and applications of engines (gas, diesel, and steam)
- Identify the major components of an engine.
- Describe the operating principles of engines.
- Describe safety and environmental hazards associated with engines.
- Identify typical procedures associated with engines.
- Describe the process technician's role in operation and maintenance of engines.
- Identify potential problems associated with engines.

Key Terms

Camshaft—a driven shaft fitted with rotating wheels of irregular shape (cams) that open and close the valves in an engine.

Compression ratio—the ratio of the volume of the cylinder at the start of a stroke compared to the smaller (compressed) volume of the cylinder at the end of a stroke.

Connecting rod—a component that connects a piston to a crankshaft.

Coolant—a fluid that circulates around or through an engine to remove the heat of combustion. The fluid may be a liquid (e.g., water or antifreeze) or a gas (e.g., air or freon).

Crankshaft—a component that converts the piston's up-and-down or forward-and-backward motion into rotational motion.

Engine block—the casing of an engine that houses the pistons and cylinders.

Exhaust port—a chamber or cavity in an engine that collects exhaust gases and directs them out of the engine.

Exhaust valve—valve in the head at the end of each cylinder. It opens and directs the exhaust from the cylinder to the exhaust port.

Head—the component of an engine on the top of the piston cylinders that contains the intake and exhaust valves.

Intake port—an air channel that directs fuel gases to an intake valve.

Intake valve—a valve located in the head, at the top of each cylinder, that opens and allows fuel gases to enter the cylinder.

Oil pan/sump—a component that serves as a reservoir for the oil used to lubricate internal combustion engine parts.

Piston—a component that moves up and down or backward and forward inside a cylinder.

Rod bearing—a flat steel ring or sleeve coated with soft metal and placed between the connecting rod and the crankshaft.

Rotating blade—a component in a gas turbine attached to the shaft of the turbine. The steam or gas causes the turbine to spin by impinging on the rotating blades. The steam or gas slows and is redirected as it transfers energy to the rotating blades.

Spark plug—a component in an internal combustion engine that supplies the spark to ignite the air/fuel mixture.

Stage—a set of nozzles or stationary blades plus a set of rotating blades.

Stationary blade—a component in a gas turbine attached to the case that does not rotate. Stationary blades change the direction of the flow of the steam or combustion gas, and redirect it to the next stage of rotating blades.

Valve cover—a cover over the head that keeps the valves and camshaft clean and free of dust or debris, and keeps lubricating oil contained.

Introduction

An engine is a machine that converts chemical (fuel) energy into mechanical force. Originally engines were considered any type of mechanical device that produced work. Most of the applications discussed here generate rotational forces, or torque. Torque is used to operate other machinery. The most common types of engines used in process facilities are internal combustion engines (e.g., gasoline and diesel engines, gas-fired turbines) and steam-driven turbines. In the process industries, engines may be referred to as drivers.

Common Types of Engines

Engines are typically one of two types: internal combustion or external combustion. Within these two types are several subtypes that include gasoline and diesel engines and steam turbines.

INTERNAL COMBUSTION ENGINES

Internal combustion engines are engines in which the combustion (burning) of fuel and an oxidizer (usually air) occurs in a confined space called a combustion chamber. As the fuel is combusted, an exothermic (heat-producing) reaction occurs that creates high temperatures. High-pressure gases expand and act on various parts of the engine (e.g., pistons or rotors), causing movement.

Examples of internal combustion engines include gasoline and diesel engines. Gasoline and diesel engines are similar in their operation, but they differ in how they ignite the fuel.

In both diesel and gasoline engines, fuel and air are introduced into the cylinders above the pistons. This fuel is then burned, producing heat and pressure. These engines differ, however, because gasoline engines use a spark (e.g., from a spark plug) to ignite the fuel, and diesel engines use compression (instead of a spark) to ignite the fuel.

In the case of gas turbines (which are discussed in Chapter 8), natural gas or other fuel is combined with compressed air to create high-energy (high-pressure and high-temperature) combustion gases. Figure 10-1 shows an example of a gas turbine.

EXTERNAL COMBUSTION ENGINES

External combustion engines are engines in which an internal working fluid is heated (usually by an external source like a heat exchanger) until it expands. This expanding fluid then acts on an engine mechanism to provide motion and usable work.

External combustion engines differ from internal combustion engines because they use an external combustion chamber (as opposed to a combustion chamber inside the engine) to heat a separate working fluid which, in turn, does work (e.g., moving pistons or turbines). An example of an external combustion engine is a steam turbine.

The output of a steam turbine is similar to other engines. However, the method for producing that rotational power is different. For example, in a steam turbine, the rotational force is produced by high-pressure steam that is produced in a boiler.

Did You Know?

Gasoline engines were invented in 1876 by Nicolaus Otto.

Rudolf Diesel developed the idea for the diesel engine and obtained the German patent in 1892.

(Nicolaus Otto)   (Rudolf Diesel)

FIGURE 10-1 Gas Turbine

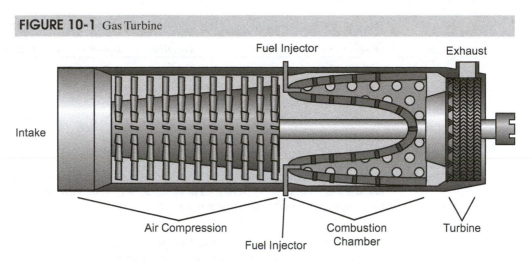

PROS AND CONS OF DIFFERENT TYPES OF ENGINES

Steam turbines can be an efficient resource in the process industries because the steam required to drive them can be produced through the combustion of waste products. As these waste products are combusted in a furnace or other combustion device, heat is produced. This heat is then used to generate the steam necessary to power the turbine. Steam turbines may also be selected when a powerful engine is needed to drive large equipment (e.g., pumps, compressors, or generators) because they are smaller and more compact than similarly sized electric motors.

Another benefit of steam turbines over motors is it is easier to vary their speed and energy output. They are also more efficient at energy conversion. A disadvantage of steam turbines is that they typically require a boiler and all its auxiliaries to produce steam.

One benefit of a gas turbine is it can be used as a backup electrical generating source because it is powered by natural gas instead of electricity. Gas turbines can also be configured as combined-cycle engines. In a combined-cycle engine, energy is extracted from the combustion exhaust gas to improve efficiency. That is, while the turbine extracts most of the energy, the gases exiting the turbine are still very hot and can be used to boil water and produce steam. Table 10-1 provides a comparison of different engine types and their uses.

Uses of Engines in the Process Industry

Engines are critical for daily operations in the process industries. They are used to drive pumps, fans, compressors, and electric generators. Engines are used for many of the same tasks as electric motors, but they may be selected instead of electric motors for a variety of reasons.

Did You Know?

The original locomotives of the 1800s were steam-driven, piston engines.

The majority of modern locomotives are diesel-driven.

TABLE 10-1   Comparison of the Different Engine Types Used in the Process Industries

Engine Characteristic	Gasoline	Diesel	Gas Turbine	Steam Turbine
Remote use	×	×		
Mobile use	×	×		
Stationary use		×	×	×
Intermittent/standby use	×	×	×	×
Continuous use			×	×
Emergency electric generation		×	×	×

Note: Although diesel engines and gas turbines are capable of continuous use, this is not common in most process facilities. However, diesel engines are used continuously on drilling rigs, ships, and remote electric generators.

Engines are sometimes used in place of motors because they provide a backup source of power. For example, diesel or gasoline-powered backup generators are often used during power outages to continue operations or to allow for the safe shutdown of the facility. They can also be used to power a backup fire pump in the event a fire occurs simultaneously with power outage.

Diesel and gasoline engines can also be used to power portable equipment or to service remote locations. For example, if a service pump is needed infrequently and at different locations around a facility, it may be preferred that the pump be driven by an engine. Another example is a large welding machine that has integral engines and generators so it can be used in remote locations where it is impractical to run large power cables. Such portable devices are frequently required during turnarounds when the normal electrical power may be unavailable.

Gasoline and diesel engines can perform many of the same tasks. However, diesel engines tend to be better suited for large tasks or tasks that require continuous operations, while gasoline engines are better suited for smaller tasks and intermittent operation. Diesel engines are also more appropriate in areas where gasoline (which is more flammable than diesel) poses an undue risk.

Turbines can also be used in place of engines for some tasks (e.g., driving pumps, compressors, and electric generators). Like gasoline and diesel engines, turbines provide a diverse power source apart from electricity. Turbines are almost always stationary and are typically used where a large engine is required. While reciprocating engines are most efficient at a set speed and are used as constant speed drivers, gas turbines can be efficient at several speeds and are used as variable speed drivers.

MAJOR COMPONENTS OF INTERNAL COMBUSTION ENGINES

The major components of a gasoline engine are discussed below and are labeled in Figure 10-2. An internal gasoline engine has many components:

The **camshaft** is a driven shaft fitted with rotating wheels of irregular shape (cams) that open and close the valves in an engine.

Did You Know?

Diesel is heavier and oilier than gasoline. It also has a different odor.

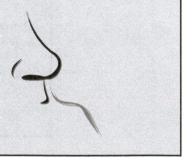

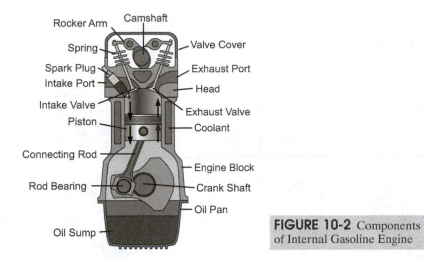

FIGURE 10-2 Components of Internal Gasoline Engine

The **connecting rod** connects the piston to the crankshaft. It swivels at both ends on connecting pins so that its angle can change as the piston moves and the crankshaft rotates.

Coolant is a fluid that circulates around or through an engine to remove the heat of combustion. This fluid may be a liquid, like water or antifreeze, or a gas, like air.

The **crankshaft** converts the piston's up-and-down or forward-and-backward motion into rotational motion.

The **engine block** is the casing of the engine that houses the pistons and cylinders.

The **exhaust port** is a cavity or chamber that collects exhaust gases from the exhaust valves and directs it to the exhaust system.

Exhaust valves are valves in the head at the end of each cylinder that open and direct the exhaust from the cylinder to the exhaust port.

The **head** is the component of an engine on top of the piston cylinders that contains the intake and exhaust valves.

The **intake port** is an air channel that directs fuel gases to the intake valves.

Intake valves are located at the top of each cylinder and open to allow fuel gases to enter the cylinder.

The **oil pan/sump** is a component that serves as a reservoir for the oil used to lubricate internal combustion engine parts.

Pistons are components that move up and down or backward and forward inside a cylinder. The pistons are usually equipped with seal and wear rings to isolate the combustion chamber from lubricant oil.

Rod bearings are normally flat, steel rings or sleeves coated with soft metal that are placed between the connecting rod and the crankshaft. Pressurized oil prevents metal-to-metal contact between the bearing surface and the crankshaft. Some small engines use splash lubrication instead of pressurized oil due to the light loads placed on small engines.

Spark plugs supply the spark that ignites the air/fuel mixture.

A **valve cover** keeps the valves and camshaft clean and free of dust or debris, and it keeps lubricating oil contained.

OPERATING PRINCIPLES OF TWO- AND FOUR-CYCLE ENGINES

Internal combustion engines can be classified as two-cycle or four-cycle. While both engine types contain the same four processes, the way they accomplish them differs.

Two-Cycle Internal Combustion Engines

Some engines are two-cycle engines (as opposed to the more typical four-cycle). Two-cycle engines complete the same four processes as four-cycle engines do, but they

FIGURE 10-3 Stages of a Two-Cycle Engine

Stroke 1: *Intake and Compression*   **Stroke 2:** *Power and Exhaust*

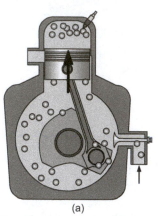
(a)

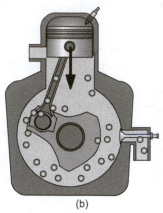
(b)

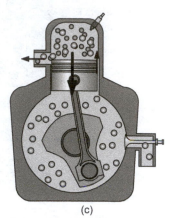
(c)

The momentum of the flywheel raises the piston and compresses the fuel mixture in the upper part of the cylinder, while another intake stroke is occurring beneath the piston.

The vacuum created during the upward stroke of the piston draws the fuel/air mixture into the lower part of the cylinder.

A spark plug ignites the fuel mixture when the piston reaches the top of the stroke. This causes the fuel to combust and force the piston downward to complete the cycle.

During the downward stroke the valve at the bottom of the cylinder is forced closed and the fuel/air mixture is forced into the upper part of the cylinder.

The intake port is exposed toward the end of the stroke. This allows the compressed fuel/air mixture to escape around the piston and into the upper part of the cylinder. This helps force exhaust gases (along with some of the fresh fuel mixture) out of the exhaust port.

complete them in two strokes instead of four. Figure 10-3 illustrates the sequence of events in a two-cycle engine.

Since a two-stroke engine fires on every revolution of the crankshaft, it is typically more powerful than a four-stroke engine of equivalent size. In addition, it is also lighter and has a simpler construction. For this reason, two-stroke engines are good for portable, lightweight applications (e.g., chainsaws, outboard motors, and some motorcycles), as well as large-scale industrial applications (e.g., locomotives). The down side to two-stroke engines is that they can be less efficient than other types of engines, and they tend to pollute more because of the unspent fuel that escapes through the exhaust port.

Four-Cycle Internal Combustion Engines

Gasoline and diesel engines function using similar four-stage engine cycles:

1. Intake
2. Compression and ignition
3. Expansion and work
4. Exhaust

Each of these cycles has slight differences and different thermodynamic names. The gasoline engine cycle is called the Otto cycle, and the diesel engine cycle is called the Diesel cycle, each named for their inventor. Each stage of these cycles corresponds with the motion of the engine pistons up and down or backward and forward. The complete cycles have two up and two down (or two backward and two forward) strokes, thus the term *four-cycle engine*. Figure 10-4 illustrates the sequence of events in a four-cycle engine.

The first stroke is the compression stroke. During this stroke, the piston compresses the gas in the cylinder. When a gas is compressed, the temperature increases.

The **compression ratio** is the ratio of the volume of the cylinder at the start of a stroke compared to the smaller (compressed) volume of the cylinder at the end of a stroke. Gasoline engines have a compression ratio of 8:1 to 12:1, while diesel engines have a compression ratio of 14:1 to 25:1. The higher compression of the diesel engine results in better efficiency. When additional workload is required, an engine may contain multiple cylinders.

FIGURE 10-4 Stages of a Four-Cycle Engine

Starting position

The cycle begins with the piston located at top dead center (the point where the piston is the furthest away from the crankshaft).

Intake

The piston moves to the bottom of the chamber, thereby reducing the pressure in the cylinder. This forces a mixture of air and fuel into the cylinder though the intake port.

Stroke 1: Compression

The piston moves upward into the chamber and compresses the fuel/air mixture.

Stroke 2: Ignition

In a gasoline engine, a spark plug ignites the fuel/air mixture. In a diesel engine the fuel/air mixture is ignited by the heat and pressure of compression.

Stroke 3: Power Stroke

When the fuel/air mixture is ignited it combusts (burns) and produces a powerful reaction that forces the piston back down to the bottom of the cylinder.

Stroke 4: Exhaust

The piston moves upward into the cylinder and forces the exhaust out of the cylinder through an exhaust valve.

In a diesel engine, only air is in the cylinder during the compression stroke. The fuel is injected into the cylinders at the top of the compression stroke and is carefully timed so that it ignites at the right moment. A high-pressure fuel pump is used to directly pump the fuel into the cylinder through an injector that sprays the fuel in as a mist.

In a gasoline engine, ignition timing is controlled by timing the spark. The fuel can be mixed with the air at low pressure when air enters the cylinder, and the fuel and air are both compressed together during the compression stroke. Older gasoline engines use an external Venturi device (the carburetor) to premix the fuel and air in the proper ratios for optimum combustion. Today the gasoline is metered by fuel injectors similar to those found in diesel engines. This results in less air pollution caused by gasoline vapors escaping from the carburetor.

The remaining stages are essentially the same. In the second stage, the fuel burns and its energy is converted to a much higher pressure and temperature in the cylinder. The expanding combustion gases force the piston back and transfer power to the crankshaft.

In the third stage, the exhaust valves in the head at the top of the cylinder open, and the piston forces the exhaust gas out of the cylinder. In the final stage, the exhaust valves close, the intake valves open, and the piston draws air into the cylinder as it moves back.

AUXILIARY SYSTEMS OF INTERNAL COMBUSTION ENGINES

All engines have a lubrication system for the bearings and moving parts. Because one of the functions of the lubrication system is to remove excess heat, there is a cooling system for the oil. Internal combustion engines also have a cooling system for the engine block. Another component of the internal combustion engine is the fuel and air delivery system (called the injection system).

MAJOR COMPONENTS OF COMBUSTION GAS TURBINES

The major components of a combustion gas turbine are discussed below and are displayed in Figure 10-5. A combustion gasoline turbine has many components:

The casing is the housing component of a turbine that contains the high-pressure air and gas and holds the stationary blades.

The combustion chamber is located between the compressor and the turbine. It is where the compressed air is mixed with fuel and the fuel is burned, thereby increasing the temperature and pressure of the combustion gases.

Compressors increase the pressure of the combustion air.

The nozzle directs the flow of fluids to the turbine blades. Nozzles are similar to blades, but they cause the velocity of the steam or gas to increase.

Rotating blades are attached to the shaft of the turbine. The steam or gas causes the turbine to spin by impinging on the rotating blades. The steam or gas slows and is redirected when transferring energy to the rotating blades.

Shafts are metal rotating components (spindles) that hold the rotor and all rotating equipment in place.

A **stage** is a set of nozzles or stationary blades, plus a set of rotating blades. Each set is shaped like a disk composed of many narrow fan blades that rotate on the shaft.

Stationary blades are attached to the case and do not rotate, but they change the direction of the flow of the steam or combustion gas and redirect it to the next stage of rotating blades.

FIGURE 10-5 Components of a Combustion Gas Turbine

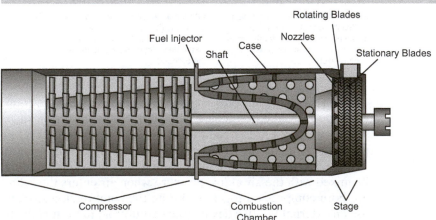

OPERATING PRINCIPLES OF A COMBUSTION GAS TURBINE

During the initial startup of a gas turbine, the compressor is started by an external starting mechanism. When the compressor starts, it delivers pressurized air to the combustion chamber. Inside the combustion chamber, fuel and air are ignited. This ignition produces

FIGURE 10-6 Diagram of Gas Turbine Combustion Stages

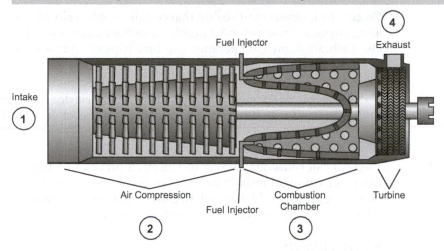

expanding combustion gases that drive the turbine. Once the turbine is operational, the external starting mechanism can be removed from the compressor. This thermodynamic operating principle is called the Brayton cycle. Figure 10-6 shows the gas turbine cycle.

In the first stage of the Brayton cycle (see Table 10-2), the hot compressed air enters the combustion chamber, where fuel is added. In the second stage, the fuel is burned. This further increases the temperature and pressure, and thus the energy of the combustion gas. In the third stage, the high-pressure, high-temperature gas applies force to the turbine blades, which does work by turning the shaft. The gas then exits the turbine at a reduced pressure and temperature.

In the final stage, gas exits through the exhaust and heat recovery takes place. An example of a heat recovery system is a waste heat boiler. A waste heat boiler is a heat exchanger that uses turbine exhaust gases to generate steam. This is called a combined cycle and it greatly improves efficiency. However, when a gas turbine is used to drive a backup emergency power generator, a combined-cycle application is not practical.

TABLE 10-2 The Stages of a Combustion Gas Turbine

Cycle Name	Cycle Name: Brayton Cycle; Open Cycle
Working fluid	Air and combustion gas.
First stage	Pressure is increased by compression.
Second stage	Combustion (increases temperature and pressure).
Third stage	Work performed by turbine rotation; pressure and temperature drop.
Fourth stage	Gas exits through exhaust; heat recovery can take place at this point.

The turbine extracts work from the combustion gas using several sets of rotating and stationary blades and nozzles. In a clockwise-rotating turbine, if you were looking down from the top, the gas goes out at an angle to the right. It then pushes against a set of blades, moving the blades and shaft clockwise. After pushing on the rotating blades, the gas changes direction to an angle to the left. A set of stationary blades, or another set of nozzles, then redirects the flow angle to the right again, and onto another set of blades.

Each pair of stationary and rotating elements in a turbine is called a stage. In most combustion gas turbines, each stage consists of a set of nozzles and a set of rotating blades, or a set of stationary blades and a set of rotating blades. Nozzles accelerate the gas to a higher velocity and result in a pressure drop. Rotating blades convert velocity energy into mechanical work. Stationary blades just change the direction of the gas.

The turbine is shaped like a cone. The blades of each stage increase in size along the shaft. Gas enters at the small end and exits at a lower pressure at the large end. When nozzles convert some of the pressure energy into velocity energy, the pressure drops and the gas expands.

Potential Problems

Process technicians must realize that certain problem situations may occur when operating engines. These potential problems include mechanical driver problems, high levels of vibration, high temperatures, turbine trips, or equipment failure.

MECHANICAL DRIVERS

If the driver is operating normally and the driven device is not performing, check the shaft coupling to see if it is connected properly or check for clutch slippage.

HIGH VIBRATION

Engines can cause high levels of vibration that can damage the machine. Common causes of vibration include an unbalanced or bowed shaft, bearing failures, misalignment, or excessively high speed. Because excessive vibrations are extremely detrimental, large engines have multiple vibration shutdown systems.

MISALIGNMENT

Misalignment is an incorrect alignment between mating components. Misalignment can be prevented through proper initial alignment.

HIGH TEMPERATURES

Typical causes of high temperatures include bearing failures, loss of lubrication, overload from driven machinery, or loss of cooling water or fan airflow. Because high temperatures can damage equipment, many large drivers have high-temperature alarms or shutdowns.

TURBINE TRIPS

A turbine trip is when a turbine has shut down due to abnormal conditions. Large drivers typically have several shutdown activation sensors. The reasons for shutdown may include high speeds, loss of oil pressure, excessive vibration, or other conditions. Some of these sensors reset to a "permit" state after re-establishing safe conditions. However, others might require a manual reset before the "run permit" state is activated. Documentation and training on the operation of shutdown interlocks must be thoroughly understood to avoid delays in restarting the equipment. For example, process technicians should understand what interlock protection trips the driver, the reasons for this protection, and the recovery procedures from the tripped state. Recovery procedures are specific to pieces of equipment and the process facility. Process technicians should review the standard operating procedures for the piece of equipment prior to attempting any type of repair.

EQUIPMENT FAILURE

Equipment design and strict adherence to operating procedures play a substantial role in preventing equipment failure. In many processes, the operation of the driver represents a significant portion of production levels and costs, as well as safety considerations. For example, failure of the engine on an emergency firewater pump could cause disastrous consequences, including spread of fire, explosions, and emissions, possibly extending to other process facility units and the surrounding community. Failure of the engine on a large single-train system could cause a unit shutdown, resulting in substantial production loss, possible flaring, or increased emissions to the environment.

Safety and Environmental Hazards

Hazards associated with normal and abnormal engine operations can affect personal safety, equipment, and environmental aspects of the process industries. Table 10-3 lists some of the equipment, safety, and environmental concerns associated with gas turbines. Because engines contain rotating parts, it is important for process technicians to

TABLE 10-3   Equipment, Safety, and Environmental Concerns Associated with Gas Turbine Engines

Personal Safety Hazards	*Equipment*	*Environment*
• Burn hazards • Fumes from exhaust gases can cause incapacitation if they are not ventilated correctly • Noise hazards • Hand and arm safety due to rotating equipment • Moving parts that can pull, tear, or rip clothing or skin	• Overspeed, which causes structure failure and flying debris • Extremely high temperatures due to combustion of gases • High pressures • Possible high levels of vibration	• Exhaust emissions from internal combustion engines and lubrication oil leaks • Fumes from exhaust gases caused by failure of the exhaust duct • Spills of fuels or lubrication oils and/or process releases from sudden engine failure

pay attention to warning signs (see Figure 10-7) and to keep loose articles (e.g., clothes, hair, and jewelry) properly secured at all times.

FIGURE 10-7 Moving Parts Danger Symbol

Typical Procedures

Typical procedures associated with engines include startup, placing in service, normal operation shutdown, lockout/tagout, and emergency procedures. Operating procedures should be closely followed when starting, operating, and shutting down an engine. Process technicians should suggest changes to these procedures if they appear to be incomplete.

STARTUP

Startup procedures vary considerably from engine type to engine type. However, the following steps apply to most engines:

1. When starting an engine, always inspect the lubrication level (this can be done with a dipstick in most internal combustion engines or oil bubblers in many gas turbine engines).
2. Ensure that the engine is clear of debris or foreign objects that could get caught in rotating parts.
3. Observe the exhaust for smoke because it can be indicative of degraded engine performance.
4. Monitor any abnormal phenomena and be prepared to shut down the engine immediately.

Small gasoline and diesel engines have electric starters, and the startup is similar to an automobile engine. Large engines typically use air-driven starters, which use high-pressure air from tanks to provide the motive force to get the engine moving.

Many engines contain interlocks that automatically shut the engine down if the lubrication oil pressure or coolant flow is low or if excessive vibration is detected. Often, the startup process includes the following:

1. Getting the engine turning by using an electric or air starter.
2. Adding gas to start the combustion process.

ENGINE IN SERVICE

Procedures for placing an engine in service vary with the engine's function. It is very important for process technicians to understand the purpose of the engine and the equipment it drives, as well as how that equipment affects the process system. Before placing the engine and its equipment in service, remember the following:

1. Establish in your mind what you expect the response to be for both the engine and the process system.
2. Check the response of the engine, the driven equipment, and the process system after placing the engine in service to verify they are responding as anticipated.

NORMAL OPERATIONS

Normal operation requires monitoring key process variables such as oil temperature, oil pressure, engine speed, vibration levels, engine temperature, and other variables. The process technician must ensure that these variables are controlled, as defined in the operating procedures.

SHUTDOWN

Shutdown of an engine involves a few important steps:

1. Cool down the engine by running it under no load or a light load for a brief period (e.g., a few minutes).
2. Monitor, record, and trend the coast downtime for some turbines because this can indicate bearing problems. Coast downtimes may also be monitored for turbochargers of large diesel engines.

LOCKOUT/TAGOUT

Lockout/tagout of an engine is similar to other lockout/tagout procedures. During lockout/tagout, process technicians should isolate the starting motive force (e.g., a battery or compressed air), external cooling sources (e.g., circulating water), and the fuel or energy source. It is always necessary to isolate the driven equipment (e.g., a generator, pump, or compressor) before performing maintenance.

EMERGENCY PROCEDURES

Emergency procedures are specific to the process facility and, in most instances, include quick isolation of the engine's energy source.

Process Technician's Role in Operation and Maintenance

Scheduled maintenance activities are an essential part of keeping engines in good condition and preventing their failure. Process technicians must know how the equipment works and how to maintain it in partnership with maintenance personnel.

MONITORING

The process technician is the key person for detecting an abnormal condition in an engine. Table 10-4 describes routine checks that are necessary for the upkeep of engines.

During operation, abnormal sounds are often the first indication of problems. These sounds can be detected if the process technician listens closely to the sounds of the engine each time it is run. Because process technicians spend hours monitoring equipment, abnormal engine sounds should be easy to detect. Process technicians

TABLE 10-4   Process Technician's Role in Operation and Maintenance

Look	*Listen*	*Feel*
• Check for loose covers and shrouds • Look for leaks • Look for excessive vibration	• Listen for abnormal noise • Listen to and check the cooling fan	• Feel for excessive heat • Check for excessive vibration

should also check operating temperatures, pressures, and flows (e.g., coolant, exhaust manifold, oil, and bearings) for abnormalities and check for vibration.

A manual vibration check is often required for small and medium-size internal combustion engines, while separate monitoring instrumentation is used for large turbines. The output of an engine-driven device should also be checked (e.g., voltage and current for generators, or pressure and flow for pumps and compressors).

LUBRICATION

Lubrication is critical to the proper operation of all engines. Routine oil sampling, on-site inspection, and off-site analysis are critical parts of a good lubrication program. On-site checks are used to identify substantial problems, such as water content, other contaminants, odor, discoloration, foam, and emulsion state. Off-site analysis reports indicate problems such as elemental materials in the oil, and other problems such as bearing wear. Off-site analysis also indicates the condition of the oil and its additives, which should be graphed to ensure the oil is not degrading faster than expected.

If makeup lubrication oil is needed, it is imperative that the proper lubricant is used, as specified in the standard operating procedures or by the manufacturer. Many locations label the required lubricant on the equipment to reduce potential error. Using the wrong lubricant can result in serious and costly equipment damage and, in some cases, catastrophic failure, fire, explosion, and injury to personnel.

OTHER MAINTENANCE ITEMS

In addition to maintaining the equipment components and lubrication, it is also important to keep the fuel system clean so the engine will continue to run smoothly. This is particularly true of diesel engines. Fuel should be sampled and inspected upon delivery, and fuel filters should be regularly inspected, maintained, and changed.

To be able to perform its function, an engine must respond reliably when called upon to do work. Process technicians should also check battery charge and condition or starting air tank pressure for large diesel engines and conduct regular performance checks for engines that provide emergency backup service.

Summary

Engines are machines that convert energy into useful work. They are used to drive pumps, compressors, electric generators, and other types of equipment. They are also useful in emergencies because they work independently of the facility's electric supply.

Gasoline and diesel engines can provide power in remote and mobile locations. Gas turbines can be used to conserve energy by utilizing waste heat. Gas turbines are capable of driving large equipment.

Common internal combustion engines are gasoline and diesel, both of which are piston engines. The internal combustion engine cycle is called the Otto cycle, and the diesel engine cycle is called the Diesel cycle. Each stage of the cycle corresponds to the motion of the engine's pistons. In a four-cycle engine, a complete cycle consists of two up and down strokes. Two-cycle engines complete the same four processes as four-cycle engines, but they complete them in two strokes instead of four. Internal combustion engines deliver work by converting the linear piston motion into high-speed rotational motion through the use of a crankshaft.

Gas turbines, another type of combustion engine, deliver work through high-energy combustion gases that drive turbine blades, causing a shaft to turn at high speeds.

Potential problems related to engines may include incorrect fuel or fuel mixtures, lack of lubrication, bearing wear or failure, drivers operating but the driven device does not, high levels of vibration, misalignment, high temperatures, overspeed, loss of coolant, or trips.

Safety and environmental hazards may be caused due to equipment malfunction or failure.

Process technicians must be aware of specific procedures associated with engines as they relate to their job requirements. These procedures include startup, shutdown, lockout/tagout, and emergency procedures. When process technicians are involved with the operation and maintenance of equipment and engines located in their operating unit, they must be familiar with required procedures and know how to access them in a time of emergency.

Checking Your Knowledge

1. Define the following terms:
   a. Engine
   b. Diesel engine
   c. Gasoline engine
   d. Combustion chamber
   e. Internal combustion engine
   f. Spark plug
2. The fuel that is most commonly used in a gas turbine engine is:
   a. diesel fuel
   b. jet fuel
   c. gasoline
   d. natural gas
3. In an _____ engine, fuel is combusted (burned) outside the engine.
   a. internal combustion
   b. external combustion
4. *(True or False)* The primary difference between a gasoline and diesel engine is that a gasoline engine uses compression ignition while a diesel engine uses spark ignition.
5. The correct order for the stages of an internal combustion engine are:
   a. compression, exhaust, intake, power
   b. power, exhaust, compression, intake
   c. exhaust, intake, compression, power
   d. intake, power, exhaust, compression
   e. intake, compression, power, exhaust
6. *(True or False)* The Otto cycle is the term used to describe a four-cycle internal combustion engine.
7. *(True or False)* Compression gas turbines can drive only small pumps, not large pumps.
8. Turbine trips are caused by (select all that apply):
   a. high speed
   b. loss of oil pressure
   c. high vibration
   d. other conditions
9. A typical cause of high lubrication oil temperature is:
   a. bearing failure
   b. blown fuse
   c. faulty start/stop
   d. misalignment

Student Activities

1. Review typical procedures associated with engines. Develop an engine procedure using the following as procedure sections. Describe when and why they are performed, the possible hazards, and precautions associated with performing them.
   a. Monitoring
   b. Lockout/tagout
   c. Routine and preventive maintenance
   d. Startup/shutdown
   e. Emergency

2. Describe the potential hazards associated with normal and abnormal operations. These hazards may include the following:
   a. Moving parts
   b. Burns
   c. Shock or electrocution
3. Using a cutaway or drawing of an engine, work with a classmate to identify additional safety precautions that process technicians may be required to take for routine and preventive maintenance.
4. Prepare a two-page report that explains the types of equipment-related operational hazards associated with engines and auxiliary systems during normal and abnormal operations.
5. With a classmate, prepare to exhibit or demonstrate different types and grades of lubricants, lubricating devices, and/or typical lubricating procedures or requirements.

Heat Exchangers

Objectives

After completing this chapter, you will be able to:

- Explain the purpose of heat exchangers in the process industry.
- Identify the common types and applications of heat exchangers.
- Identify the components of heat exchangers and explain the purpose of each.
- Describe the operating principles of heat exchangers.
- Describe safety and environmental hazards associated with heat exchangers.
- Identify typical procedures associated with heat exchangers.
- Describe the process technician's role in heat exchanger operation and maintenance.
- Identify potential problems associated with heat exchangers.
- Identify heat exchanger symbols.

Key Terms

Backwashing—a procedure in which the direction of flow through the exchanger is reversed to remove solids that, for example, have accumulated in the inlet tubes of a heat exchanger.

Baffle—a metal plate that is placed inside a vessel or tank and is used to alter the flow of chemicals, facilitate mixing, or cause turbulent flow.

British thermal unit (BTU)—a measure of energy in the English system, referring to the heat required to raise the temperature of 1 pound of water 1 degree Fahrenheit at sea level.

Calorie—the amount of heat energy required to raise the temperature of 1 gram of water by 1 degree Celsius at sea level.

Co-current flow—flow that occurs when two fluids are flowing in the same direction.

Condenser—an exchanger used to convert a substance from a vapor to a liquid.

Conduction—the transfer of heat from one substance to another by direct contact.

Convection—the transfer of heat as a result of fluid movement.

Cross flow—occurs when two streams flow perpendicular (at 90-degree angles) to each other.

Exchanger head—located on the end of a heat exchanger, it directs the flow of fluids into and out of the tubes.

Fouling—accumulation of deposits (such as sand, silt, scale, sludge, fungi, and algae) built up on the surfaces of processing equipment.

Heat transfer—the transfer of energy from one object to another as a result of a temperature difference between the two objects.

Interchanger—a process-to-process heat exchanger. Interchangers use hot process fluids on the tube side and cooler process fluids on the shell side. Also known as cross exchanger.

Laminar flow—a streamline flow that occurs when the Reynolds number is low (at very low fluid velocities).

Parallel flow—occurs in a heat exchanger when the shell flow and the tube flow are parallel to one another.

Preheater—a heat exchanger that adds heat to some substance prior to a process operation.

Radiation—the transfer of heat through electromagnetic waves (e.g., warmth emitted from the sun).

Reboiler—a tubular heat exchanger, placed at the bottom of a distillation column or stripper, that is used to supply the necessary column heat.

Reynolds number—a number used in fluid mechanics to indicate whether a fluid flow in a particular situation will be smooth or turbulent.

Shell—the outer casing, or external covering of a heat exchanger.

Temperature—the measure of the thermal energy of a substance (i.e., the "hotness" or "coldness") that can be determined using a thermometer.

Tube bundle—a group of fixed, parallel tubes though which process fluids are circulated.

Tube sheet—a formed metal plate with drilled holes that allows process fluids to enter the tube bundle.

Turbulent flow—flow that occurs when a dimensionless number (Reynolds number) is above 10,000 (at high velocities).

Introduction

Heat exchangers are devices that are used to transfer (exchange) heat from one substance to another without the two process fluids physically contacting each other. Without heat exchangers, many processes could not occur properly. Heat can be transferred through conduction, convection, or radiation. However, the primary modes of heat transfer in the process industries are conduction and convection.

When there is a temperature difference between two fluids, heat is transferred from the fluid with the higher temperature to the fluid with the lower temperature. In other words, the heat travels from hot to cold.

Heat exchangers come in a variety of types, including double-pipe, spiral, plate and frame, and shell and tube. The most common type, however, is the shell and tube exchanger. While the designs of the different heat exchangers may vary, all have similar components and use the same principles of heat transfer. When working with heat exchangers, process technicians should be aware of the operational aspects of the exchanger and the factors that could affect heat exchange.

Heat Transfer Overview

When working with heat, it is important to know that heat and temperature are not the same thing. **Temperature** is the measure of the thermal energy of a substance (i.e., the "hotness" or "coldness"), which can be determined using a thermometer. **Heat transfer** is the transfer of energy from one object to another as a result of the temperature difference between the two objects.

According to the principles of physics, heat energy always moves from hot to cold. Heat also cannot be created or destroyed; it can only be transferred from one object to another.

Heat is typically measured in British thermal units or calories. A **British thermal unit (BTU)** is the amount of heat energy required to raise the temperature of 1 pound of water by 1 degree Fahrenheit. A **calorie** is the amount of heat energy required to raise the temperature of 1 gram of water by 1 degree Celsius. One BTU is equivalent to 252 calories. (The term *Calorie*, with a capital "C," found on food labels and in diet books is really 1,000 of the calories referred to in this chapter. That means one BTU is actually equivalent to 252,000 of the Calories we eat!)

Heat is commonly transferred by three methods: conduction, convection, and radiation. **Conduction** is the transfer of heat from one substance to another by direct contact (e.g., transferring heat energy from a frying pan to an egg or from heat tracing attached to a pipe to a process fluid). **Convection** is the transfer of heat as a result of fluid movement (e.g., warm air circulated by a hair dryer). **Radiation** is the transfer of heat by electromagnetic waves (e.g., warmth emitted from the sun). Figure 11-1 shows examples of the different types of heat transfer.

Theory of Heat Exchanger Operation

In the process industries, many chemical reactions generate heat or require the addition of heat. Heat can also be produced by friction in rotating equipment (e.g., pumps and compressors). Some of this heat is transmitted to process fluids and must be

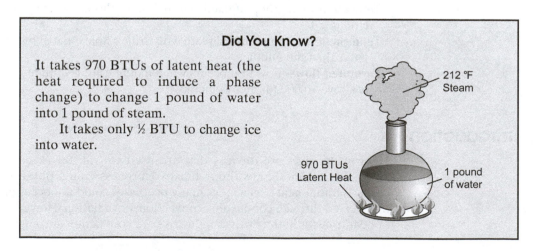

Did You Know?

It takes 970 BTUs of latent heat (the heat required to induce a phase change) to change 1 pound of water into 1 pound of steam.

It takes only ½ BTU to change ice into water.

212 °F Steam

970 BTUs Latent Heat

1 pound of water

FIGURE 11-1 Heat may be Transferred Through Conduction, Convection, or Radiation

Conduction	Convection	Radiation
(frying pan)	(hair dryer)	(sunlight)

removed. The only way to remove heat from a process stream is to transfer it to another process stream or to some other heat-absorbing material.

To improve efficiency and reduce energy costs, heat in the process industries is often recycled. For example, the material leaving a distillation column (tower) can be used to heat the feed coming to the column (tower) by means of a heat exchanger. Another way to reduce energy costs is to use the most efficient heat exchangers possible. The more efficient the exchanger, the less energy is required for heat transfer.

Heat exchangers can be used for a variety of applications. The amount of heat transfer that occurs in a heat exchanger is a function of several factors, including the temperature difference between the streams, the surface area available for heat transfer, and the overall heat transfer coefficient.

HEAT TRANSFER COEFFICIENTS

The overall heat transfer coefficient is a calculated variable that indicates the rate at which heat can be transferred. The temperature difference between the two streams, the surface area of the exchanger, and the overall heat transfer coefficient are the three variables that determine the heat transfer rate. The heat transfer rate increases when the following occur:

- The temperature difference between the fluids increases.
- The surface area increases.
- The heat transfer coefficient increases.

Did You Know?

Heat exchange systems also occur in nature. Penguins and wading birds use a heat exchange system to reduce the amount of body heat lost through their legs as they stand on the ice or move through cold water. The heat exchangers in the body are composed of blood vessels that run parallel to each other, but the blood flows in opposite directions.

Warm blood flows from the body into the legs. Cooled blood runs from the legs up into the body alongside the warm blood. As the two flows move parallel to one another, the cooler blood is warmed by the warmer blood, thereby conserving body heat. In other words, the blood vessels act like a countercurrent heat exchanger. Such an arrangement can prevent hypothermia, or excessive loss of body heat to the environment.

There are also factors that decrease the heat transfer rate. One of those is heat exchanger fouling.

FOULING

Fouling is the accumulation of deposits (such as sand, silt, scale, sludge, fungi, and algae) built up on the surfaces of processing equipment. Fouling can be caused by a variety of factors, including mineral deposits or other impurities (e.g., corrosion products) in the fluid. These contaminants or impurities form a thin layer or scale on the walls or tubes of the exchanger. This scale acts like an insulating barrier that increases the resistance to heat transfer and reduces the effectiveness of the exchanger. It also has the ability to reduce the interior diameter of the tubing, causing a drop in pressure across the exchanger.

A symptom of fouling is an undesirable change in temperature (ΔT, or delta T) and pressure (ΔP, or delta P). When the buildup (fouling) becomes too thick, the exchanger must be taken out of service and its surfaces cleaned.

TYPES OF FLOWS

In a shell and tube exchanger, liquids or gases of varying temperatures are introduced in the shell and tubes to facilitate heat exchange. For example, hot process fluids can be pumped into the tubes while cool water is pumped into the shell. As the cool water circulates around the tubes, heat is transferred from the tubes to the water in the shell.

A practical example of a heat exchanger is a car radiator. In this type of exchanger, hot liquid flows through the radiator tubes. As the car moves forward, cool air is drawn into the radiator grill and over the radiator tubes. As the air passes over the tubes, the heat from the liquid is transferred to the air via conduction and convection.

Each fluid in an exchanger has an inlet and an outlet flow path. The flow paths may be countercurrent, co-current, parallel, or cross flow. Flow may also be characterized as turbulent (disrupted or unorganized) or laminar (smooth and streamlined). These different flow types are important because they affect heat exchanger operations.

Countercurrent flow occurs when the tube side stream and the shell side stream flow in opposite directions. Because it is the most efficient flow type, most shell and tube exchangers are designed for countercurrent flow.

Figure 11-2 shows an example of countercurrent flow.

Co-current flow (shown in Figure 11-3) occurs when the streams on both the shell and tube sides flow in the same direction. Co-current flow is most often used with heat-sensitive fluids that must be kept below a certain temperature.

Parallel flow (shown in Figure 11-4) occurs when countercurrent and co-current streams flow parallel to each other. When fluids flow across a surface, a slow-moving film exists at the tube wall that inhibits heat transfer.

Cross flow (shown in Figure 11-5) occurs when two streams flow perpendicular (at 90-degree angles) to each other. In a steam condenser, the tubes with cooling water run parallel to the ground. Steam enters from the top, flows perpendicular and across the outside of the tubes, is condensed into its liquid state, and then exits out the bottom.

As fluids move through a pipe or heat exchanger, the flow can be either laminar or turbulent. **Laminar flow**, or streamline flow (shown in Figure 11-6), occurs when the fluid flow velocities are low, resulting in regular flow patterns. In other words, the fluid flow is smooth and unbroken, like a series of laminations or thin cylinders of fluid slipping past one another inside a tube. Laminar flows occur when the **Reynolds number**

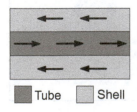

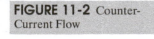

FIGURE 11-2 Counter-Current Flow

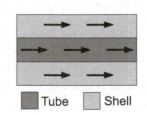

FIGURE 11-3 Co-current Flow

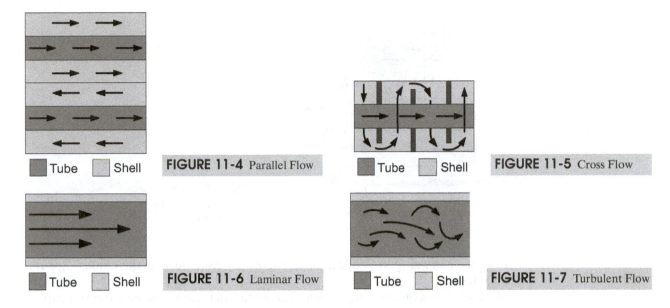

FIGURE 11-4 Parallel Flow

FIGURE 11-5 Cross Flow

FIGURE 11-6 Laminar Flow

FIGURE 11-7 Turbulent Flow

(a number used in fluid mechanics to indicate whether a fluid flow in a particular situation will be smooth or turbulent) is low.

Turbulent flow (shown in Figure 11-7) is a condition in which fluid flow patterns are disturbed so there is considerable mixing. Turbulent flow occurs when the fluid flow velocities and the Reynolds number are high.

The ideal flow type in a heat exchanger is turbulent flow because it provides more mixing and better heat transfer for heating and cooling. That is why heat exchangers contain **baffles**, which are metal plates used to alter the flow of chemicals or facilitate mixing. Baffles support the tubes, increase turbulent flow, increase heat transfer rates, and reduce hot spots.

Types of Heat Exchangers

The basic purpose of all heat exchangers is to heat or cool fluids. However, the design of the exchangers may differ. For example, heat exchangers can be double-pipe, spiral, plate and frame, and shell and tube, and the heads can be fixed or floating. They can also be classified by their function (e.g., condenser, air cooler, intercooler, and after-cooler).

DOUBLE-PIPE HEAT EXCHANGERS

Double-pipe heat exchangers (shown in Figure 11-8) consist of a pipe within a pipe. This double-pipe configuration provides extra strength in the pipe walls and reduces the likelihood of rupture.

The diameter of the larger pipe is reduced by a flange or other transition piece and is welded to the smaller pipe to create an annular (shaped like a ring) space between the two pipes. Nozzles attached to each end of the outer pipe allow a stream to pass

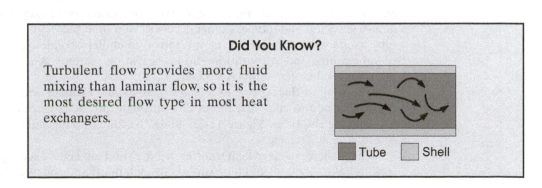

Did You Know?

Turbulent flow provides more fluid mixing than laminar flow, so it is the most desired flow type in most heat exchangers.

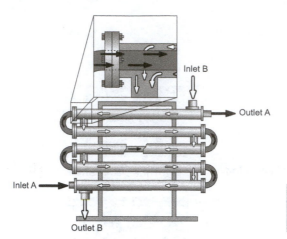

FIGURE 11-8 Double-Concentric Tube Heat Exchanger

through the annular space, while another stream is passed through the smaller pipe for heat exchange.

In some cases, the inner pipe has louvers or fins to increase the heat transfer capacity. Finned tubes improve heat exchange for low-density fluids, such as gases. Normally, double-pipe heat exchangers are used only when a small amount of heat transfer is necessary between streams.

SPIRAL HEAT EXCHANGERS

Spiral heat exchangers (shown in Figure 11-9) consist of spiral plates or helical tube cores fixed in the shell. Spiral heat exchangers offer high reliability and are good for high fouling services such as slurries. Because there are no dead spaces, the helical flow pattern creates a high-turbulence area that is self-cleaning.

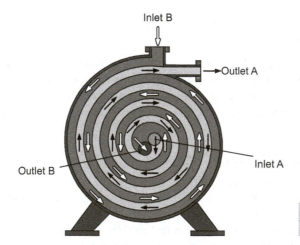

FIGURE 11-9 Spiral Heat Exchanger

SHELL AND TUBE HEAT EXCHANGERS

Shell and tube heat exchangers are the most common type of exchangers in the process industries. Shell and tube exchangers consist of a tube bundle inside a shell. Shell and tube exchangers are used for heat removal in chillers, condensers, reboilers, process steam cooling, and evaporation and refrigeration systems. A drawing of a shell and tube heat exchanger is shown in Figure 11-10.

The shell in this type of exchanger consists of a cylinder with flanged, removable heads attached to ends that direct flow through the exchanger and allow easy access for cleaning the tubes. Figure 11-11 shows the tube bundles and baffles inside a shell and tube heat exchanger.

The main methods of heat transfer in a shell and tube heat exchanger are conduction (through the tube wall and tube surface, to shell fluid) and convection (fluid movement

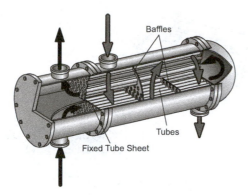

FIGURE 11-10 Shell and Tube Heat Exchanger

FIGURE 11-11 Shell and Tube Heat Exchanger Bundles and Baffles

Courtesy of Brazosport College

within the shell and the tubes). For example, hot process fluids can be passed through the tubes while cool water is pumped into the shell. As the cool water circulates around the tubes, heat is transferred from the fluid in the tubes to the water in the shell.

A practical example of a heat exchanger is a car radiator. In this type of exchanger, hot fluids flow through the radiator tubes. As the car moves forward or a fan system is in operation, cool air is drawn into the radiator grill and over the radiator tubes. As the air passes over the tubes, the heat from the fluid is transferred to the air via conduction and convection.

Shell and tube exchangers can come in several different designs. One is a multipass design that contains baffles (shaped metal dividers) in the heads and shells. These baffles help multipass exchangers increase fluid velocity, thereby increasing the overall amount of heat transferred. This results in an exchanger that requires less surface area. Multipasses also make more efficient use of the cooling medium and reduce fouling rates. Figures 11-12 and 11-13 show examples of a multipass heat exchanger.

U-tube (hairpin) heat exchangers (shown in Figure 11-14) are a type of shell and tube heat exchanger that uses U-shaped tubes. In this type of heat exchanger, baffles route the shell side back and forth across the tubes. One benefit to this type of exchanger is that the tubes save space and can be removed or replaced; however, a number of tubes may have to be removed before replacing the innermost tube.

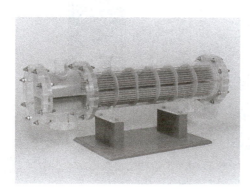

FIGURE 11-12 Picture of a Multi-Pass Heat Exchanger

Courtesy of Design Assistance Corporation (DAC)

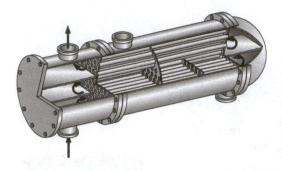

FIGURE 11-13 Diagram of a Multi-Pass Heat Exchanger.

FIGURE 11-14 Picture of U-tube Showing the Boundless Inside the Cover

Courtesy of Design Assistance Corporation (DAC)

Another benefit is that the tubes are better able to deal with thermal expansion because the tubes start and end on the same side of the exchanger.

PLATE AND FRAME HEAT EXCHANGERS

Plate and frame heat exchangers (shown in Figure 11-15) consist of a series of thin plates that are mounted on a frame with alternating hot and cool fluids. The plates are sealed with gaskets. The high fluid velocities and the thin wall of the plates result in a very high overall heat transfer rate (two to three times that of a shell and tube exchanger). This allows a typical plate and frame to save considerable space in a process plant.

Because of their design, plate and frame exchangers are usually used with clean (free of solids), nontoxic materials, and they operate at fairly low pressures. Because of their efficiency, plate and frame exchangers are used where very high overall heat transfer rates are required.

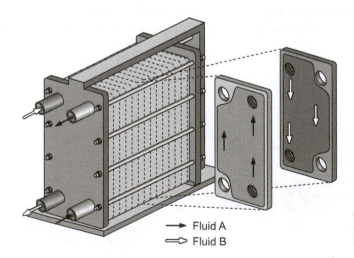

→ Fluid A
⇨ Fluid B

FIGURE 11-15 Plate and Frame Heat Exchanger

FIGURE 11-16 Photo of Tube Bundle, Tubes, Tube Sheets
Courtesy of Brazosport College

EXCHANGER COMPONENTS

The components of a heat exchanger may vary based on the design and purpose of the exchanger. However, there are some commonalities among exchangers. For example, the components of a typical shell and tube heat exchanger include the tubes, tube sheets, segmental baffles, and tie rods that make up the tube bundle. Figure 11-16 shows a photo of tubes, a tube bundle, and tube sheets.

A **tube bundle** is a group of fixed or parallel tubes through which process fluids are circulated. Within the tube bundles are numerous small tubes used to increase the surface area available for heat transfer. These tubes are housed inside the shell. The **shell** of a heat exchanger is the outer covering that retains the shell-side fluid. The tube bundle is sealed in the shell to prevent leakage between the tube side and the shell side. Fluid moves into and out of the shell and tubes through a set of inlets and outlets. Figure 11-17 shows the tube bundles, the shell, and other components of a shell and tube heat exchanger. The **exchanger head** (also called the channel head) is located on the end of a heat exchanger and directs the flow of fluids into and out of the tubes.

Plain tubing is used in shell and tube exchangers. Plain tubing consists of a tube bundle inside a shell. Finned tubing (shown in Figure 11-18) consists of thin plates of metal that are attached to the outside of a tube to provide additional surface area for heat transfer. This increased surface area allows more heat transfer to occur through a finned tube than an equivalent length of plain tubing. Your radiator is a typical finned tube application.

The **tube sheet** is a flat plate to which the tubes in a heat exchanger are fixed. The tube sheet consists of a formed metal plate with holes drilled through it. Tubes are then fitted into the holes and attached to the tube sheet by welding or flaring. This group of tubes is called a tube bundle.

FIGURE 11-17 Components of a Counter Current Flow Shell and Tube Heat Exchanger

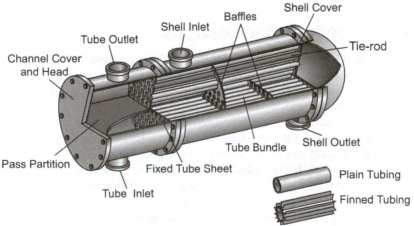

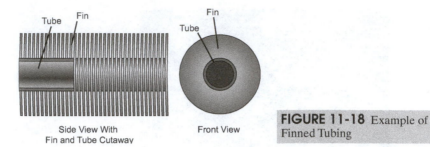

Tube   Fin

Fin
Tube

Side View With
Fin and Tube Cutaway

Front View

FIGURE 11-18 Example of
Finned Tubing

The **tube bundle** consists of a group of fixed or parallel tubes, a tube sheet, baffles, and tie rods. The channel is the flanged section of the heat exchanger head that contains flow nozzles and any baffles used to create a multipass exchanger.

The channel cover is a plate that is bolted to the outside end of the channel and retains exchanger fluid and provides access for cleaning. (*Note:* Some exchangers are built with shell bonnets instead of channel covers. Bonnets are built in one piece, with a flanged end bolted onto the shell so that they can be removed easily.) A gasket is compressed between the flanged metal pieces of the channel cover and the exchanger to prevent leakage to the atmosphere.

Baffles, or dividers, are metal plates spaced along the tube bundle that hold and support the tubes in place and divert the flow of fluid across the tubes. Baffles contain the same tube-hole pattern as the tube sheet, with a section cut off to form an opening for fluid flow. The baffles cause fluid to flow back and forth across the tube bundle to improve heat transfer.

Baffles are fastened together by tie-rods. Tie-rods are small-diameter metal rods along the tubes. Tube vibration results from fluid flow, condensing vapors, and piping pulsations. Spacing the baffles properly and using tie rods to hold them in position minimizes vibration.

Some tube bundles must be allowed to expand when they are heated. These bundles either have expansion joints in the shell, with fixed tube sheets (tube sheets that are welded to the shell) on both ends, or they have one fixed tube sheet and a floating head on the other end. Expansion joints allow the tube bundle to expand without touching the shell. Packing is used to prevent leakage from the shell on the floating end of the exchanger.

A floating head is an arrangement in which the process fluid coming out of the tubes at one end is captured and directed by a head that is not fixed to the shell of the heat exchanger. Floating heads are used to cope with thermal stress due to thermal expansion. A U-tube heat exchanger also solves the problem of thermal stress.

A weir is a flat or notched dam that functions as a barrier to flow. The weir is located at the top of the tube bundle and creates an equal flow of liquid to each tube by acting as a dam to help contain the liquid at a constant level within the exchanger. When excess flow enters the exchanger, it flows over the weir and is circulated back into the system.

Exchanger Applications and Services

Heat exchangers are used to heat or cool process fluids. They can be used alone or coupled with other heat-exchanging devices such as cooling towers. Heat exchanger applications include reboilers, preheaters, after-coolers, condensers, chillers, intercoolers, or interchangers.

A **preheater** is a heat exchanger that adds heat to fluids prior to a process operation. A preheater can be used to heat a liquid near the boiling point prior to entering a distillation tower, or to heat the feed before entering a reactor.

A **reboiler** is a tubular heat exchanger, usually placed near the base of a distillation column or stripper, that is used to supply necessary column heat. The purpose of a reboiler is to convert a liquid to a vapor and to control temperature and product quality. The heating medium for a reboiler can be steam or hot fluids from other parts of the facility. Common types of reboilers are thermosiphon, forced circulation, and kettle.

A **condenser** is a heat exchanger used to convert a vapor to a liquid. The design of a condenser can be the same as a preheater (i.e., a typical shell and tube exchanger). The difference, however, is the temperature of the fluid being used for heat exchange (i.e., cool instead of warm). An example of a condenser is the coil on the back of a refrigerator.

Figure 11-19 shows an example of a preheater, a reboiler, and a condenser attached to a distillation column.

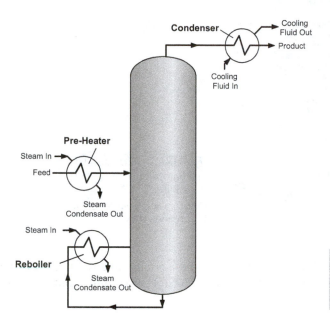

FIGURE 11-19 Pre-Heater, Reboiler, and Condenser Attached to a Distillation Column

An after-cooler is a shell and tube heat exchanger, located on the discharge side of a compressor, that is used to remove excess heat created during compression. An inter-cooler is used between the stages of a process to cool fluid before it reaches the next stage.

An **interchanger** (cross exchanger) is a process-to-process heat exchanger. Interchangers use hot process fluids on the tube side and cooler process fluids on the shell side. Interchangers are used on columns or towers where the hot bottom fluids are pumped through the tube side of an exchanger, and the feed to the column is put through the shell side. This action preheats the feed to the column and cools the bottom material.

Chillers (coolers) are devices used to cool a fluid to a temperature below ambient temperature. Chillers generally use a refrigerant as a coolant.

Air coolers consist of banks of finned tubes connected by an external tubesheet. The finned tubes transfer heat from process streams flowing through the tube banks to the atmosphere. Heat transfer is assisted by fans that push or pull a high volume of air across a bank of tubes. The finned tubes provide additional heat transfer capacity by increasing the heat transfer surface. When finned tubes and fans are added to air coolers, they are called fin-fan coolers.

A heater is a heat exchanger used to heat process materials as needed. A vaporizer (a device that converts a liquid into a vapor) is constructed by combining a heat exchanger with a separate vapor/liquid separator. Figure 11-20 shows a photograph of a distillation column with a heat exchanger in the column.

A kettle reboiler has a vapor disengaging space built into the shell. When more complete separation of liquid from vapor is needed, a mist eliminator or coalescer is installed in the vapor space.

Thermosiphon reboilers facilitate circulation by using the difference in the densities of the fluids. The circulation depends on the difference in density between the very

FIGURE 11-20 Distillation Column with Heat Exchanger in Column

Courtesy of Design Assistance Corporation (DAC)

hot material in the exchanger and the cooler material in the distillation tower. A transfer line exchanger (TLE) is used to cool gas coming from a furnace.

Potential Problems

Heat exchangers are designed to perform a specific function. Deviations from normal operating conditions can cause process upsets. For example, when a distillation column reboiler provides insufficient heat, it often results in improper separation of the materials in the column and thus poor product quality and higher production cost.

CONSEQUENCES OF DEVIATION FROM NORMAL OPERATION

A temperature differential across a heat exchanger that is too low can result in insufficient heat transfer. Excessive pressure drop may cause loss of flow, lower the amount of heat transfer, and increase the likelihood of fouling.

Process technicians should be familiar with the causes and solutions of the following troubleshooting situations: excessive chest pressure, system upsets, leaks, fouling, corrosion, erosion, reduced fluid velocity, back-flushing and vapor-locked conditions. Table 11-1 lists potential problems and their solutions when troubleshooting heat exchangers.

Safety and Environmental Hazards

Hazards associated with normal and abnormal heat exchanger operations can affect personnel safety and the environment. For example, opening and closing valves too quickly can be hazardous operations when high pressure, high temperature, and hazardous materials are involved. Also, draining and cleaning an exchanger for maintenance exposes the process technician to the potential for burns and contact with chemicals. During process upsets and emergencies, additional hazards, such as chemical contact with the skin and eyes, are present. Burns are also possible when working around steaming piping.

Potential Problem	Cause/Solution	Description of Cause/Solution
Fouling	Cause:	Cause: Plugging of the tube side caused by a buildup of contaminants at the tube sheet inlet and inside the tubes. Scaling and carbon buildup (coking) are also common forms of fouling.
	Solution:	Solution: In the case of contaminate buildup such as scaling and sludge buildup, back-flushing or chemical cleaning is usually required to clean tube deposits. If shell slide fouling is caused by coking, burning off the accumulation in a furnace may clean the bundle.
Excessive chest pressure	Cause:	In heat exchangers that use steam as the heating medium, an increase in chest pressure (increasing shell side pressure) indicates reduced heat transfer. The high pressure is typically due to a decrease in the rate of steam condensing on the shell side. Common causes of excessive chest pressure are: • Low liquid levels • Malfunctioning steam traps • Tube fouling • Noncondensable compounds in the steam • Excessive steam temperature • Closed valve on the steam outlet line
	Solution:	Monitor the chest pressure for any loss of heat transfer and watch for tube fouling and polymer formation through sample results. Monitor liquid level, maintain proper steam temperatures, and check for proper steam trap operation.
Leaks	Cause:	Leaks around flange gaskets and valve packing can be the results of worn gaskets, corrosion, loose bolts, or normally occurring wear. Leaks across tube sheets or through ruptured tubes are caused by erosion, corrosion, or substandard construction. These structural abnormalities result in the contamination of the heating/cooling medium or the product, depending on relative pressure levels in the tubes/shells of the exchanger. Leaks can also be caused by thermal shock, high pressure differential, and/or incorrect design.
	Solution:	Frequent inspection of the packing material around flanges and valve gaskets for corrosion. Replace any materials that appear to be damaged from erosion, corrosion, or substandard construction. Inspect for and tighten any loose bolts. Repair the tube sheet by plugging leaking tubes or rolling tubes.
Fouling, corrosion, erosion	Cause:	Fouling is the deposit of solids such as scale or tar-like materials on heat transfer surfaces. Corrosion is the removal of metal as a result of a chemical reaction within a process fluid or other material. Erosion is the removal of metal by impingement caused by an abrasive and/or high-velocity fluid stream.
	Solution:	A fouled tube bundle is designed to allow tubes to be cleaned easily by a machine with a rotating rod-out device or by a high-pressure water spray (called hydroblasting). Tube corrosion should not occur if the proper material of construction is employed in the heat exchanger tube bundle. Erosion may not occur if suspended solids are removed from the fluid stream or the velocity of the stream can be lowered without significantly affecting the heat transfer.
Reduced fluid velocity	Cause:	Reduced fluid velocity can occur as a result of fouling (scaling), restrictions (partially opened valves or debris restricting the flow), or improper pump operations (internal wear on a pump that reduces the performance, the pump screens plugging, or a low suction head to a pump).
	Solution:	To increase fluid velocity due to fouling, restrictions, or improper pump operations, process technicians should clean screens on the pump, switch pumps or repair failing pumps, and adjust levels to provide proper suction head.
Vapor-locked conditions	Cause:	Vapor-locked conditions can occur when vapors become trapped in spaces where heat transfer normally occurs. For example, a thermosiphon reboiler can become vapor-locked when the liquid level is above the return nozzle of the reboiler.
	Solution:	Necessary adjustments to steam and process flows are needed to correct the vapor-lock condition.

Personal protective equipment (PPE) should be worn during normal and abnormal heat exchanger operations. PPE includes the following:

- Goggles when sampling around exchangers.
- Breathing protection if toxic gases may be present.
- Face shields when using chemicals, steam, or high-pressure water.

Thermal shock and overheating of fouled tubes can result in ruptured and/or melted tubes that can cause injury to personnel and damage to equipment.

Environmental effects are also important hazards to keep in mind. Many chemicals are toxic and harmful to the environment if leaked to the atmosphere. Because of this, the government regulates the amounts that can be released. Exceeding the regulated amounts requires reporting, and fines are possible if reporting and regulations are not followed. Process fluids can also leak into other systems like cooling tower water causing corrosion, foaming, and release to the atmosphere.

There are many hazards associated with improper heat exchanger operation. Table 11-2 lists some of these hazards and their effects.

TABLE 11-2   Hazards Associated with Improper Heat Exchanger Operation

Improper Operation	*Possible Effects*			
	Individual	*Equipment*	*Production*	*Environment*
Attempting to place the exchanger in service with the bleed valves open	Exposure to chemicals or steam	Adverse effects on adjacent equipment	Loss of production	Spill to the environment
Applying heat to the exchanger without following the proper warm-up procedure	Exposure to chemicals or steam if the tubes rupture	Tube rupture due to thermal shock	Downtime due to repair	Spill to the environment if the tubes rupture
Improper alignment of the inlet and outlet valves		Overheating, tube fouling, or melting	Downtime due to repair	
Operating a heat exchanger with ruptured tubes	Exposure to chemicals or steam	Possible explosive condition resulting in severe damage	Product ruined by contamination; downtime due to repair	Potential environmental release
Opening a cold fluid to a hot heat exchanger	Exposure to chemicals or steam	Overpressurizing the heat exchanger due to rapid expansion of the liquid; potential for thermal shock	Heat exchanger damage resulting in downtime due to repair	Potential environmental release

Typical Procedures

Process technicians should be aware of typical heat exchanger procedures, which include monitoring, inspecting for leaks, lockout/tagout, routine and preventive maintenance, sampling, startup, shutdown, and emergency procedures. Failure to perform proper maintenance and monitoring could affect the process and result in equipment damage or injury.

STARTUP

The following steps represent a general startup procedure for a shell and tube heat exchanger that was taken out of service for maintenance activities. In this example, our procedures require that the heat exchanger be filled from the low side first. It is important to note that facility personnel must always follow specific operating procedures developed for individual facilities.

Filling the Low Side

1. Inspect the heat exchanger to see if it is ready for startup.
2. Verify that all the drains and vents are shut.
3. Establish the cold stream flow to the heat exchanger.
4. Open the cold side vent valve to allow any trapped vapors to escape and partially open the cold side inlet valve.
5. Shut off the vent valve when the cold side is completely filled and open the cold side inlet valve fully.
6. Open the cold side outlet valve.

Lining Up the Hot Side

The next step in this procedure is to line up the hot side of the exchanger:

1. Open the hot side vent valve.
2. Partially open the hot side inlet valve to allow the hot side to fill with process fluid and to remove any air trapped in the tubes.
3. After the hot side is filled, close the vent valve and open the inlet valve fully.
4. Open the outlet valve on the hot side slowly until it is fully open.
5. As the exchanger comes up to normal operating conditions, monitor and check for leaks or abnormal conditions.

SHUTDOWN

The following steps represent a general procedure for shutting down a shell and tube heat exchanger. Individual facilities have specific procedures that may vary from this procedure. In this example, the side with the hot stream is usually shut down first.

1. Shut off the hot side inlet valve.
2. Shut off the hot side outlet valve.
3. When the hot side has cooled, open the hot side vent and the drain valves.
4. Shut off the cold side inlet valve.
5. Shut off the cold side outlet valve.
6. Open the cold side vent and drain valves.

GENERAL MAINTENANCE TASKS

General maintenance tasks performed on heat exchangers may include backwashing, water blasting, sand blasting, acidizing, and high point bleeding.

Backwashing (back-flushing) is a procedure in which the direction of flow through the exchanger is reversed to remove solids that have accumulated in the heat exchanger. Heat exchangers are equipped with special valves so back flushing can be performed.

Water blasting is a surface cleaning technique used on heat exchangers that uses high-pressure jets of water to remove scale or other deposits from the inside of exchanger tubes.

Sand blasting is a surface cleaning technique involving entraining sand into a stream of high-pressure air to remove scale or other deposits from accessible surfaces.

Acidizing is a surface cleaning technique that uses acid to remove scale or other deposits from accessible surfaces.

High point bleeding is the process of venting gases from piping and equipment that are normally filled with liquid. Particularly with large piping systems, vent valves are put at the high points of these systems to allow them to vent. Vapors can be trapped in equipment, and vents allow the process technician to vent gas from the equipment.

TABLE 11-3	Process Technician's Role in Operation and Maintenance	
Look	*Listen*	*Feel*
• Inspect equipment for external leaks. • Check for internal tube leaks by collecting and analyzing samples. • Look for abnormal pressure changes (could indicate tube plugging). • Check temperature gauges for high temperature readings. • Inspect insulation.	• Listen for abnormal noises (e.g., rattling or whistling). High-pressure leaks are often invisible but very audible and highly dangerous.	• Feel for excessive vibration (vibration can loosen the tubes in the tube sheet, or wear through tube walls). • Feel inlet and outlet lines for unusually high or unusually cold temperatures.

Process Technician's Role in Operation and Maintenance

The role of the process technician is to ensure the proper operation of heat exchangers, including routine and preventive maintenance. Process technicians should always monitor heat exchangers for abnormal noises, leaks, excessive heat, pressure drops, temperature drops, or other abnormal conditions, and they should conduct preventive maintenance as needed. Table 11-3 lists some of the items a process technician should monitor when working with heat exchangers.

Summary

Heat exchangers are devices used to transfer heat from one process to another without the two processes fluids physically contacting each other. Heat exchangers accomplish this heat transfer through conduction and convection.

There are many types of heat exchangers, including double-pipe, spiral, shell and tube, U-tube, and plate and frame. The most common type of heat exchanger is the shell and tube exchanger.

Double-pipe heat exchangers consist of one pipe within a larger diameter pipe. The double-pipe provides extra strength in the pipe walls, reducing the likelihood of rupture.

Spiral heat exchangers consist of spiral plates, or helical tube cores fixed in the shell.

In a shell and tube exchanger, liquids or gases of varying temperatures are pumped into the shell and tubes. As the fluids move through the shell, baffles cause turbulence (mixing) in the fluid. Turbulence is desired because it facilitates heat transfer and creates more surface area for heating and cooling.

U-tube (hairpin) exchangers are a type of shell and tube heat exchanger that uses U-shaped tubes. The tubes start and end on the same side of the exchanger and thus are able to absorb thermal expansion. Baffles create the turbulence necessary for heat transfer to occur.

The basic components of a heat exchanger vary based on design and purpose. However, there are commonalities among exchangers. The basic components of a shell and tube heat exchanger include tube bundles, tube sheets, baffles, tube inlet and outlet, shell, shell inlet and outlet, exchanger cover, and pass partition.

Heat exchangers, including reboilers, preheaters, after-coolers, condensers, chillers, or interchangers, can be used in a variety of applications. An exchanger service is the actual functional application of the exchanger such as cooler, heater, condenser, vaporizer, thermosiphon reboiler, transfer line, and air cooler.

Flow characteristics affect heat exchanger operations through the inlet and outlet flow path. Flow paths may include countercurrent, co-current, parallel, or cross. Flow may also be characterized as turbulent or laminar.

Improperly operating a heat exchanger can lead to safety and process issues, including possible exposure to chemicals or hot fluids, or equipment damage due to thermal shock.

Process technicians should always monitor heat exchangers for abnormal noises, leaks, excessive heat, pressure drops, temperature drops, or other abnormal conditions. They should conduct preventive maintenance as needed.

Checking Your Knowledge

1. Define the following terms:
   a. Backwashing
   b. Baffle
   c. BTU
   d. Fouling
   e. Countercurrent flow
   f. Cross flow
   g. Laminar flow
   h. Parallel flow
   i. Turbulent flow
   j. Laminar flow
   k. Convection
   l. Conduction
   m. Radiation
   n. Heat exchanger

2. Heat exchangers are used to transfer heat between:
   a. separation equipment
   b. hot and cold streams
   c. chemical reactions
   d. temperature and pressure

3. Which of the following is the flow characteristic that occurs when the tube side stream and the shell side stream flow in the opposite directions?
   a. Cross flow current
   b. Countercurrent flow
   c. Parallel flow
   d. Turbulent flow

4. Select the components that are found in a heat exchanger (select all that apply).
   a. Channel cover
   b. Baffle/divider
   c. Expansion joint
   d. Tube sheet

5. A _____ is the piece of equipment that is bolted to the outside end of the channel to retain exchanger fluid and provide access for cleaning.
   a. tube sheet
   b. channel cover
   c. tie rod
   d. bonnet

6. Which of the following pieces of equipment is used to convert a vapor to a liquid?
   a. Preheater
   b. Reboiler
   c. Chiller
   d. Condenser

7. Which of the following is *not* a potential problem in heat exchangers?
   a. Chemical balance
   b. Plugging of the U-tubes
   c. Fouling, erosion, and corrosion
   d. Low heat transfer in the thermosiphon reboiler

Match the services on the left with the correct description on the right.

8. Thermosiphon reboiler
9. Cooler
10. Transfer line exchanger
11. Condenser

a. Used to cool gas from a furnace.
b. Uses cooling to convert a substance from a vapor to a liquid.
c. Used to cool hot liquids.
d. Provides circulation based on density differences.

Student Activities

1. Using a diagram or a cutaway of a multipass heat exchanger, identify the various components and explain the purpose of each.
2. In a lab setting, pressure test a shell and tube exchanger and locate the leak source. Explain how you would correct the leak.

3. With a group, determine how the heat transfer coefficient is calculated and make a short presentation to the class.
4. List the three variables that determine the heat transfer rate.
5. Describe the different applications of a typical shell and tube exchanger. Be sure to include the following:
   a. Reboiler
   b. Thermosiphon
   c. Preheater
   d. After-cooler
   e. Condenser
   f. Interchanger

CHAPTER 12

Cooling Towers

Objectives

After completing this chapter, you will be able to:

- Explain the purpose of cooling towers in the process industry.
- Identify the common types of cooling towers.
- Identify the components of cooling towers and explain the purpose of each.
- Describe the operating principles of cooling towers.
- Describe safety and environmental hazards associated with cooling towers.
- Identify typical procedures associated with cooling towers.
- Describe the process technician's role in cooling tower operation and maintenance.
- Identify potential problems associated with cooling towers.

Key Terms

Ambient air—any part of the atmosphere that is accessible and breathable by the public.

Approach range—a range that describes how close to the dew point a cooling tower can cool the water.

Basin—a reservoir at the bottom of the cooling tower where cooled water is collected so it can be recycled through the process.

Biocide—a chemical agent that is capable of controlling undesirable living organisms in a cooling tower.

Blowdown—the process of removing small amounts of water from the cooling tower to reduce the concentration of impurities.

Circulation rate—the rate at which cooling water flows through the tower and through the process exchangers.

Cooler—a heat exchanger that may use cooling tower water to lower the temperature of process materials.

Cooling range—the difference in the temperature between the hot water entering the tower and the cooler water exiting the tower.

Cooling tower—a structure designed to lower the temperature of water using latent heat of evaporation.

Counterflow—the condition created when air and water flow in opposite directions.

Dew point—the temperature at which air is completely saturated with water vapor (100 percent relative humidity).

Drift—carrying of water with the air stream.

Drift eliminator—devices that prevent water from being blown out of the cooling tower; the main purpose of a drift eliminator is to minimize water loss.

Dry bulb temperature—the actual temperature of the air that can be measured with a thermometer.

Duty—the amount of heat energy a cooling tower is capable of removing. Usually expressed in MBTU/hour.

Erosion—the degradation of tower components (fan blades, wooden portions, and cooling water piping) by mechanical wear or abrasion by the flow of fluids (often containing solids). Erosion can lead to a loss of structural integrity and process fluid contamination.

Evaporation—a process in which a liquid is changed into a vapor through the latent heat of evaporation.

Fill—material inside the cooling tower, usually made of wood or plastic, that breaks water into smaller droplets and increases the surface area for increased air-to-water contact.

Foaming—the formation of a froth generated by water contaminants mixing with air. Foam causes impairment of water circulation and may cause pumps to loose suction or cavitate.

Forced draft cooling towers—cooling towers that contain fans or blowers at the bottom or on the side of the tower that force air through the equipment.

Humidity—moisture content in ambient air measured as a percentage of saturation.

Induced draft cooling tower—a cooling tower in which air is pulled through the tower internals by a fan located at the top of the tower.

Louver—a moveable, slanted slat that is used to adjust the flow of air.

Natural draft cooling tower—a cooling tower in which air movement is caused by wind, temperature difference, or other nonmechanical means.

Relative humidity—a measure of the amount of water in the air compared with the maximum amount of water the air can hold at that temperature.

Tube leak—a leak in a tube that can result in process chemical entering the circulating water, possibly resulting in a fire, explosion, or environmental or toxic hazard.

Water distribution header—a pipe that provides water to a distributor box located at the top of the cooling tower so the water can be distributed onto the fill.

Wet bulb temperature—the lowest temperature to which air can be cooled through the evaporation of water.

Introduction

These towers usually flow water over internal components made of plastic or wood. These internal components are designed to break the water into tiny droplets, thereby increasing surface area and promoting maximum water-to-air contact. There are many different types of cooling towers in use today. While their designs differ, all have similar components and use the same principles of heat transfer.

The main purpose of cooling towers is to remove heat from process cooling water so the water can be recycled and recirculated through the process. Cooling towers reduce the temperature of water from heat exchangers and other devices through convection and evaporation. Cooling tower water can be supplied from municipal water systems or from other bodies of water, such as the ocean or a cooling pond.

Types of Cooling Towers

Cooling towers are structures designed to lower the temperature of water by evaporation and sensible heat loss. Cooling towers (shown in Figure 12-1) come in many shapes, sizes, and air flow classifications (e.g., natural draft, induced draft, forced draft, single-cell, and multi-cell).

— **Natural draft cooling towers**, which are most frequently seen in nuclear and coal-fired power plants, use temperature differences inside and outside the stack to facilitate air movement. The chimneys on natural draft cooling towers are very tall (typically 300 feet) and do not contain fans. Instead of forced movement, air flow occurs naturally as a result of density differences between the warm, moist air inside the tower and the cooler, drier air outside the tower. These cooling towers require massive amounts of heat duty to work. In these types of towers, a cloud forms right above the tower's most narrow point. As the water vapor goes from the inlet of the narrow section (the vena contracta) to the outlet, it drops in pressure. This pressure drop causes the water vapor to release most of its water via rain. Figures 12-2 and 12-3 show examples of a natural draft cooling tower.

—**Induced draft cooling towers** (shown in Figure 12-4) have fans at the top of the tower that pull air through the tower. In this type of cooling tower, the fan at the top induces (pulls) hot moist air out of the tower and into the atmosphere. The result is cool air that is entering at a low velocity and warm air exiting at a high velocity. This is usually the most efficient arrangement for cooling towers.

FIGURE 12-1 Types of Cooling Towers

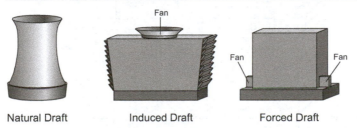

Natural Draft   Induced Draft   Forced Draft

FIGURE 12-2 Natural Draft Cooling Tower

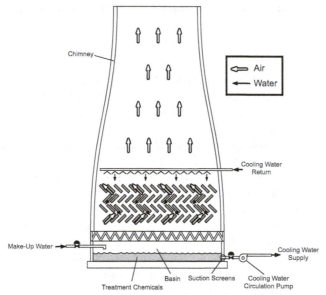

FIGURE 12-3 Diagram of Natural Draft Cooling Tower

FIGURE 12-4 Diagram of a Induced Draft Tower Showing the Flow Inside the Equipment

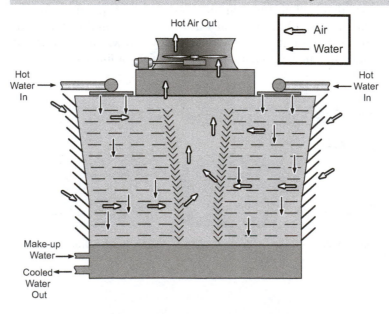

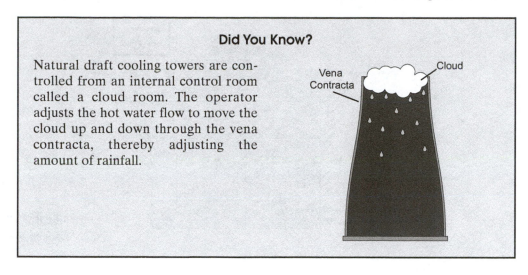

Did You Know?

Natural draft cooling towers are controlled from an internal control room called a cloud room. The operator adjusts the hot water flow to move the cloud up and down through the vena contracta, thereby adjusting the amount of rainfall.

FIGURE 12-5 Diagram of a Forced Draft Tower

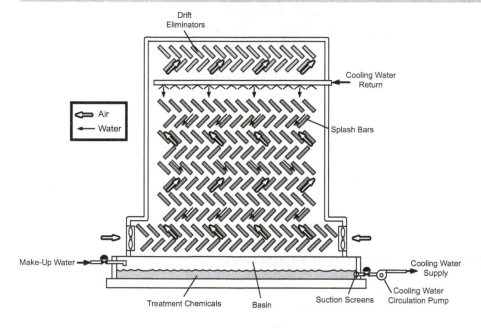

━ **Forced draft towers** (shown in Figure 12-5) have fans or blowers at the bottom side of the tower that forces air through the equipment. The pressure produced by the fan helps move the air through the tower and out of the top. This results in cool air entering at a high velocity, and warm air exiting at a low velocity. Forced-draft towers are usually built in units or cells. Several cells can be connected together to create the appearance of a single unit. However, each cell has its own fans and water feed lines.

Single-cell and multi-cell towers are both commonly used in the industry. In a multi-cell cooling tower, individual cells can be taken out of service or isolated for maintenance and control purposes. Multi-cell towers are used most often when the BTU removal demand increases.

Cross flow and counterflow are two types of air flow patterns used in cooling towers. In a cross flow cooling tower (shown in Figure 12-6) water flows downward and air is forced horizontally across the water flow path. In a **counterflow** cooling tower (shown in Figure 12-7), air and water flow in opposite directions. The water flows downward while the air is forced upward by induced, forced, or natural draft.

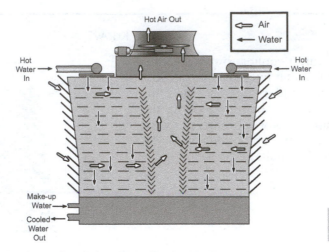

FIGURE 12-6 Crossflow Cooling Tower

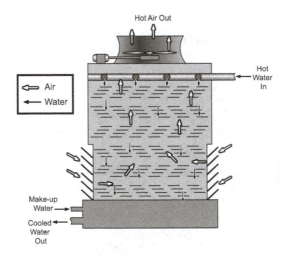

FIGURE 12-7 Counterflow Cooling Tower

Components and Their Purposes

While cooling tower designs vary, most cooling towers are made of plastic and wood and have similar components (shown in Figure 12-8), which include the following:

- Fans and motors
- Fan shrouds or chimney
- Water distribution system
- Fill (slats, splash bars)
- Louvers
- Drift eliminators
- Basin
- Pumps and sumps

At the top of the cooling tower is a **water distribution header** that provides returned (hot) cooling water to a distributor box located at the top of forced and induced draft cooling towers. The water flows out of the distributor box through a series of nozzles and begins to fall on a set of splash bars called fill.

Fill is the material that breaks the water into smaller droplets as it falls through the cooling tower. The fill materials direct the flow of water and maximize the falling path. Breaking the water into smaller droplets helps increase the surface area of the water. This, coupled with an increased falling path, helps facilitate the process of evaporation and increase the amount of cooling.

FIGURE 12-8 Induced Draft, Cross-Flow Cooling Tower Components

Motor · Drift Eliminators · Gear Box · Fan · Fan Shroud or Chimney · Distribution Header · Cooling Water Return · Distribution Nozzles · Air Flow · Water Box · Air Intake Louvers · Fill · Make-Up Water · Cooling Water Supply · Blow Down · Basin · Suction Screens · Cooling Water Circulation Pump · Chemical Treatment Injection Site · Manual Drain Valve

Louvers are installed on the outer sides of the cooling tower to adjust the quantity and direction of the air flow into the tower. In an induced draft cooling tower, air flow is facilitated by a large motor-driven fan located at the top of the tower. This fan is usually equipped with speed-reducing gear boxes that are located at the top of each cell.

In an induced draft tower, there is a shroud surrounding the fan. Air is drawn through the cooling tower and exhausted through this shroud. In a forced draft cooling tower, the fan is located at the base of the tower. This fan forces air upward through the tower. In a hyperbolic, natural draft cooling tower, there is no fan. Instead there is a chimney that induces air flow through natural air currents (draft).

On the interior surfaces of the cooling tower (the surfaces closest to the center where the air movement from the fan is the greatest), **drift eliminators** prevent water from being blown out of the cooling tower. The main purpose of a drift eliminator is to minimize water loss. However, some water (5 to 10 percent) is always lost due to **drift** (carrying of water with the air stream). To compensate for this water loss, makeup water is added to the basin. The addition of makeup water may require increased chemical treatment.

The **basin** is a reservoir (sump) at the bottom of the cooling tower where cooled water is collected so it can be recycled. The basin is also where chemicals are injected into the cooling water to reduce bacterial growth and adjust water chemistry.

Once inside the basin, cooling water circulation pumps pump the water out of the cooling tower and back into the heat exchangers and other process components. After the water has passed through the exchangers and heat transfer has occurred, the heated water is returned to the top of the cooling tower to be cooled and recycled.

Did You Know?

Cooling water may contain chemicals that are unhealthy. If you are exposed to these chemicals, you should remove contaminated clothing immediately, wash the contaminated skin thoroughly, and dispose of the contaminated clothes or put them in an industrial wash.

Applications of Cooling Towers

Cooling tower water can be used in a variety of heat exchanger applications, including process condensers and coolers. Cooling water can also be used to heat and vaporize cryogenic (very low temperature) liquids such as anhydrous ammonia or liquid propane.

Condensers are heat exchangers that cool hot vapors and convert them into a liquid. As the vapors flow through the heat exchanger, the cool water is heated and the hot vapors are cooled. As the vapors cool, they condense and are pumped out for storage or are returned to the process. The cooling water (which is now hotter) is returned to the cooling tower to be cooled.

Coolers are heat exchangers that may use cooling tower water to lower the temperature of process materials. In lubricating systems (found in rotating equipment), coolers are used to cool lubrication oils.

Theory of Operation

Cooling towers work primarily based on the process of **evaporation**, a process in which a liquid is changed into a vapor. When water changes from a liquid into a vapor, it absorbs a considerable amount of heat. This evaporative process is what cools the water.

The process of evaporation is much more efficient at removing heat than the process of convection (the transfer of heat through the circulation or movement of liquid or gas). Water is an excellent cooling medium because it is 50 to 100 times more efficient at conducting heat than air.

OVERVIEW OF THE COOLING PROCESS

The following is an overview of the cooling process in a cooling tower:

1. Cool water is pumped from the cooling tower basin to the process heat exchangers.
2. In the heat exchangers, heat is transferred to the circulating water.
3. The heated water is returned to the cooling tower distribution header for cooling.
4. The distribution header fills the distribution boxes, and the hot water flows through spray nozzles downward onto the fill (splash bars).
5. The fill provides good air-to-water contact by breaking the water into smaller droplets, thereby increasing the surface area of the water and promoting heat exchange.
6. As the water falls through the tower, it is exposed to air that removes heat through latent heat of evaporation and convection.
7. As the cool water reaches the bottom of the tower, it is collected in the basin.
8. The water in the basin is then pumped back to the process heat exchangers and steps 2 to 8 are repeated.

Factors That Affect Cooling Tower Operations

Many factors affect cooling tower operations, including relative humidity, outside air temperature, wind velocity, and tower design. Other factors include tube rupture or inadequate circulation due to pump malfunction. Water temperature can also be a problem if the cooling tower is located in an area with extremely low temperatures. For this reason, cooling towers in cold areas are usually equipped with temperature control devices to protect from possible freezing.

HUMIDITY

Humidity is the moisture content in the air. Local humidity levels must be considered when designing a cooling tower because evaporation is much more difficult in areas with high levels of humidity (e.g., a cooling tower in the hot, dry conditions of Arizona could be significantly smaller than one in the hot, humid climate of south Texas and still provide the same amount of cooling). The reason evaporation is more difficult in humid areas is because the ambient air is already partially saturated with moisture, so less water can be absorbed.

Relative humidity is the amount of water in a given amount of air at a given temperature, compared to the maximum amount of water that the air can hold at that same temperature. Relative humidity and the temperature of the air determine the amount of heat that can be removed.

If the humidity is high, the air can hold only a small amount of water vapor, so the evaporation process will be slow. However, if the humidity is low, the air can hold a much larger amount of water vapor, so the evaporation rate will be much quicker.

Dew point is the temperature at which air is completely saturated with moisture. As the temperature in the cooling tower approaches the dew point, the evaporation rate declines and reduces the cooling rate. Once the dew point is reached, the amount of evaporation drops to zero.

The **approach range** refers to the cooling tower's ability to evaporate and cool water. For example, a 5-degree approach means that, if everything is working as designed, the cooling tower should cool the water temperature to within 5 degrees of the dew point. The approach range depends on the design of the tower, the relative flow rates of the air and the water, and the contact time.

TEMPERATURE

Air temperature affects cooling towers by increasing or decreasing the rate of evaporation. If the temperature is cold, the molecules of air move much more slowly, so the rate of evaporation is decreased. If it is hot, the molecules of air move much more rapidly, so the process of evaporation occurs more quickly.

Air temperature can be evaluated through wet or dry bulb measurements. **Dry bulb temperature** is the actual temperature of the air that can be measured with a thermometer. **Wet bulb temperature** is the lowest temperature that air can be cooled to through the evaporation of water. If the humidity in the air is less than 100 percent (i.e., the air is not completely saturated with water), the process of evaporation will take place as the water comes into contact with the air.

Wet and dry bulb temperatures, measured with a psychrometer (see Figure 12-9), are used to evaluate the effectiveness of a cooling tower at current atmospheric conditions. Because the evaporation of water from the surface of the thermometer has a cooling effect, the temperature indicated by a wet bulb thermometer is less than the temperature indicated by a dry bulb thermometer. The rate of evaporation from the wet bulb thermometer depends on humidity since the rate of evaporation is slower when the air is already full of water vapor. For this reason, the difference in the temperatures indicated by the two thermometers (wet and dry bulb) gives a measure of atmospheric humidity.

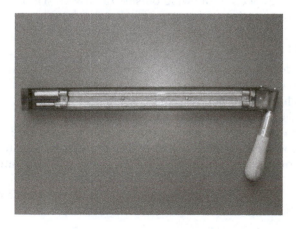

FIGURE 12-9 Psychrometer

High cooling water temperature occurs as a result of improper system balance or ambient air conditions. **Ambient air** is any part of the atmosphere that is accessible and breathable by the public. Worn-out fan blades, nonworking fans, improper water/air distribution, foaming, fouling, scaling, and corrosion can all lead to the improper heat balance that results in inadequate cooling in the cooling tower and reduces the heat transfer efficiency.

Low cooling water temperature can be the result of improper system balance or ambient air temperature. For example, cold weather can make it difficult to control the temperature of the tower. Sometimes controlling the number of fans in operation helps alleviate this problem, and sometimes it does not. In geographic areas where cold weather is more prevalent, cooling towers may be equipped with bypasses that divert return water to the basin in order to control temperature. Also during cold weather, fans may be shut down and the towers closed to prevent ice accumulation.

Many factors affect cooling tower performance:

- Humidity and ambient temperature
- Cooling tower design
- Wind speed and direction (in natural draft cooling towers)
- Contamination of cooling water

Table 12-1 lists how each of these variables affects cooling tower performance.

TABLE 12-1 Variables That Affect Cooling Tower Performance

Variable	*How Each Variable Affects Cooling Tower Performance*
Temperature	High ambient temperature = high cooling tower loads
Tower design	High air flow = high cooling
Humidity	High humidity = high water vapor in the air = low cooling
Wind velocity and direction	High wind velocity = high air flow = high cooling
	Wind direction opposite of cooling tower orientation = decreased evaporation
	(*Note:* Induced or forced draft towers have greater air flow than similar-size natural draft towers and do not depend on wind currents).
Water contamination	High contamination (e.g., algae buildup) = low cooling (e.g., due to fouling)

As evaporation occurs, the air absorbs water. The air continues to absorb water until there is either no more water for evaporation or the air becomes saturated (i.e., it cannot hold any more water). Energy is required to change this water from a liquid to a vapor. This energy comes in the form of heat.

WIND VELOCITY

Wind velocity is the speed of the wind outside the tower. Wind velocity affects cooling towers by increasing or decreasing the rate of evaporation.

TOWER DESIGN

Cooling tower design can affect heat transfer in a variety of ways. For example, the design of the fill can increase or decrease heat transfer based on the amount of water-to-air contact. The amount and type of draft can also increase or decrease the amount of cooling based on the amount of water-to air contact.

The **cooling range** is the difference between the temperature of the entering water and the exiting water. **Circulation rate** is the rate at which water flows through the tower. Circulation rate can affect heat transfer (increased circulation equals increased cooling).

Duty is the amount of heat energy a cooling tower is capable of removing, usually expressed in MBTU/hour.

Did You Know?

Natural draft cooling towers are commonly used in nuclear plants. These types of towers use natural draft instead of fans to facilitate movement.

CORROSION, EROSION, FOULING, AND SCALE

Other factors, such as corrosion, erosion, fouling, and scale can also affect cooling tower operations. For example, corrosion and erosion can lead to a loss of structural integrity, and fouling and scale can prevent water from flowing properly.

Corrosion

Corrosion is the deterioration of a metal by a chemical reaction (e.g., iron rusting). Over time, corrosion can cause a phenomenon known as stress corrosion cracking in some stainless steel materials. In cooling systems, this can result in decreased heat transfer and loss of structural integrity, and it may allow process fluids to leak into the cooling water if the corrosion in the heat exchanger is severe enough.

Erosion

Erosion is the mechanical degradation or wearing away of tower components (e.g., fan blades, wooden portions, and cooling water piping). Erosion can also lead to a loss of structural integrity and process fluid contamination.

Fouling

Fouling is the accumulation of deposits (such as sand, silt, scale, sludge, fungi, and algae) built up on the surfaces of processing equipment. As fouling increases, tubes become blocked so water flow rate decreases. As the flow rate decreases, heat exchange decreases. Fouling of pumps, which is commonly caused by plugged suction screens, can lead to low pump discharge pressure and can prevent water from circulating properly through the cooling water system.

Scale

Scale occurs as a result of dissolved solids depositing on the inside surfaces of hot equipment. As scale is deposited, an insulating layer is formed. This layer reduces the efficiency of the equipment and reduces heat transfer capability.

FOAMING

Foaming is a continuous formation of bubbles that remain on the surface of a liquid or some other substance. If foaming is not prevented, it can result in a frothy mixture (foam) that reduces the effectiveness of the product, causes poor equipment performance, and overflows tanks or sumps.

Foaming can result from excessive agitation, improper fluid levels, air leaks, cavitation, or contamination. Foaming is visible at the top of the tower cells where the distribution takes place and in the bottom basin. Foaming can be controlled or eliminated with anti-foam agents. The potential for foaming is increased when impurities are present. Foaming can cause an impairment in water circulation and may cause pumps to lose suction.

Safety and Environmental Hazards

Many hazards are associated with normal and abnormal cooling tower operations. These hazards can affect the equipment, unit operations, the environment, and the personal safety of the process technician, Table 12-2 lists some of these hazards and their possible effects.

TABLE 12-2   Hazards Associated with Improper Cooling Tower Operation

| Improper Operation | Possible Effects | | | |
	Individual	*Equipment*	*Production*	*Environment*
Failure to chemically treat cooling water	Exposure to harmful microorganisms (e.g., Legionella bacteria, which causes Legionnaires' disease)	Algae growth or sludge buildup can foul the exchanger	Production is reduced; products created are off spec due to improper heat exchange	Extremely high chemical residuals in the blowdown
Improper cooling tower line-up	Eye and skin irritation can result if exposure occurs	Pump damage due to blocked discharge or suction.	Products created are off spec due to improper heat exchange	Wrong valves open could bypass effluent water treatment facilities, resulting in excursions (digressions from the intended path)
Circulating cooling water through a heat exchanger with ruptured tubes	Exposure to chemicals, steam, or bacteria	Tower contamination; depletion of treatment chemicals; possible fire or explosive conditions; possible damage to tower components	Product ruined by contamination	Hazardous chemicals spilled to the environment
Exposing skin to or ingesting chemically treated cooling water	Eye and skin irritation and/or death			
Chemical addition to cooling tower	Eye and skin irritation and/or death	Corrosion; erosion	Loss of tower performance	Hazardous chemicals spilled to the environment

In addition to the hazards caused by improper operation, hazards are also associated with normal operations. For example, slip hazards exist around towers because of water leakage and moisture collection caused by drift. Typical personal protective and safety equipment used during normal and abnormal cooling tower operations includes goggles, rubber gloves, slicker suits, respirators, and hydrocarbon or gas detectors (used to detect tube leaks).

Because of the potential for injury or environmental hazards, all precautions and safeguards should be taken to ensure that cooling towers are in proper working condition to prevent potential damage to individuals and the environment.

HEAT EXCHANGER TUBE LEAKS

Heat exchanger **tube leaks** are a potential problem that can result in process chemicals entering the circulating water, possibly resulting in a fire, explosion, or toxic hazard. In some cases, air-monitoring instruments are added to detect toxic or flammable chemicals entering the cooling tower through leaks in heat exchanger tubing.

Did You Know?

In addition to degrading tower components, improperly treated cooling tower water can also lead to serious illnesses because cooling towers and evaporative condensers may contain Legionella and other microorganisms.

Legionnaires' disease is a form of pneumonia. It is called Legionnaires' disease because the first known outbreak occurred in 1976 at a hotel that was hosting an American Legion convention.

During that outbreak, 221 people contracted this previously unknown type of bacterial pneumonia, and 34 of those people died. The source of the illness was later found to be Legionella bacteria in the cooling water for the hotel's air conditioning system.

This is why it is important for cooling tower water to be properly treated and maintained at all times.

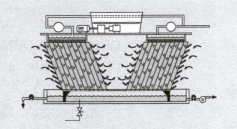

FIRES

Many cooling towers are large wooden structures that can act like kindling (thin, dry wood used to start a fire) in certain situations. For example, if a heat exchanger is being used to cool a flammable substance and a tube leak develops, the cooling tower may be saturated with flammable gas in a very short time. Static electricity, lightning, and other forms of ignition can occur. This, coupled with the fans providing an extremely large amount of air flow, turns the tower into a potentially huge fire hazard. Thus, cooling towers are generally equipped with sprinkler and trip systems that shut down the fans in the event of a fire.

Particularly dangerous is the situation that occurs when an operating unit is down for a particularly long period of time. During this downtime, the wooden structure can dry out completely and becomes susceptible to ignition sources. To prevent fire, regular water hoses with lawn sprinklers are often placed on top of the tower to keep it wet at all times.

Process Technician's Role in Operation and Maintenance

Typical procedures associated with cooling towers include monitoring, adjusting to weather conditions, temperature control, blowdown, adding makeup water, testing, adding chemicals, and maintenance. A process technician working in a cooling tower unit must understand the operating procedures associated with the unit. In the event of an emergency or shutdown of the unit, the process technician must be able to troubleshoot the problem and communicate with the control room on the activity in the field.

COMPONENT AND CONCENTRATION MONITORING

Cooling tower components must be monitored during normal operation. This monitoring includes checking for overloading of electric motors, cycles of concentration, vibration, water basin level, and proper operation of the liquid level water addition system.

When water passes through the cooling tower, some water is evaporated and minerals are left behind in the recirculating water. As evaporation continues, the water becomes more concentrated than the original makeup water. This can eventually lead to saturated conditions. If the cooling tower contains an excessive amount of solids, then the process technician must perform blowdown operations to reduce the concentration.

TEMPERATURE MONITORING

Generally, cooling towers are designed with some excess capacity in the initial design phase of the facility because additional heat loads are often added to the process unit without additional cooling tower capacity. However, conditions can affect water temperature that, in turn, affects the capacity for heat removal.

The capacity of a tower during the long, hot summer is less than on a cold winter day. Any decline in cooling tower capacity limits the manufacturing capacity of the facility. Because of this, cooling water temperature is monitored and controlled by the number of cooling tower cells and fans operating, by changing the number of pumps running, and by adjusting the throttle valves to ensure the water is circulating effectively.

CHEMICAL TREATMENT

Chemical treatment is very important in cooling towers because they are open systems. An untreated cooling tower system is susceptible to algae growth, sludge, scale buildup, and fouling of exchangers and cooling equipment.

To determine the proper chemical treatment, the circulating system water chemistry must be monitored on a regular basis for parameters such as conductivity, pH, total dissolved solids, and the presence of process chemicals. The most common chemical additives are biocides, corrosion inhibitors, dispersants, and pH control chemicals.

Biocides control biological growth (e.g. bacteria) in the tower and in circulating water. Biocides are often added at night when evaporation is lowest, giving the biocides the maximum effect.

Corrosion inhibitors are chemicals added to cooling tower water to prevent corrosion in piping and heat exchange equipment. Corrosion inhibitors are added to reduce the corrosive effect of the cooling water, both in the cooling tower and in the facility equipment where cooling water is used.

Dispersants are used to clean the cooling tower and other equipment by keeping foreign material dispersed in the water until the material can be removed by filters.

A cooling tower requires pH adjustment for the health of the tower and equipment. These pH controls aid in corrosion control and suppress organic growth (e.g., algae).

Be aware that as water is lost to evaporation, the dissolved solids accumulate, changing the pH and the chemical concentrations in the water. Additives and/or blowdown are required to counteract these effects.

BLOWDOWN

Blowdown is the process in which a certain amount of water is discharged and removed from the system and is replaced with fresh makeup water. Fresh makeup water must also be added to compensate for losses due to evaporation and leaks. Generally this is an automatic process, but process technicians should still monitor these levels to make sure the system is working properly.

MAINTENANCE

Some work on a cooling tower requires that lockout/tagout procedures be implemented. When a lockout/tagout procedure is implemented, unit technicians must know the unit in detail to be able to completely isolate the unit from all energy sources and thus ensure the safety of themselves and other employees.

Wooden cooling towers, whether in or out of service, are notorious fire hazards. The natural oils in cedar planks, for example, can quickly ignite. The cribbed design of the fill adds to the likelihood of flame propagation (spread). The hazard increases exponentially if tower components are allowed to dry out during outages. Because of this, hot work around cooling towers should be approved, limited, and closely monitored.

When working with cooling towers, process technicians should be aware of problems that can affect exchangers and be able to perform preventive maintenance. Typical scheduled maintenance activities for process technicans include:

- Sampling for chemical control (daily)
- Performing blowdown for solids control and removal (as required)
- Pulling and cleaning pump suction screens and washing the stairway and deck (as required)
- Disinfecting the cooling tower (as required)

Table 12-3 lists some additional monitoring and maintenance tasks that process technicians must perform.

TABLE 12-3   Process Technician's Role in Operation and Maintenance

Look	*Listen*	*Feel*
• Check for leaks • Check basin water levels to make sure they are adequate • Check chemical balance (pH and conductivity) • Check filter screens for plugging • Check temperature differentials • Look for broken fill materials (to fix at next turnaround) • Look for ice buildup in cold climates (*Note:* The weight of ice buildup can collapse the internal components of a cooling tower.) • Check for proper water distribution on top of the tower • Check cycles of concentration	• Listen for abnormal noises (e.g., grinding sounds associated with pump cavitations, or high pitched sounds associated with improperly lubricated fan bearings)	• Feel for excessive heat in fan and pump motors • Feel for excessive vibration in fans and pumps

Cooling towers require routine preventive maintenance, like other elements in a process system. Mechanical and chemical cleaning is required per manufacturer and unit specifications. The cooling tower must be disinfected periodically, and biocide treatments must be maintained on a regular basis to ensure that the system is within its chemical range.

In addition to maintaining biocide levels, process technicians should also routinely sample cooling water to ensure that the pH balance is correct. If the readings are not in alignment with specs, then the process technician needs to make changes to the system as needed.

Process technicians should monitor the temperature of the cooling water supply and only run the number of fans required to produce the proper cooling water supply temperature. Technicians should also take additional safety precautions for routine and preventive maintenance, such as the use of breathing air when adding toxic materials to the cooling tower water.

Failure to perform proper maintenance and monitoring can affect the process and result in personnel or equipment damage. Equipment operational hazards are also associated with cooling towers during normal and abnormal operation. For example, cooling towers use large, motor- or turbine-driven pumps. At times, emergency cooling water is needed during power outages. In these cases, circulating pumps with other drivers, engines, or turbines are used, so hazards associated with rotating equipment are also present.

Summary

Cooling towers are structures designed to lower the temperature of a circulating water stream. They work through the thermodynamic principle of evaporation. When water changes from a liquid into a vapor, it absorbs a substantial amount of heat. This process is called latent heat of vaporization.

The main purpose of cooling towers is to remove heat from process cooling water so the water can be recycled and recirculated through the process. Cooling tower water can be used in a variety of applications, including process condensers and equipment coolers.

Cooling towers are open cooling systems (as opposed to the closed cooling water systems in your vehicle). This means that they require a considerable amount of attention for a variety of reasons, which include the potential for contact with very hot water and exposure to toxic or corrosive chemicals.

Heat exchangers remove heat from process fluids. Cooling towers remove heat from heat exchanger cooling water. Maintaining the heating and cooling relationship between exchangers and towers is a continuous process.

Cooling towers are classified as natural draft, forced draft, or induced draft. They also come in many shapes and sizes and can be single-cell or multi-cell. The two types of flow patterns in cooling towers are cross flow and counterflow. In a cross flow, air flows generally perpendicular (at 90-degree angles) to the direction of the water flowing downward. In counterflow, water and air generally flow on the same plane (e.g., vertically) but in opposite directions.

Cooling towers include a water distribution header and water boxes that distribute the water while fill redirects the flow. A basin stores water, and makeup water replaces water lost to evaporation, blowdown, and drift. Drift eliminators minimize water loss. Suction screens filter out debris. Fans with a shroud force or induce air flow in induced or forced draft towers.

Many factors affect cooling tower performance. These factors include temperature, humidity, wind velocity, water contamination, and tower design.

Potential problems associated with cooling towers include heat exchanger tube rupture and inadequate circulation due to pump malfunction. Heat exchanger tube rupture can affect cooling tower design by allowing process chemicals to enter the circulation water system and being exposed to the atmosphere in the cooling tower.

Safety and environmental hazards associated with cooling towers include normal and abnormal cooling tower operations and operation of the circulating water system. Hazards include slip hazards, chemical exposure, and fire or explosions.

When monitoring and maintaining cooling towers, process technicians must remember to check pumps for excessive vibrations, noise, or heating. They must also ensure that water levels are adequate, there are no leaks, filter screens are not plugged, and proper heat exchange is occurring. In cold climates, they should check for excessive ice buildup inside the tower because heavy ice layers can damage the internal structures of the tower. They must also monitor and adjust chemistry.

Checking Your Knowledge

1. Define the following terms:
   a. Dew point
   b. Basin
   c. Biocide
   d. Blowdown
   e. Cooling tower
   f. Drift eliminator
   g. Evaporation
   h. Fill
   i. Forced draft
   j. Induced draft
   k. Natural draft
   l. Relative humidity
   m. Wet bulb temperature/dry bulb temperature
   n. Circulation rate
   o. Louvers
   p. Cross flow
   q. Counterflow
2. *(True or False)* Cooling towers are designed to raise the temperature of a water stream.
3. *(True or False)* Natural draft towers have fans or blowers at the bottom of the tower that force air through the equipment.

4. _____ are moveable plates that adjust the quantity and direction of the air flow into the tower.
   a. Spray nozzles
   b. Fans
   c. Louvers
   d. Basins
5. Drift is water carried over with _____, from the tower.
   a. air
   b. heat transfer
   c. entrainment
   d. evaporated
6. The _____ and temperature of the air determines the amount of heat removed in a standard amount of air.
   a. steam
   b. heat transfer
   c. entrainment
   d. relative humidity
7. Local _____ levels are considered when designing a cooling tower.
   a. humidity
   b. water chemistry
   c. temperature
8. Algae can accumulate and other plant and animal life can appear if cooling tower _____ is not controlled.
   a. thermodynamics
   b. water chemistry
   c. temperature
   d. flow

Student Activities

1. Given a model or diagram of a cooling tower, identify the following components and explain the purpose of each:
   a. Water distribution header and water boxes
   b. Fill (splash bars)
   c. Basin
   d. Makeup water
   e. Suction screens
   f. Drift eliminators
   g. Fan
2. Prepare a report on the three main types of cooling towers used in industry. In the report, describe how water is cooled in the tower. Include the flow description and a drawing with the components labeled.
3. Work with a team member to prepare a presentation on the cycles of concentration, the importance of chemical treatment, water chemistry (include types of common chemical additives), and blowdown. Include in your presentation the importance of each, how they interact, and how they can improve productivity within a process facility.
4. Research and explain why nuclear power facilities use hyperbolic cooling towers.

13

Furnaces

Objectives

After completing this chapter, you will be able to:

- Explain the purpose of furnaces in the process industry.
- Identify the common furnace types and their applications.
- Identify the components of furnaces and explain the purpose of each.
- Explain the operating principles of furnaces.
- Identify typical procedures associated with furnaces.
- Describe safety and environmental hazards associated with furnaces.
- Identify potential problems associated with furnaces.

Key Terms

Air register—an air intake device used to adjust air flow to a burner in a furnace.

Assisted draft furnace—a system that uses electric motor, steam turbine driven rotary fans, or blowers to push combustion gases into the furnace (forced draft) or draw flue gas from the furnace to the stack (induced draft).

Balanced draft furnace—a system that uses induced draft and forced draft to force air through a furnace.

Box furnace—a square, box-shaped furnace designed to heat process fluids or to generate steam.

Burner—a mechanical device where fuel is burned in a controlled manner to produce heat.

Cabin furnace—a cabin-shaped furnace designed to heat process fluids or to generate steam.

Convection section—the upper portion of a furnace where heat is transferred by convection.

Convection tube—a furnace tube, located above the shock bank, that receives heat primarily through the process of convection.

Firebox—the portion of a furnace where burners are located and radiant heat transfer occurs.

Forced draft furnace—a type of furnace that uses a fan to push air flow through the furnace.

Fuel lines—provide the fuel required to operate the burners.

Induced draft furnace—a type of furnace that uses a fan located at the top of the furnace to draw flue gas from the furnace body into the stack to draw the flow of air through the furnace.

Natural draft furnace—a type of furnace that does not use fans to create air flow. Natural draft is created by the difference in density between the hot combustion gases, the cooler outside air, and the height of the stack.

Nitrogen oxides (NO$_x$)—undesirable air pollution produced from reactions of nitrogen and oxygen. The primary NO$_x$ species in process heaters are nitrogen monoxide (NO) and nitrogen dioxide (NO$_2$).

Pilot—a device used to ignite burner fuel.

Radiant section—the portion of a furnace firebox where heat transfer is primarily through radiation.

Radiant tubes—tubes located along the walls of the radiant section that receive radiant heat from the burners.

Radiation—the transfer of heat by electromagnetic waves.

Refractory lining—a brick-like form of insulation used inside a furnace operating at high temperatures.

Shock bank tubes—tubes that receive both radiant and convective heat and protect the convection section from direct exposure to the radiant heat of the firebox.

Stack—a cylindrical outlet at the top of a furnace; it removes flue gas from the furnace.

Vertical furnace—provide even temperature control to process fluids.

Introduction

A furnace, also referred to as a process heater, is an apparatus in which heat is liberated by burning fuel and is transferred directly or indirectly to a fluid mass for the purpose of increasing the temperature of fluids flowing through tubes. Furnaces are an important part of many processes because they provide the heat necessary to facilitate chemical reactions and physical changes, and they can also be used to incinerate unwanted waste streams.

Did You Know?

Burning municipal waste in an incinerator produces heat that can be used to make steam. This steam can be used to power turbines that drive generators that produce electricity.

Metals that might be present during the incineration process can also be reclaimed and sent for recycling.

The most common types of industrial furnaces are box, vertical, and cabin. While all of these furnace types have similar principles of operation, they differ in shape, heat source, function, process cycle, draft type, mode of heat application, and the atmosphere inside the furnace.

Applications

Furnaces can be used in a wide variety of applications. For example, in mining and metallurgical operations, furnaces can be used to extract metal from ore, facilitate the casting and shaping of metal, and improve the properties of metal through heat treating and hardening. In glass and plastics manufacturing, furnaces can be used to melt glass or plastic so it can be shaped into containers, formed into light bulbs, extruded into fibers, pressed into sheets, or molded into other shapes. In the waste treatment industry, furnaces can be used to incinerate (burn at a high temperature and convert to ash) medical and municipal solid wastes so they take up less space and are less hazardous. In other industrial applications, like oil and gas refining or chemical processing, furnaces are used to heat fluids in order to change the physical properties (e.g., lower the viscosity or convert from a liquid to a gas), and facilitate chemical reactions. These are just a few of the many applications of furnaces.

Common Furnace Designs

Furnace designs vary with regard to function, heating duty, type of fuel, and method of introducing combustion air. However, the most common types of industrial furnaces are box furnaces, vertical furnaces, and cabin furnaces. Figure 13-1 through Figure 13-3 show examples of each of these types of furnaces.

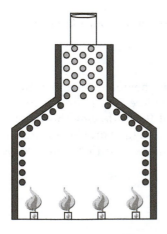

FIGURE 13-1 Cabin Furnace

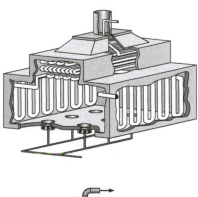

FIGURE 13-2 Box Furnace

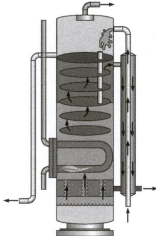

FIGURE 13-3 Vertical Furnace

Cabin furnaces (shown in Figure 13-1) are used in high-temperature processes. Cabin furnaces are so-called because they are shaped like a log cabin with a chimney. **Box furnaces** (shown in Figure 13-2) are used to heat process fluids and generate steam. They can also be used for heat treating applications such as tempering, hardening, or firing. **Vertical furnaces** (shown in Figure 13-3) are designed to provide even temperature control for process fluids. These types of heaters, which are also referred to as vertical cylindrical heaters (VCs), are similar to box furnaces but are cylindrical (tube-like) in shape and stand upright (vertical), so they require less space.

Furnace Draft Types

Like cooling towers, air flow (draft) inside furnaces can be provided by natural air currents (natural draft) or mechanical air currents (assisted draft).

NATURAL DRAFT

Natural draft furnaces (Figure 13-4) have no mechanical draft or fans. Because they lack mechanical air movers, the stacks on natural draft furnaces must be taller than on other furnaces in order to achieve proper draft.

Because hot furnace gases are not as dense as the ambient air, they rise, creating a differential pressure between the top and bottom of the furnace. This differential pressure (referred to as the chimney effect, furnace draft, or thermal head) creates a slight vacuum that pulls atmospheric air into the furnace and causes draft. As hot flue gas rises through the stack, a negative pressure is created inside the firebox. This negative pressure causes air to enter the burner air registers at the bottom of the furnace.

ASSISTED DRAFT

Assisted draft furnaces use rotary blowers, driven by electric motors or steam turbines, to push combustion gas into the furnace (forced draft), draw flue gas from the furnace to the stack (induced draft), or both (balanced draft). The fan speed can be either fixed

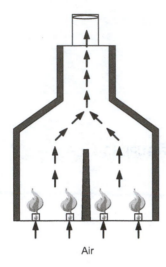

Air

FIGURE 13-4 Natural Draft Furnace (Cabin-Type)

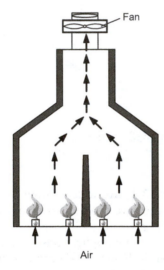

Fan

Air

FIGURE 13-5 Induced Draft Cabin Furnace

or variable, and it can be controlled manually or automatically from a control panel. While assisted draft systems require more energy to operate than natural draft systems because of the fans, they are actually more efficient because they provide more control and use smaller burners for the same firing capacity.

There are three main types of air flow in assisted draft furnaces: induced draft, forced draft, and balanced draft. All three of these draft types are shown in Figures 13-5 through 13-7.

Induced draft furnaces (shown in Figure 13-5) use a fan located at the top of the furnace to draw flue gas from the furnace body into the stack. As the gas moves from the body to the stack, the pressure in the firebox is decreased, thereby inducing draft.

Forced draft furnaces (shown in Figure 13-6) use fans or blowers located at the air inlet of the furnace to force air flow. During combustion in a forced draft furnace, positive pressure is created by the fan or blower. This positive pressure forces air into the burner air registers through a header or plenum. Rising combustion gases inside the furnace create draft or negative pressure.

Balanced draft furnaces (shown in Figure 13-7) use two fans to facilitate air flow. One fan forces air flow into the burner registers, while the other fan pulls (induces) air out of the stack.

Furnace Sections and Components

Furnaces are divided into two sections: the radiant section, and the convection section. Both of these sections are identified in Figure 13-8.

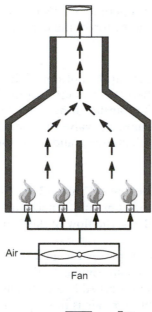

FIGURE 13-6 Forced Draft Cabin Furnace

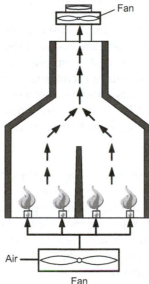

FIGURE 13-7 Balanced Draft Cabin Furnace

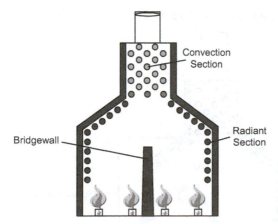

FIGURE 13-8 Cabin Furnace Sections

The **radiant section** is the furnace section located at the bottom of the unit, closest to the heat source. This section is called the radiant section because the primary method of heat transfer is **radiation** (the transfer of heat by electromagnetic waves). The **convection section** is the upper portion of a furnace where heat is transferred by convection. To better understand furnace operations, let's follow the flow of heat from the bottom of the radiant section through the top of the convection section.

RADIANT SECTION

The radiant section (Figure 13-9) is located at the bottom of the furnace and is closest to the heat source. This section is called the radiant section because the primary method of heat transfer is radiation (the transfer of heat by electromagnetic waves). Located at or near this section are fuel lines, fuel valves, burners, draft gauges, the firebox, the purge system, the bridgewall, radiant tubes, and refractory lining.

Fuel Lines and Burners

At the base of the radiant section are **fuel lines** (Figure 13-9). These fuel lines provide the fuel required to operate the burners. Common furnace fuels include natural gas, fuel oil, process oil, process gas, and fuel gas. The flow of the fuel through these lines is controlled by fuel valves.

As the fuel gas leaves the fuel line, it passes through the burners where a **pilot** ignites the burner fuel. The **burners**, which are similar to the burners on your kitchen stove, are mechanical devices where fuel is burned in a controlled manner to produce heat. As the fuel is transferred to the burners, air is introduced through **air registers**. This air helps facilitate the combustion process.

During the combustion process, draft gauges are used to measure the differential pressure between the outside of the furnace and the inside of the furnace. Draft gauges are devices, calibrated in inches of water, that measure the differential pressure between the outside of the furnace and the flue gas contained inside the furnace.

The heat that is produced by the burners is radiated into the main body of the furnace called the **firebox**. Because so much heat is generated in this section, the firebox must be lined with a special refractory lining in order to protect the furnace structure.

Purge System

Before lighting a furnace, it is important to remove any combustible materials from the firebox because failure to do so can cause an explosion. The system that is used to remove combustible materials from the firebox is called a purge system.

In forced, induced, or balanced draft furnaces, fans are used to purge combustibles from the firebox with air. In natural draft furnaces, steam is used. Steam purge systems can also be used to extinguish uncontrolled fires in the firebox should they occur (this type of steam is referred to as snuffing steam).

FIGURE 13-9 Natural Draft Cabin Radiant Section

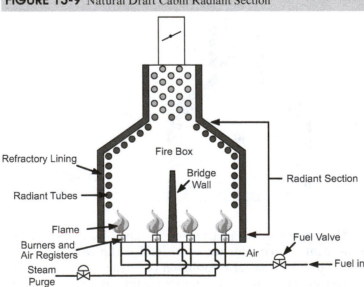

Bridgewall

Also contained in the firebox of some furnaces is a bridgewall. A bridgewall is a wall or vertical partition in the fire chamber of the furnace that is used to deflect the heat. The bridgewall compartmentalizes the radiant section and helps redirect the heat toward the tubes located along the outer wall. This redirected heat allows for radiant heating of the process fluids passing through the tubes.

Radiant Tubes

The tubes located along the walls of the radiant section are called **radiant tubes** (Figure 13-10) because they receive radiant heat from the furnace burners. Radiant tubes can be mounted either vertically or horizontally, and they can be placed in different locations or arrangements depending on the type of furnace.

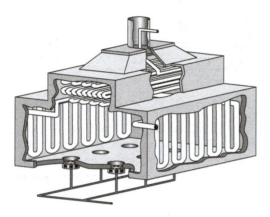

FIGURE 13-10 Furnace with Radiant Tubes Surrounding the Inside of the Furnace

Refractory Lining

To protect the outside metal walls of the furnace from excessive heat, a refractory lining is installed. A **refractory lining** (Figure 13-11) is a brick-like form of insulation used inside high-temperature furnaces. The purpose of this lining is to radiate heat back into the firebox and to protect the steel structure of the furnace. The refractory lining also helps maintain a uniform temperature on tube walls, minimize heat loss, and keep the outside wall from becoming too hot (excessive heat on the exterior walls could cause personal injury or damage to the firebox).

Refractory linings are typically composed of specialized heat-resistant materials such as firebrick, castable refractories, and ceramic fiber. Ceramic fibers and firebricks are typically used on the ceiling and walls of the furnace, while the furnace floor is typically composed of castable refractories (a stronger material that can be molded) because it must be hard enough to walk on during maintenance. Each of these insulation types is held together with heat-resistant mortar and attached to the furnace wall by studs welded to the outside steel structure wall.

FIGURE 13-11 Examples of Insulating Firebrick

CONVECTION SECTION

As the heat moves up through the furnace, it eventually leaves the radiant section and moves into the convection section (Figure 13-12). The convection section is located at

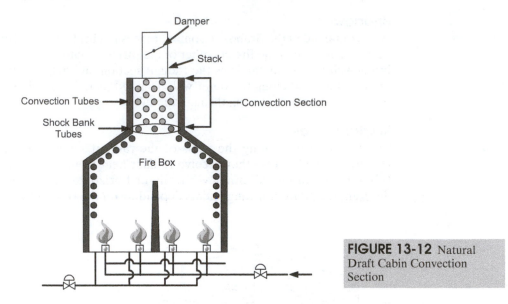

FIGURE 13-12 Natural Draft Cabin Convection Section

the top of the furnace firebox and is further away from the heat source. This section is called the convection section because the primary method of heat transfer is convection (the transfer of heat through the circulation or movement of a fluid). Within this section are convection tubes. Above this section are the stack and damper.

Shock Bank

The convection section has relatively few components. At the base of the convection section is a row of shock bank tubes. **Shock bank tubes** receive both radiant and convective heat, and they protect the convection section from direct exposure to the radiant heat of the firebox. Shock bank tubes are normally made of materials that have greater heat resistance than the tubes in the convection section.

Convection Tubes

Above the shock bank tubes is a series of tubes called convection tubes. **Convection tubes** receive most of their heat through the process of convection because they are protected from direct exposure to radiant heat by the shock bank. Convection tubes may be made of the same material as the radiant tubes and may be finned to increase heat transfer.

Stack and Damper

At the top of the furnace is the **stack**, a cylindrical outlet at the top of a furnace that removes flue gas from the furnace. Within the stack is a valve or movable plate called the stack damper. The damper is a movable plate that regulates the flow of air, draft, or flue gases in a furnace.

Furnace Operating Principles

Furnaces burn fuel inside a containment area (firebox) to produce heat. Process fluids are pumped through tubes located in this heated firebox. The heat from the firebox is then transferred through the tube walls by conduction and through the process fluid by convection. The products of combustion (flue gas) then flow from the firebox up through a stack and to the atmosphere, where they are released.

FUEL SUPPLY

The fuel supply flow is determined by temperatures inside the furnace system. In many process furnaces, the temperature of the stack gas and the product exiting the tubes indicate how much fuel is needed.

FIGURE 13-13 Fuel and Steam Leading into a Set of Furnaces

Corbis Royalty free Image

Fuel flow is typically controlled by adjusting the fuel pressure with some type of regulating valve. The regulator may be manually or automatically adjusted. When heavier liquid fuels are used (e.g., No. 6 fuel oil), heat must be applied to reduce viscosity. This is accomplished by raising the temperature of the fuel oil to ensure proper atomization. The temperature control of the liquid fuel is commonly accomplished by attaching electrical or steam-jacketed heat tracing to the fuel supply line. Figure 13-13 shows an example of fuel and steam lines leading into a set of furnaces.

COMBUSTION AIR

All furnaces require a certain amount of air for complete combustion to occur. Without sufficient air, the furnace could become too fuel-rich and could result in inefficient burning, smoke production, high carbon monoxide (CO) emissions, or furnace explosion. However, operating with more air than is required wastes energy and increases the amount of hot flue gas exiting the furnace.

To maintain proper ratios and prevent hazardous situations, online analyzers are used to monitor the oxygen (O_2) content of the flue gas. If oxygen (O_2) levels are too low, then CO emissions may increase. If oxygen (O_2) levels are too high, then energy efficiency decreases and nitrogen oxide (NO) emissions increase. **Nitrogen oxides (NO_x)** are forms of undesirable air pollution produced from reactions of nitrogen and oxygen. NO_x is more often produced by too hot a furnace, not by excess O_2.

Did You Know?

Carbon monoxide (CO) is extremely poisonous. It is a colorless, odorless gas that is produced during the incomplete combustion of carbon-based fuels (e.g., gasoline, wood, or oil).

When carbon monoxide enters the body, it binds with the hemoglobin in the blood (the place where oxygen normally binds). This prevents the blood from carrying oxygen to the cells. The end result can be serious injury or death.

Because of this, process technicians should always be careful when working in or around combustion equipment.

FLAME TEMPERATURE

Auto-ignition is the term used to describe the temperature at which a mixture of fuel and air automatically ignites. The auto-ignition temperature of most fuels used in the process industries is in the 800-degree to 1300-degree F range. For the fuel to auto-ignite, the furnace's burner region must operate at a temperature exceeding the auto-ignition temperature of the fuel mixture.

Since burner flames can exceed 3000 degrees F, peak flame temperature must be controlled using a correct fuel/air ratio to the burners. Lowering the flame temperature also significantly decreases the production of nitrogen oxides (NO_x), an undesirable byproduct of furnace operation.

FURNACE PRESSURE CONTROL

Process heaters are operated at negative pressure, not positive pressure, and the furnace pressures fluctuate with the burner firing rate (e.g., the pressure is lowest at the lowest firing rate and highest at the highest firing rate). To compensate for this constantly changing condition, furnaces are equipped with fuel pressure control systems. These control systems consist of a stack damper that is automatically controlled to maintain the desired pressure in the combustion chamber. As the burner fire rates decrease, the damper throttles the flow out of the stack to hold the pressure constant.

Furnace pressure controllers regulate and stabilize the pressure in the combustion chamber. A pressure gauge in the furnace chamber or duct regulates the air flow to maintain the target pressure. Pressure controls can be manual or automatic.

INTERLOCK CONTROLS

Interlock controls are safety devices used to protect the furnace system from dangerous operating conditions. Instrument control systems use sensors and logic control to shut off the supply of furnace fuel or process feed and to shut down the furnace under certain conditions. Some examples of items that can trip an interlock and shut down a furnace include the following:

- Loss of flame at the burners
- High temperature at the process tube exit
- Low or high pressure in the burner fuel supply
- Low speed on the draft fan, or loss of draft fan (trip out)
- Loss of process flow through the tubes trips out the fuel gas supply
- High firebox pressure
- High stack temperature
- Excessive O_2 in the stack
- Excessive NO_x in the stack
- Excessive CO in the stack
- Smoke in the stack

Did You Know?

In homes throughout the United States, the household furnace is the most common major appliance.

These furnaces provide heat through air, steam, or hot water.

Safety and Environmental Hazards

Hazards are associated with normal and abnormal furnace operations. Hazards can affect the personal safety of the process technician, the equipment, facility operations, and the environment. Process technicians must always follow proper safety precautions when working around furnaces. Table 13-1 lists some of the hazards associated with improper furnace operations.

TABLE 13-1   Hazards Associated with Improper Furnace Operations

| *Improper Operation* | *Possible Effects* | | | |
	Individual	*Equipment*	*Production*	*Environment*
Failure to follow flame safety procedure (e.g., lighting off burners without purging the firebox)	Burns, injuries, or death	Explosion in the firebox or flashbacks	Lost production due to downtime for repairs	Exceeding Environmental Protection Agency (EPA) capacity limits for opacity
Poor control of excess air and draft control	Burns, injuries, or death	Flame impingement and/or tube rupture and explosion	Lost production due to downtime for repairs; loss of efficiency	Exceeding EPA CO, NOx, and capacity limits
Bypassing safety interlocks	Burns, injuries, or death	Explosion in the firebox	Lost production due to downtime for repairs	Fines from operating above environmental limits
Opening furnace inspection ports when the firebox pressure is greater than atmospheric pressure	Burns, injuries, or death caused when hot flames or combustion gases are forced out of the inspection port	Sudden drop in box pressure	Cooling of the furnace	Hot combustion gases blown into the environment
Failure to wear proper protective equipment (e.g., gloves and face shield) when opening furnace inspection ports	Burns, injuries, or death if the pressure in the firebox is greater than atmospheric pressure			

FIRES, SPILLS, AND EXPLOSIONS

Because process furnaces usually contain combustible gases and liquids under high pressure, there is always the potential for uncontrolled fires, spills, or explosions. Tube failures, for example, can cause large and sudden fires. Furnace leaks that allow flammable product to escape from the furnace also create a fire hazard.

To prevent dangerous conditions from occurring, it is absolutely critical that all instrument and interlock problems be corrected immediately. Interlocks should never be bypassed.

Always be aware that an accumulation of flammable liquids in the furnace firebox is an explosion hazard, and startup operations are especially hazardous. Each type of furnace has its own specific startup and emergency procedures that must be closely followed. To prevent flashbacks or explosions when preparing to light off the burners, make sure that the firebox has been properly purged and that adequate draft is provided.

Responding to Uncontrolled Fires

Every process technician must know how to extinguish an uncontrolled fire, either inside the furnace or underneath it. In most cases, the proper response is to shut off all fuels, close the damper, and flood the firebox with snuffing steam. However, technicians should always be familiar with plant-specific procedures before attempting to extinguish any type of fire.

HAZARDOUS OPERATING CONDITIONS

Hazards are present during both normal and abnormal furnace operations. For example, contact with hot pipelines or equipment can cause burns. Excessive smoke caused by incomplete combustion creates a potential ignition or overpressure hazard in the firebox. Failure to properly control the balance of the fuel and air supplies cause symptoms such as "puffing," flame, or smoke emission from the furnace stack.

If electrical power is lost, the furnace is usually automatically shut down by failsafe equipment. When this occurs, make sure the fuel supply is positively shut and then check all the interlock actions for proper activation.

PERSONAL SAFETY

In addition to wearing standard personal protective equipment, process technicians should observe the following precautions when performing routine preventive maintenance on furnace equipment:

- When inspecting a natural draft furnace or lighting the burners, always wear a face shield.
- When opening the inspection door, never assume that the firebox is at negative pressure, and minimize the amount of time that the inspection door is open. During upsets, there is usually the possibility of positive pressure in the firebox.
- Always stay in contact with the control room operator.
- Be aware of hot surfaces and hot flue gases (e.g., at the entrance and exit points around the tubes).
- Always wear long sleeves and gloves to prevent contact burns.
- Wear flame-retardant clothing when required.

ENVIRONMENTAL IMPACT

It is important for furnaces to operate properly because excessive smoke, SO_x (sulphur oxide), CO (carbon monoxide), or NO_x (nitrogen oxide) emissions from furnace operations can have a negative impact on the environment and can result in citations and/or fines.

Nitrogen Oxide (NO_x) Emissions

Nitrogen oxides (NO_x) are byproducts of the combustion of fuel at high temperatures or with fuel containing nitrogen (N_2) compounds. High combustion temperatures oxidize atmospheric nitrogen and produce nitric oxide (NO), which can be further oxidized to nitrogen dioxide (NO_2). Both compounds are atmospheric pollutants and are collectively referred to as NO_x. In recent years, many industries have been required to comply with stricter standards for NO_x emissions.

The amount of NO_x formation depends on combustion characteristics such as temperature, residence time, and the concentrations of oxygen and nitrogen in the flame zone. Of these, peak flame temperature is the most important parameter in determining potential NO_x formation.

Atmospheric pollution by nitrogen oxides can be controlled by reducing the amount of source emissions and by treating the exhaust gas. Low-NO_x burners modify the combustion process by staging the air or fuel (i.e., having the correct air/fuel ratio), or by providing premixed fuel and flue gas into the burner throat. These measures are designed to reduce peak flame temperature by distributing the flame over a larger space than is provided by traditional burners. Some companies have recently developed ultra-low-NO_x burners that combine premixing and staging to achieve lower NO_x emissions. These burners are more difficult to operate and more operator attention is required.

To keep NO_x emissions under control, ensure that the burner air registers and fuel valves are always operating properly. Monitor burner flame appearance and adjust burner conditions as needed.

Carbon Monoxide Emissions

Carbon monoxide (CO) typically forms from the incomplete combustion of a fuel containing carbon. CO is an asphyxiant that can cause death in high enough concentrations. It is a regulated pollutant that can be controlled by ensuring there is adequate air available for combustion. This is why some amount of excess air should be supplied to account for imperfect mixing of the fuel and combustion air. Excessive CO emissions are an indication of reduced thermal efficiency because CO is a fuel that becomes CO_2 (carbon monoxide) when it has been combusted.

Smoke Emissions

In addition to creating a hazard in the firebox, smoking creates an emissions violation. Smoking is caused by incomplete combustion of fuels. Understanding and controlling burner operations provides the right balance for complete combustion and controls smoke emissions.

Typical Procedures

Furnace procedures are specific to each process unit. Process technicians are required to have a clear understanding of site- and unit-specific procedures, and are responsible for knowing how to locate the procedures in the event of an emergency.

Potential Problems

Furnace operations have a direct and immediate impact on the operation of other process equipment such as reactors or separation systems. The consequences of deviating from normal furnace operation can result in loss of production, loss of product quality, and the high cost of repairing or replacing failed or damaged components. The following are some of the problem conditions associated with furnaces:

- Burner flames can impinge on tubes, causing coking (the deposit of carbon) inside the process tubes and thus the local metal to overheat and fail.
- A change in the flow or composition of the process fluid flowing through the tubes can create or increase coke deposits in the tubes, create excessively high temperatures on the tube walls, produce an extreme pressure drop in the tubes, and ultimately cause tube failure.
- Poor flame distribution can cause overheating and tube failure.
- Liquid entrained in the fuel gas can cause a flame failure or temperature shock in the furnace.
- Improper warm-up or cool-down of furnaces can result in the refractory materials peeling away from the furnace walls.
- Poor atomization of liquid fuels may result in pooling of liquids inside the firebox, which can ignite.
- Ignition of excess unburned fuel inside the firebox can cause furnace explosions.

Many other factors can contribute to problems in furnace operations, including the design, normal wear over time due to the age of the equipment, instrument problems, fouling, too high or low tube temperatures, external factors, improper equipment maintenance, and failure to follow proper procedures during furnace startup and shutdown.

EQUIPMENT AGE AND DESIGN

Over time, equipment can age or wear out. Rotating equipment, such as forced air fans, are included among the types of furnace equipment that are most subject to failure due to normal wear over time. Air feed control linkages can wear out or jam up as a result of inadequate lubrication or from the collection of dust and grit, or fail due to corrosion.

Freezing and earth movements can crack concrete equipment foundations and cause external corrosion of structural components. Tube supports and spring hangers that have not been properly adjusted and maintained can create stress in the furnace tubes due to sagging or coming into contact with other tubes.

INSTRUMENT PROBLEMS

Instrument malfunctions can contribute to a number of problems, including an improper fuel/air ratio or a complete trip out of the furnace system. Faulty sensors can also activate furnace interlocks, causing unnecessary shutdowns.

FOULING

Some furnace problems are caused by fuel or process-related fouling. Some fuels contain impurities that produce corrosive materials and cause fouling in the cooled flue gas. In some cases, the lower limit of the flue gas temperature in the furnace stack is set by the dew point of these acidic materials in the stack gas. Operating below the dew point temperature corrodes the tubes and refractory in the upper sections of the furnace. Other fuels contain trace metal impurities such as vanadium pentoxide (V_2O_5). Vanadium ash collects on upper furnace tubes, causing oxidation failures and critically attacking refractory insulation.

Sometimes feeds to the furnace decompose from the high furnace temperatures. This decomposition causes the formation of coke deposits on the inside surface of the furnace tubes. Some furnaces are designed to operate for long periods of time by controlling coke deposits through the use of proper temperatures, upstream purification, and the addition of steam. Other furnaces, such as ethylene crackers, are designed for periodic decoking (offline removal of coke deposits with steam and air), essentially burning the coke deposits off.

TUBE LIFE AND TEMPERATURE

Several factors determine the length of time that a tube can be in service before it must be repaired or replaced. These factors include the uniformity of burner firing, internal fouling, tube support hangers, and carburization (a process by which carbon is introduced into a metal in order to make the surface harder and more abrasion-resistant). In addition, tubes can be damaged by direct flame impingement. Furthermore, if the fuel contains sulfur and water, the heat from the flame causes the formation of sulfuric acid that can damage the tubes.

EXTERNAL FACTORS

Other factors that contribute to furnace problems include the loss of steam supply, electrical failure, cold weather, and upstream process upsets. Cold weather can freeze the water in process, utility, and instrument lines. Upsets in the upstream processing units can cause off-spec feed to be sent to the furnace.

IMPROPER MAINTENANCE

Failure to properly maintain the furnace and burners can lead to potentially serious problems. For example, burner tips (fuel injectors) must be clean and in good condition to ensure proper combustion. If the tips become plugged, they could cause the flame to temporarily go out, leading to an explosion when the fuel re-ignites on a hot surface in the furnace.

If the furnace is improperly sealed, air can leak into the furnace, preventing a significant portion of the combustion air from coming through the burners. This prevents some of the fuel from being combusted in the radiant section and leads to afterburning in the convection section when the unburned fuel contacts air that has leaked into the furnace. This could damage the convection section as well as reduce thermal efficiency.

Figure 13-14 displays examples of gas flames that need service. *Note:* These flames are also shown in the color section located in the back of this textbook.

FURNACE STARTUP AND SHUTDOWN

Furnaces must be heated up and cooled down gradually to prevent damage to the furnace firebox insulation and refractory materials because these components are susceptible to damage from thermal shock. Specific time and temperature guidelines in furnace

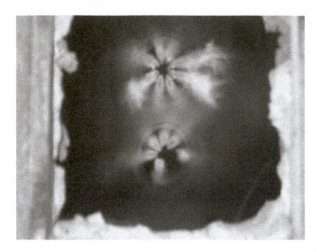

FIGURE 13-14 Gas Flames that Need Service; Plugged Fuel Tips Causing the Flames to be Non-Uniform

startup and shutdown procedures should always be followed because startups have the greatest risk for spills and fires. When using combination burners, firebox temperature must be near the operating range before the fuel oil is introduced. The protection of instrument safety interlocks is also essential during startup operations and during normal operations.

Summary

A furnace is an apparatus in which heat is liberated by burning fuel and transferred directly or indirectly to a fluid mass for the purpose of increasing the temperature of a fluid. The most common types of industrial furnaces are box, vertical, and cabin.

Cabin furnaces are used in high-temperature processes that require frequent cleaning of the furnace combustion chamber. Cabin furnaces are called cabins because they are shaped like a log cabin with a chimney.

Box furnaces are used to heat process fluids and generate steam. They can also be used for heat treating applications such as tempering, hardening, or firing.

Vertical furnaces are designed to provide even temperature control for pumping process fluids. They are similar to box furnaces, but they are cylindrical in shape and stand upright, so they require less space.

Air flow inside furnaces can be provided by natural draft or assisted draft. Natural draft furnaces use natural air currents instead of fans to create air flow. Assisted draft furnaces use an electric motor, steam turbine driven rotary fans, or blowers. There are three main types of air flow in assisted draft furnaces: induced draft, forced draft, and balanced draft.

Furnaces are divided into two sections: the radiant section and the convection section. The radiant section is located at the bottom of the furnace and is closest to the heat source. Within this section are fuel lines, fuel valves, burners, draft gauges, the firebox, the purge system, the bridgewall, radiant tubes, and the refractory lining.

The convection section is located at the top of the furnace and is further away from the heat source. Within this section are convection tubes, the stack, and a damper.

Furnaces burn fuel inside a firebox to produce heat. Process fluids are pumped into this heated firebox through a series of tubes. The heat from the firebox is then transferred through the walls of the tubes to the process fluid via conduction. The flue gas then flows from the firebox up through a stack and to the atmosphere.

Interlock controls help protect the furnace system from dangerous operating conditions by shutting off the supply of furnace fuel or process feed and shutting down the furnace. Improperly operating a furnace can lead to safety and process problems, including injury, explosion or rupture, lost production, and environmental exposure to harmful gases.

Process technicians should always monitor furnaces for puffing (excessive smoke), flames, pressure drops, temperature drops, or other abnormal conditions. They should conduct preventive maintenance as needed.

Checking Your Knowledge

1. Define the following terms:
   a. Pilot
   b. Stack
   c. Burner
   d. Firebox
   e. Draft gauge
   f. Induced draft
   g. Natural draft
   h. Air register
   i. Vertical furnace
   j. Damper
   k. Shock bank
   l. Flue gas
   m. Radiant tube
   n. Cabin furnace
   o. Radiant section

2. Furnaces are used to:
   a. increase the temperature of a fluid
   b. release harmful gases to the environment
   c. decrease temperature and pressure

3. Which of the following is *not* a common type of furnace?
   a. Cabin
   b. Spherical
   c. Box
   d. Vertical

4. Select the components that are found in a furnace (select all that apply).
   a. Radiant tubes
   b. Bridgewall
   c. Fuel valve
   d. Tube sheet

5. A _____ is the piece of equipment that is bolted to the outside end of the channel to retain exchanger fluid and provide access for cleaning.
   a. tube sheet
   b. channel cover
   c. tie rod
   d. bonnet

6. Which of the following pieces of equipment is used to convert a vapor to a liquid?
   a. Pre-heater
   b. Reboiler
   c. Chiller
   d. Condenser

7. Which of the following is *not* a potential problem in heat exchangers?
   a. Chemical balance
   b. Plugging of the U-tubes
   c. Fouling, erosion, and corrosion
   d. Low heat transfer in the thermosiphon reboiler

Match the services on the left with the correct description on the right.

8. Thermosiphon reboiler   a. Used to cool gas from a furnace
9. Cooler   b. Uses cooling to convert a substance from a vapor to a liquid
10. Transfer line exchanger   c. Used to cool hot liquids
11. Condenser   d. Provides circulation based on density differences

Student Activities

1. Given a diagram or a cutaway of a furnace, identify the various components and explain the purpose of each.
2. Describe the importance of operating a furnace properly. Be sure to include emissions concerns, environmental impact, and fines or citations.

14

Boilers

Objectives

After completing this chapter, you will be able to:

- Explain the purpose of boilers in the process industry.
- Identify the common types and applications of boilers.
- Identify the components of boilers and explain the purpose of each.
- Explain the operating principles of boilers.
- Identify typical procedures associated with boilers.
- Describe the process technician's role in boiler operation and maintenance.
- Identify potential problems associated with boilers.

Key Terms

Coagulation—a method for concentrating and removing suspended solids in boiler feedwater by adding chemicals to the water, which causes the impurities to cling together.

Deaeration—a method for removing air or other gases from boiler feedwater by increasing the temperature and aeration time.

Demineralization—a process that uses ion exchangers to remove mineral salts. Also known as deionization. The water produced is referred to as deionized water.

Desuperheated steam—superheated steam from which some heat has been removed by the reintroduction of water. It is used in processes that cannot tolerate the higher steam temperatures.

Desuperheater—a temperature control point at the outlet of the boiler steam flow that maintains a specific steam temperature. The temperature is maintained by using boiler feedwater injection through a control valve.

Downcomer—a tube, located in the firebox, that transfers water from the steam drum to the mud drum.

Draft fan—a fan used to supply combustion air to the burners.

Economizer—the section of a boiler located in the flue gas stream, used to preheat feedwater before it enters the upper steam drum.

Filtration—a method for removing suspended matter and sludge from boiler feedwater and condensate return systems.

Fire tube boiler—a device that passes hot combustion gases through the tubes to heat water on the shell side.

Igniter—a device (similar to a spark plug) that automatically ignites the flammable air and fuel mixture at the tip of the burner.

Impeller—a fixed, vaned device that causes the air/fuel mixture to swirl above the burner. An impeller in this instance is not the same as an impeller in a turbine or pump.

Knockout pots—devices designed to remove liquids and condensate from the fuel before it is sent to the burners.

Pilot—a device used to light a burner.

Premix burner—a device that mixes fuel gas with air before either exits the burner face.

Raw gas—gas that has not been premixed with air

Reverse osmosis—a process that uses pressure to force a solvent through a membrane and retains the dissolved solids on one side of the membrane while allowing the solvent to pass to the other side of the membrane.

Riser tubes—tubes that allow water or steam from the lower drum to move to the upper drum.

Saturated steam—steam in equilibrium with water (i.e., steam that holds all of the moisture it can without condensation occurring).

Softening—the treatment of water that removes dissolved mineral salts such as calcium and magnesium, known as hardness in boiler feedwater.

Spiders—devices used to inject fuel into a boiler.

Spuds—devices used to inject fuel into a boiler.

Steam drum—the upper drum of a boiler where all of the generated steam is collected.

Steam trap—a device used to remove condensate or liquid from steam systems.

Superheated steam—steam that has been heated to a temperature above its saturated temperature.

Superheater—a set of tubes located in or near the boiler flue gas outlet that increases (superheats) the temperature of the steam flow.

Waste heat boiler—a device that uses waste heat from a process to produce steam.

Water tube boiler—a type of boiler that contains water-filled tubes that allow water to circulate through a heated firebox.

Introduction

Steam has many applications in the process industries. For example, it is used to heat and cool process fluids, power and purge equipment, fight fires, facilitate distillation, and induce other physical and chemical reactions.

Boilers are an important source of energy in the process industries because they supply steam energy to process equipment and produce the steam used throughout the process facility. Examples of process equipment that use steam energy include turbines, reactors, distillation columns, stripper columns, and heat exchangers.

General Components of Boilers

Boilers are devices in which water is boiled and converted into steam under controlled conditions. Boiler components can vary with the different types. However, the most common components include a firebox, burners, drums, tubes, an economizer, a steam distribution system, and a boiler feedwater system.

FIREBOX

Like other process furnaces and direct-fired heaters, boiler fireboxes have a refractory lining, burners, a convection-type section, a radiant section, fans, air flow control, a stack, and a damper. The boiler firebox is insulated to reduce the loss of heat and enhance the heat energy being transferred to the boiler's internal components.

BURNERS

Burners inject air and fuel through a distribution system that mixes them in proper concentrations so combustion can occur. Most boilers use natural gas, fuel oil, or coal burners to provide heat to the boiler.

Did You Know?

A new innovation in burners is low NO_x formation burners called spuds.

Spuds contain unique gas injectors that promote better combustion than traditional burners.

Courtesy of John Zink Co. LLC

The key components of a natural gas burner include knockout pots, dampers, impellers, spuds, and igniters. **Knockout pots** remove liquids and condensate from the fuel before it is sent to the burners. Dampers regulate the flow of air to the burner. **Impellers** are fixed, vaned devices that cause the air/fuel mixture to swirl above the burner. An impeller in this instance is not the same as an impeller in a turbine or pump. **Spuds** (Figure 14-1) or **spiders** (Figure 14-2) are used to inject fuel into the boiler, and **igniters** are used to automatically ignite the flammable air and fuel mixture at the tip of the burner.

DRUMS

The drums inside a water tube boiler resemble a large water distribution header connected by a complex network of tubes. The lower drum (mud drum) and water tubes are filled completely with water, while the upper drum (water and steam drum) is only

FIGURE 14-1 Spud

FIGURE 14-2 Spider

partially full. Maintaining this vapor space in the upper drum allows the saturated steam to collect and pass out of the header.

Feedwater to the boiler inlet is treated to achieve the required chemical composition. Water lost in the boiler is replaced by a makeup water line. Sediment accumulates in the bottom of the mud drum and is removed through blowdown. Blowdown is the process of removing small amounts of water from the boiler's lower drum to reduce the concentration of impurities.

ECONOMIZER

The economizer section is used to increase boiler efficiency by preheating the water as it enters the boiler system. This heat exchanger transfers heat from the stack gases to the incoming feedwater and is usually located close to the stack gas outlet of the boiler. Economizers can be supported from overhead or from the floor. The boiler feedwater line that serves the boiler is piped and controlled to the economizer. No additional feedwater control valves or stack gas dampers are required.

An economizer is similar to the convection section in a direct fired heater. Both operate under the energy saving concept of recovering some of the heat from the hot flue gases before they are lost out of the stack. The typical improvement in efficiency of a boiler is 2 to 4 percent.

STEAM DISTRIBUTION SYSTEM

The steam distribution system consists of valves, fittings, piping, and connections suitable for the pressure of the steam being transported. Steam exits the boiler at sufficient pressure required for the process unit or for electrical generation. For example, when steam is used to drive steam turbine generators to produce electricity, the steam must be produced at a much higher pressure than that required for process steam. The steam pressure can then be reduced for the turbines that drive process pumps and compressors that require lower pressure steam.

Most steam used in the process facility is ultimately condensed to water. This condensation takes place in various types of heat exchangers. Condensate is typically

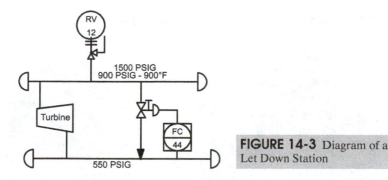

FIGURE 14-3 Diagram of a Let Down Station

reused as boiler feedwater and returned through condensate return systems. This saves the plant considerable amounts of money because the boiler feedwater, which has been treated with chemicals, need not be repurchased.

In most cases, when a facility contains a steam system, this system is typically two or three piping systems with different pressure steam headers. One method of maintaining several steam headers is through a letdown valve or a reducing station. Typically, a pressure control valve is used. In this type of arrangement, a pressure drop across the valve equalizes with the pressure contained in the lower steam header. In many instances, the steam header itself has another letdown valve or reducing station that performs the same function, supplying steam to yet another, lower pressure steam header.

Another method of supplying steam to headers of different pressures is piping the exhaust of steam turbines into that header. For example, a steam turbine that operates with 1,500 psig pressure may exhaust into a 550 psig steam header. The turbine that uses the 550 psig steam may then exhaust into a 50 psig header. A letdown valve or reducing station (see Figure 14-3) offers the advantage of maintaining a steam header at an interim pressure that may not be suitable for all steam turbines.

Another component used in steam systems is a steam trap. **Steam traps** are used to remove condensate or liquid from steam systems. There are several steam trap designs. The most common, however, are the mechanical trap (Figure 14-4) and the thermostatic trap (Figure 14-5).

Mechanical traps operate based on the amount of condensation present. Inside most mechanical traps are internal floats attached to mechanical linkages. As the condensate levels rise, the linkage causes the valve to open. As the levels drop, the valve closes. Inverted bucket and float traps are both common examples of mechanical traps.

Thermostatic traps operate based on temperature change. These traps contain valves that are opened or closed by thermal expansion and contraction. They differ from other types of traps because they must retain condensate for a period of time until it cools and the valve opens. Because of the way they operate, thermostatic valves are inappropriate for situations that require the condensate to be removed as quickly as it is formed. Bimetallic and bellows traps are examples of temperature-operated traps.

FIGURE 14-4 Examples of Mechanical Steam Traps

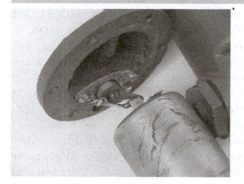

Courtesy of Brazosport College

FIGURE 14-5 Examples of Thermostatic Steam Traps

Courtesy of Brazosport College

BOILER FEEDWATER SYSTEM

The boiler feedwater supply is a critical part of steam generation. There must always be as many pounds of water entering the system as there are pounds of steam leaving.

The water used in steam generation must be free of contaminants such as minerals and dissolved impurities that can damage the system or affect its operation. Suspended materials such as silt and oil create scale and sludge and must be filtered out. Dissolved gases such as carbon dioxide and oxygen cause boiler corrosion and must be removed by deaeration and other means of treatment. Because dissolved minerals cause scale, corrosion, and turbine blade deposits, boiler feedwater must be treated with lime or soda ash to precipitate these minerals from the water, and recirculated condensate must be deaerated to remove dissolved gases.

Depending on the individual characteristics of the raw water, boiler feedwater may be treated by clarification, sedimentation, filtration, ion exchange, deaeration, membranes, or by several of these methods. Boiler feedwater treatment is discussed in greater detail in the following section.

Water Tube, Waste Heat, and Fire Tube Boilers

Many factors are considered when selecting a boiler. These factors include the pressures and temperatures required, total capacity, number of generating tubes, number of drums, type of circulation (forced), superheating and desuperheating, tube configuration, and cost. The most common types of boilers used in the process industries today include water tube, waste heat, and fire tube boilers.

WATER TUBE BOILERS

Water tube boilers (shown in Figure 14-6) are one of the most common types of boilers. These boilers are called water tube because they contain water-filled tubes that allow water to circulate through a heated firebox. Water tube boilers have an upper and lower drum connected by tubes. The upper drum is called the steam drum, and the lower drum is called the mud drum. Chemicals are added to the boiler feedwater that enters these drums in order to prevent fouling and corrosion.

There are several types of tubes inside the boiler. Generating tubes are attached to the upper and lower drums. Water flows through the tubes and back up to the water and steam drum. The downcomer tube is the cold water line between the upper and lower drums. The riser tube is the hot water line between the upper and lower drums. Superheated tubes are where steam is removed from the water drum and heated without an increase in pressure.

The water level in a water tube boiler is controlled in the steam drum. It is important for the water level in this drum to be maintained for safety reasons and for compliance with standard operating procedures. Loss of water level can damage boiler equipment,

FIGURE 14-6 Diagram of a Water Tube Boiler

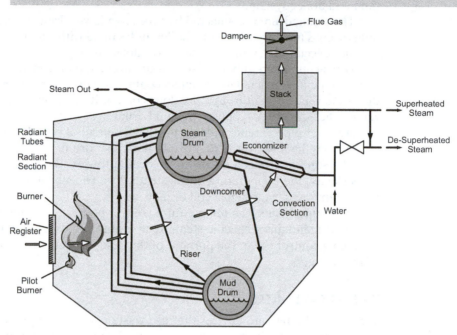

and excessively high water levels can result in carryover that causes the steam to become saturated with water containing chemicals that can cause fouling in the steam system.

Heat is generated in the boiler through a direct-fired heater with a natural gas, oil, or combination burner. The heat from the burner is transferred to the water tubes. As water flows through the tubes, the combustion gases boil the water and produce steam. This steam is collected in the upper drum, while combustion gases exit the boiler stack as flue gas.

Water Tube Boiler Components

The heating portion of a boiler is very similar to a furnace. Like furnaces, boilers contain a firebox. The **firebox** is where the burners are located and radiant heat transfer occurs. A special **refractory lining** (a brick-like form of insulation) is used to reflect heat back into the box and protect the structural steel in the boiler.

Located in the firebox area of the boiler are radiant tubes. **Radiant tubes** and riser tubes both contain boiler feedwater that is heated by radiant heat from the burners and boiled to form steam that is returned to the steam drum.

Burners are devices that introduce, distribute, mix, and burn a fuel (e.g., natural gas, fuel oil, or coal) for heat. **Pilots** are used to light burners. Burners can be **premix** (they mix the fuel gas with the air before either exits the burner face), **raw gas** (gas that has not been premixed with air), or a combination.

Associated with burners are air registers. **Air registers** control the flow of air to the burners to maintain the correct fuel-to-air ratio and to reduce smoke, soot, or NO_x (nitrogen oxide) and CO (carbon monoxide) formation.

Draft fans are used to supply combustion air to the burners. Depending on the design of the boiler, the draft fan either forces the air through the boiler (forced draft), pulls the air through the boiler (induced draft), or performs a combination of the two (balanced draft).

The **stack** is an opening at the top of the boiler that is used to remove flue gas. Contained within the stack is a **damper**, a moveable plate that regulates the flow of air, draft, or flue gases in boilers.

At the top of the boiler is a steam drum. The **steam drum** is the component where all generated steam gathers before exiting the boiler. Water enters the steam drum from the **economizer**, the section of a boiler used to preheat feedwater before it enters the main boiler system.

The steam drum is connected by tubes to a lower drum, called the mud drum. The mud drum is the lower drum of a boiler and is filled with water. The mud drum is where sediment accumulates and blowdown occurs manually.

Downcomers are tubes, located in the firebox, that transfer water from the steam drum to the mud drum. Downcomers contain cooler water descending from the steam drum. As water flows through the downcomers, it picks up heat from the firebox and replenishes the water supply to the mud drum.

Another type of tube contained within a boiler is a riser. **Riser tubes** allow water or steam from the lower drum to move to the upper drum.

The **superheater** is a set of tubes located toward the boiler outlet that increases (superheats) the temperature of the steam flow. The steam drum is usually connected to the superheater through a coil or pipe.

A **desuperheater** is a temperature control point at the outlet of the boiler steam flow that maintains a specific steam temperature by using boiler feedwater injection through a control valve. The purpose of the desuperheater is to lower the temperature of the steam.

WASTE HEAT BOILERS

Waste heat boilers are devices that use waste heat from a process to produce steam. Waste heat boilers have two functions: producing steam and providing cooling for a process that would otherwise be too hot for use farther down in the process. Figure 14-7 shows a waste heat boiler that may be used to recover waste heat energy and cool the vent stream from a turbine exhaust.

Waste heat boilers improve efficiency and save money by allowing steam to be produced through the combustion of waste products. By using waste products as a fuel source, facilities can reduce the amount of money spent on burner fuels and waste disposal.

Because of the duty required of waste heat boilers, the construction is usually thick-walled and designed to withstand high pressures and temperatures. Waste heat boilers are commonly single-pass, floating-head-type heat exchangers that experience a considerable amount of expansion and contraction of the tubesheet.

In many furnaces, the waste heat boiler is located on the outlet of the furnace. This design recovers heat by generating steam and cools the vent gas stream that exits from the furnace stack. Because of this, waste heat boilers are sometimes referred to as steam generators.

FIGURE 14-7 Diagram of a Waste Heat Boiler

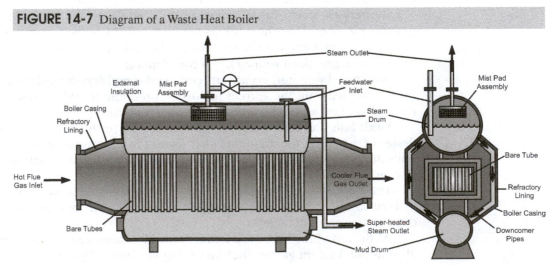

Courtesy of John Zink Co. LLC

Did You Know?

Steam locomotives and home water heaters are both examples of fire tube boilers?

FIRE TUBE BOILERS

Fire tube boilers are devices that pass hot combustion gases through the tubes to heat water on the shell side of the exchanger. In this type of boiler, combustion gases are directed through the tubes while water is directed through the shell. As the water begins to boil, steam is formed. This steam is directed out of the boiler to other parts of the process, and makeup water is added to compensate for the fluid loss. In this type of system, the water level within the shell must always be maintained so that the tubes are covered. Otherwise, the tubes could overheat and become damaged. Figure 14-8 shows an example of a fire tube boiler and its components.

FIGURE 14-8 Fire Tube Boiler

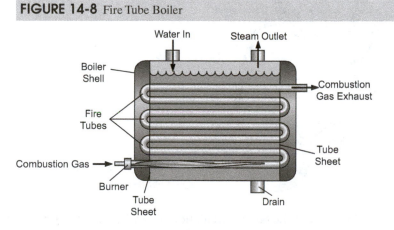

Theory of Operation

Boilers use a combination of radiation, convection, and conduction to convert heat energy into steam energy. Proper boiler operations depend on controlling many variables, including boiler feedwater quality, water flows and levels in the boiler, furnace temperatures and pressures, burner efficiency, and air flow.

To illustrate how boilers work, consider a simple boiler like the one shown in Figure 14-9. Simple boilers consist of a heat source, a water drum, a water inlet, and a steam outlet. In this type of boiler, the water drum is partially filled with water and then heat is applied. Once the water is sufficiently heated, steam forms. As the steam leaves the vessel, it is captured and sent to other parts of the process (e.g., it is used to turn a steam turbine, or it is sent to a heat exchanger to heat a process fluid). Makeup water is then added to the drum to compensate for the liquid lost as steam.

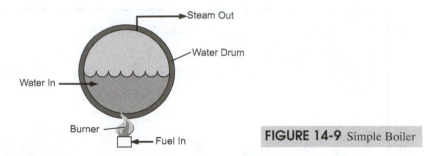

FIGURE 14-9 Simple Boiler

Boilers use the principle of differential density when it comes to fluid circulation. For boilers to work properly, they must have adequate amounts of heat and water flow. Factors that affect boiler operation include pressure, temperature, water level, and differences in water density. As fluid is heated, the molecules expand and it becomes less dense. When cooler, denser water is added to hot water, convective currents are created that facilitate water circulation and mixing.

WATER CIRCULATION

The circulation of boiler water (shown in Figure 14-10) is based on the principle of convection. When a fluid is heated, it expands and becomes less dense, moving upward through the heavier, denser fluid. Convection and conduction transfer heat through pipe walls and water currents, resulting in unequal densities. Cold water flows through the downcomer to the bottom of the mud drum and then flows upward through the riser (water wall tubes) as it heats.

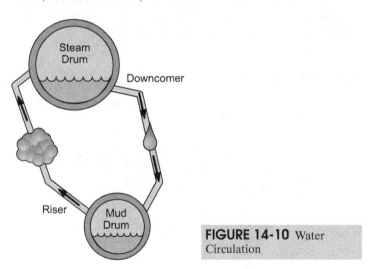

FIGURE 14-10 Water Circulation

In a water tube boiler, circulation occurs because the temperature of the fluid in the downcomer is always lower than the temperature in the boiler and generating tubes. As the liquid temperature continues to increase, steam bubbles are formed. These bubbles increase the circulation as they rise up the tubes. As the water vapor collects in the upper drum, the pressure builds. Each time the water passes through the tubes, it picks up more heat energy. As the pressure increases, the boiling point of the water increases. When the target pressure is achieved, steam is delivered to the steam header. To maintain this pressure, makeup water must be added, heat must be continually applied, and circulation must be controlled. In a fire tube boiler, the water level in the boiler shell must be maintained above the tubes to prevent overheating of the tubes.

SUPERHEATED STEAM

Saturated steam is steam in equilibrium with water (i.e., steam that holds all of the moisture it can hold and still remain a vapor). Saturated steam can be used as is to purge boilers or perform other functions, or it may be resuperheated.

As long as the steam and water are in contact with each other, the steam is in a saturated condition. Once steam is saturated, it cannot absorb additional water vapor. However, the boiler can continue to add heat energy to the steam. Steam that continues to take on heat energy or get hotter is known as superheated steam.

DESUPERHEATED STEAM

Superheated steam, which is produced downstream of the steam drum (typically in the firebox), is steam that has been heated to a temperature above its saturated temperature. Superheated steam is typically 200 to 300 degrees F hotter than the saturated steam. Typical uses for superheated steam include the following:

- Driving turbines
- Catalytic cracking
- Product stripping
- Maintenance of steam pressures and temperatures over long distances
- Producing steam for systems that require dry, moisture-free steam

Superheated steam may not be the best choice for heat transfer in some heat exchangers because the amount of energy given up by superheated steam is relatively small compared to the energy given up by saturated steam. Some facility processes cannot tolerate the high temperatures of superheated steam. The process of cooling the superheated steam is called desuperheating. **Desuperheated steam** is superheated steam from which some heat has been removed by the reintroduction of water. Typically, desuperheating steam is not performed at the boiler but at specific points in the process by injecting condensate into superheated steam.

BOILER FEEDWATER

Boiler feedwater levels and flows are critical to proper boiler operation. If feedwater runs low and the water level decreases to the point where the boiler runs dry, the tubes will overheat and fail. If the boiler water level is allowed to become too high, excess water will be carried over into the steam distribution system, which negatively affects process facility steam consumers and can damage turbines and other equipment.

During the boiling process, most solids stay in the water section of the drum while steam is sent to the superheater. As solids in the water increase, they are removed by sending a small amount of the feedwater flow through a drum blowdown pipe to a blowdown tank. This water is usually released to a waste water treatment processing unit. Boilers have continuous or intermittent blowdown systems to remove water and solids from the steam drums and limit scale buildup on turbine blades and superheater tubes.

Feedwater must be free of contaminants that could affect operations. The general rule is the higher the steam pressure, the more strict the feedwater quality requirements. Important feedwater parameters include pH (the alkalinity or acidity of the

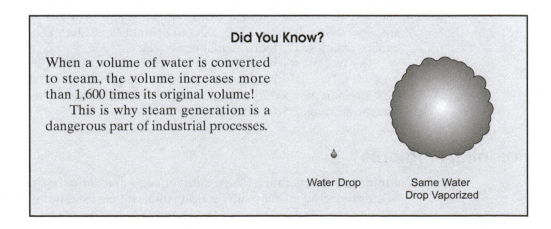

Did You Know?

When a volume of water is converted to steam, the volume increases more than 1,600 times its original volume!
This is why steam generation is a dangerous part of industrial processes.

Water Drop   Same Water
                    Drop Vaporized

water), hardness (the amount of mineral content in the water), oxygen and carbon dioxide concentration, silicates, dissolved or suspended solids, and the concentration of organics. Water treatment techniques include reverse osmosis, ion exchangers, deaeration, membrane contractors, and electro-deionization or demineralization.

WATER TREATMENT METHODS

Raw water can come from a variety of sources, such as lakes, rivers, or wells. Each water source has its own components and treatment requirements. In general, however, raw water goes through the following steps during the cleaning processes:

1. Deaeration
2. Coagulation
3. Filtration
4. Softening
5. Demineralization
6. Reverse osmosis

Deaeration removes air or other gases from boiler feedwater by increasing the temperature and aeration time.

Coagulation adds chemicals to reduce coarse suspended solids, silt, turbidity, and colloids through the use of a clarifier. The impurities gather together (coagulate) into larger particles and settle out of the chemical/water solution.

Filtration removes coarse suspended matter and sludge from coagulation or from water softening systems. Gravel beds and coarse anthracite coal are common materials used for filter beds. Special coated filters can be used to remove oil and reduce color.

Softening is the treatment of water to remove dissolved mineral salts such as calcium and magnesium, known as hardness, in boiler feedwater. There are several different softening methods, including the addition of calcium carbonate (lime soda), phosphate, and/or zeolites (crystalline mineral compounds).

Demineralization is the removal of ionized mineral salts by ion exchangers. The process is also called deionization, and the water produced is referred to as deionized water.

Reverse osmosis uses pressure to remove dissolved solids from boiler feedwater by forcing the water from a more concentrated solution, through a semi-permeable membrane, to a less concentrated solution.

BURNER FUEL

Boilers may use a single fuel or a combination of fuels, including refinery gas, natural gas, fuel oil, and powdered coal. In some complexes, scrubbed-off gases are collected from process units and combined with natural gas or liquefied petroleum gas in a fuel-gas balance drum. The balance drum serves as a constant system pressure and fairly stable BTU (British Thermal Unit) content. It also provides automatic separation of suspended liquids in the gas vapors to prevent large slugs of condensate from being carried over into the fuel distribution system.

The fuel oil system delivers fuel to the boiler at the required temperatures and pressures. The fuel oil is heated to pumping temperature, pulled through a coarse suction strainer, pumped to a temperature-control heater, and then pumped through a fine mesh strainer before being burned.

Potential Problems

Boiler system operations have a direct and immediate impact on the operation of other process equipment, such as distillation columns and turbines. The boiler must be

TABLE 14-1   Problems Associated with Boilers

Condition	Problem
Burner flame impingement or poor flame distribution	Causes overheating and failure of the tubes
A change in the fuel flow	Creates or increases soot deposits on the tubes, causing excessively high temperature on the tube walls
A change in the composition of the fuel	Creates or increases soot deposits on the tubes, causing excessively high temperature on the tube walls
Loss of boiler feedwater flow	Allows the lower drum to run dry, causing the tubes to overheat and fail, in turn causing catastrophic equipment failure and potential injuries
A high water concentration in the steam drum	Causes water to carry over into the steam distribution system, which can damage equipment downstream
Poor control of feedwater treatment	Creates the formation of scale in the tubes and may cause tube failure, resulting in a loss of boiler efficiency
Power outage	Loss of pumps and blowers

operating properly to produce steam energy at steady pressures. Failure to maintain normal boiler operation can result in loss of production, compromised product quality, and high costs to repair or replace failed or damaged components. Table 14-1 lists problem associated with boilers.

CONTRIBUTING FACTORS
Several factors can contribute to problems in boiler operations, including the design and age of the equipment, instrument problems, fouling, tube temperatures, and external factors.

Equipment Age and Design
The types of boiler equipment that are most subject to failure due to aging include boiler feedwater pumps, blowdown valves, and system piping. In addition, air feed control linkages on stack dampers can wear out from inadequate lubrication or from the collection of dust and grit. Freezing and earth movements can crack concrete equipment foundations and cause external corrosion of structural components. Worn or damaged tube supports and spring hangers can create stress problems for the boiler tubes.

Water and Other Contaminants
Instrument malfunctions can contribute to a number of problems and activate boiler interlocks. For example, water can condense in the pressure sensing leads in the firebox and cause faulty pressure readings.

Tube life in a boiler can also be significantly affected by the quality of the feedwater. For example, improper feedwater treatment can create salt deposits inside the tubes. These salt deposits cause tube corrosion and/or hot spots.

External Factors
Other factors that contribute to boiler problems include the loss of the water supply, electrical failure, cold weather, and downstream process upsets. For example, if a large steam user suddenly stops taking steam, steam header pressure can increase, suddenly causing pressure swings and instrument sensing problems faster than the boiler firing controls can respond. Another involves cold weather causing water in the process and utility lines to freeze and plug the line. Upsets in the upstream processing units can cause off-spec fuel to be sent to the boiler.

TABLE 14-2   Hazards Associated with Boiler Operations

| Improper Operation | Possible Effects | | | |
	Individual	Equipment	Production	Environment
Opening header drains, vents, and peepholes	Burns and eye injuries			
Failing to purge the firebox (startup)	Possible burns or injuries	Explosion in the firebox; damage to the internal components of the boiler	Process facility upset; lost steam production; downtime for repairs	Exceeding Environmental Protection Agency (EPA) limits for opacity (the amount of light blocked by a medium)
Poor control of excess air and draft control		Flame impingement	Reduced boiler efficiency	Exceeding EPA opacity limits
Loss of boiler feedwater		Tube rupture; loss of downstream equipment use	Lost production due to downtime for repairs	
Loss of fuel gas or oil		Loss of downstream equipment use	Lost steam production	

Safety and Environmental Hazards

Hazards associated with normal and abnormal boiler operations can negatively affect personal safety, equipment, and the environment. Process technicians must always take proper safety precautions when working around boilers. Table 14-2 lists some of the hazards associated with improper operation of boilers and their effects.

NO_x AND SMOKE EMISSIONS

Excessive smoke or nitrogen oxide (NO_x) emissions from boiler operations have a negative impact on the environment. Both are considered air pollutants and can lead to emissions violations. To keep NO_x emissions under control, ensure that the burner air registers and fuel valves are always operating properly. Understanding and controlling oil burner operations allows the right balance for complete combustion and control of smoke emissions. Process technicians should also take care when handling chemicals to reduce environmental impact.

FIRE PROTECTION AND PREVENTION

The most potentially hazardous operation in steam generation is boiler startup. During startup, a flammable mixture of gas and air can build up as a result of flame loss at the burner during light-off. Each type of boiler requires specific startup and emergency procedures, including purging before light-off and after misfire or loss of burner flame.

HAZARDOUS OPERATING CONDITIONS

Hazards are present during both normal and abnormal boiler operations. For example, there is the potential for exposure to feedwater chemicals, steam, hot water, radiant heat, and noise. Alternate fuel sources may be available in the event fuel gas is lost due to a process unit shutdown or emergency.

PERSONAL SAFETY

Process technicians should always observe safe work practices and wear appropriate personal protective equipment when working around boilers or performing process sampling, inspection, maintenance, or turnaround activities. Process technicians should also avoid skin contact with boiler blowdown because it may contain hazardous chemicals.

Typical Procedures

Process technicians should be aware of several boiler procedures, including startup, shutdown, lockout/tagout, and emergency procedures. The procedures listed below are generic in nature, so process technicians should always refer to site-specific procedures before performing any work on their unit.

STARTUP

1. Inform control room personnel that the boiler is going to be put into service.
2. Use the peepholes to inspect the inside of the boiler and verify that it is free of debris (e.g., scaffold boards, rain suits, and tools) and that the refractory lining, tubes, and tube hangers are all intact.
3. Inspect the outside of the boiler for loose flanges around inlet and outlet piping.
4. Inspect the condition of the draft fan, the damper, and the fuel gas system to ensure all are in satisfactory operating condition.
5. While in close communication and with control room personnel, open the damper and start up the draft fan per standard operating procedures.
6. Satisfy the interlock (control safety systems that have to be satisfied before boiler startup) for proper water level.
7. Once the draft in the boiler is stable and it has been purged for the required amount of time, open the fuel gas supply to the burner and the pilot gas block valves.
8. Inform control room personnel that the pilot is being ignited.
9. Once all of the pilots are lit and burning (confirmed through visual observation), open the primary and secondary air registers as required and slowly open the main burner block valves one at a time until all burners are burning.
10. Remain in the boiler area to inspect the flames for proper flame patterns and general operation (e.g., draft fan and fuel supply).

Note: Many of these steps in the process industries are automatically included in the boiler automated control system.

SHUTDOWN

With some exceptions, boiler shutdown is the reverse of the startup procedure.

1. Turn off fuel.
2. Shut off the steam.
3. Go through shutdown procedure.

During shutdown, the control room operator should gradually reduce the firing and the process flow until conditions are safe enough for the outside operator to block in all the burners, shut down the fan, and close the damper.

EMERGENCY

In an emergency, one fuel gas block valve (often referred to as the fireman) is usually designated as the main shut-off valve. Generally, the draft fan continues to run and the damper remains open. The only thing the outside operator does is block in the burners, block in the fuel, and stop all pumps.

LOCKOUT/TAGOUT FOR MAINTENANCE

Each company has its own lockout/tagout procedures. Process technicians should be familiar with these procedures before performing any maintenance.

Process Technician's Role in Operation and Maintenance

Process technicians are responsible for performing specific procedures to operate and maintain boiler system equipment. Starting up a boiler, for example, requires filling the drum with water, lighting the burner, bringing the boiler up to pressure, and then placing the boiler on line. Each of these steps requires that the process technician perform

TABLE 14-3   Process Technician's Role in Operation and Maintenance

Look	*Listen*	*Feel*
• Check firebox for flame impingement on tubes • Check burner flame color • Check for wall hotspots (external and internal) • Check draft balance (pressures) • Check temperature gradient • Check firing efficiency (CO_2, CO, and O_2 in the stack) • Check burner balance (fuel and air) • Check controlling instruments (water level, fuel flow, feedwater flow, pressure, steam pressure, and temperature) • Check the boiler stack for smoke • Inspect the burners and flame pattern • Maintain proper steam flow • Monitor the air flow and oxygen level, and adjust draft as needed • Collect boiler feedwater samples and ensure proper chemical composition • Ensure that the blowdown on the mud drum is operating efficiently • Check for burner wear, which may cause uneven flame or hotspots	• For abnormal noise (e.g., fans, burners, water leaks, steam leaks, or external alarms) • For huffing or puffing, either of which can indicate improper draft operation	• Check for excessive vibration (fans and burners) • Check atomization for uniformity (if not uniform, burner tips on the atomization gun or burner require replacement)

a number of tasks and follow specific procedures that vary according to the specific site and boiler type. When monitoring and maintaining boilers, process technicians must always remember to look, listen, and feel for all the factors listed in Table 14-3. Failure to perform proper maintenance and monitoring could affect the process and result in equipment damage.

Summary

Boilers convert water into steam, which is supplied to steam consumers throughout the process facility. The steam produced by boilers provides heat to other process equipment and supplies the steam energy to drive turbines and compressors.

Boilers use a combination of radiant, conductive, and convective methods to transfer heat. These devices consist of a number of tubes that carry the water-steam mixture through the boiler for maximum heat transfer.

The process industries use three types of boilers: water tube, waste tube, and fire tube. Water tube boilers are the most commonly used boilers in process operations. In water tube boilers, the water is circulated through tubes that run between the upper steam and water drum and the lower water collection drum (mud drum). Water circulation is created using the principles of differential density. The downcomer is the cooler water line that goes from the upper drum to the lower drum. The riser is the hotter water line that goes from the lower drum to the upper drum.

A fire tube boiler is similar to a shell-and-tube exchanger. A combustion tube equipped with a burner transfers heat through the tubes and out of the boiler. The tubes in a fire tube boiler are submerged in water and are designed to transfer heat energy to the liquid through conduction and convection.

Sediment accumulated in the bottom of the mud drum is removed by water blowdown. Steam from the upper drum is superheated before entering the steam distribution system. Desuperheaters are used to lower the steam temperature for processes that cannot tolerate the higher steam temperatures.

Steam used in the process facility is usually condensed to water in various types of heat exchangers. This condensate may be reused as boiler feedwater.

Boiler heat is provided by an oil- or gas-fired burner. Dampers are used to regulate air flow to the burner. Combustion gases exit the boiler stack as flue gas.

The boiler feedwater supply is a critical part of steam generation. Feedwater is treated to remove contaminants that can damage system equipment or compromise its efficient operation. An economizer is used to preheat the feedwater as it enters the boiler. Makeup water is added to maintain proper water levels in the system.

Boiler operations depend on controlling feedwater quality, maintaining specific water levels in the system, controlling boiler temperatures and pressures, and maintaining burner efficiency. Process technicians are responsible for monitoring boiler system operations and performing preventive maintenance on boiler system equipment. Typical tasks include checking the boiler stack for smoke; inspecting burners and flame patterns; checking burner fuel pressure, temperature, and air flow; and maintaining proper steam flows, temperatures, and pressures.

Problems associated with boiler operations include the impingement of burner flames on the tubes, loss of boiler feedwater, inadequate treatment of feedwater, or a high water level in the steam drum. A change in the flow or composition of the fuel can increase soot deposits and cause excessively high temperature on the tubes.

Some problems are a natural consequence of equipment age and design. Other problems are the result of instrument malfunctions, which can be caused by rain, dust, or moisture. Performing the required preventive maintenance, such as making sure that the air feed control linkages are properly lubricated, can prevent many problem conditions.

As is the case with all furnaces or direct-fired heaters, there is always the potential for fire or explosion in the boiler firebox. This is of greatest concern during boiler startup. Because of this, the boiler firebox must be purged of combustibles before light-off and in the event of misfire or loss of burner flame. Process technicians must always follow the specific startup and emergency procedures for the particular boiler.

Other potential hazards associated with boiler operations include the potential for exposure to feedwater chemicals, steam, hot water, radiant heat, and noise. Process technicians should always wear appropriate personal protective gear and avoid skin contact with blowdown materials, which may contain hazardous chemicals.

Checking Your Knowledge

1. Define the following terms:
   a. Boiler
   b. Damper
   c. Superheated steam
   d. Saturated steam
   e. Blowdown
   f. Fire tube boiler
   g. Water tube boiler
   h. Waste heat boiler
2. *(True or False)* Boilers provide heat to other process equipment and produce steam used throughout the process facility.
3. The component(s) on a water tube boiler is(are) (select all that apply):
   a radiant tubes
   b. burners
   c. air registers
   d. fire box
4. What type of boiler consists of a shell and a series of steel tubes that transfer heat through a combustion chamber into the horizontal fire tubes?
   a. Water tube boiler
   b. Waste heat boiler
   c. Fire tube boiler

5. Burners inject air and fuel through a distribution system. The key components of a natural gas burner include (select all that apply):
   a. dampers
   b. impellers
   c. spuds or spiders
   d. drums
6. *(True or False)* The economizer section is used to decrease boiler efficiency by reheating the water as it enters the boiler system.
7. Dissolved minerals in the boiler feedwater system include metallic salts and calcium carbonates that cause scale, corrosion, and turbine blade deposits. Boiler feedwater is treated with _____ to precipitate these minerals from the water.
   a. lime
   b. salt
   c. lemon
   d. ethanol
8. A water treatment method that uses ion exchangers to remove all ionized mineral salts is:
   a. deaeration
   b. coagulation
   c. reverse osmosis
   d. demineralization
9. *(True or False)* Burner flame impingement or poor flame distribution can cause overheating and failure of the tubes.

Student Activities

1. Write a one-page paper on the two boiler types and their functions. Include drawings of the boilers and label their components.
2. Write a one-page paper on the operating principles of boilers (i.e., how do boilers work?).
3. Within a group, select one type of water treatment method used in boiler feedwater. Research this treatment method and present what you learned to the class. Be sure to include drawings and explain how the water treatment can enhance productivity for the process unit.
4. With a partner, review the typical startup procedure mentioned in the chapter. After reviewing the procedure, decide what steps can be added to the procedure to make the procedure safer for the individual, the equipment, the production goals, and the environment. Be prepared to present your information to the class.

15

Auxiliary Equipment

Objectives

After completing this chapter, you will be able to:

- Explain the purpose of auxiliary equipment in the process industry.
- Identify the common types and applications of auxiliary equipment.
- Identify the components of auxiliary equipment and explain the purpose of each.
- Explain the operating principles of auxiliary equipment.
- Identify typical procedures associated with auxiliary equipment.
- Describe the process technician's role in the use of auxiliary equipment.
- Identify potential problems associated with auxiliary equipment.

Key Terms

Agglomerator—a mixer that feeds dry and liquid products together to produce an agglomerated (clustered together) product like powdered laundry detergent

Agitator blades—devices that act on the fluid by generating turbulence, which promotes mixing.

Agitator shaft—a cylindrical rod that holds the agitator blades.

Bernoulli principle—principle stating that, as the speed of a fluid increases, the pressure of the fluid decreases.

Centrifuge—a device that uses centrifugal force to separate solids from liquids or to extract liquid from a liquid solvent. Also called a centrifugal extractor or contactor.

Drum mixer—a screw-type device used to blend two or more materials (sometimes with small amounts of liquid) into a consistent composition.

Dynamic mixer—a piece of equipment that rotates to mix products rapidly.

Eductor—a device that transports fluid by using a Venturi design.

Hydrocyclone—a mechanical device that promotes separation of heavy and light liquids by centrifugal (rotating) force.

Jet pump eductor—a device that uses the movement of high pressure liquids to create suction (Venturi effect). Also referred to as an aspirator.

Mixer—a piece of equipment that rotates inside or outside to combine products.

Static—a mixer that operates by creating a complex flow path that results in contact between the streams. Also referred to as an inline mixer.

Steam jet eductor—a device that uses the movement of steam to create suction (Venturi effect).

Introduction

The process industries use many different types of auxiliary equipment. The term auxiliary refers to items that act in a supporting capacity. In other words, auxiliary equipment is equipment that is secondary to the main processing unit but is necessary for the process to be completed.

Auxiliary equipment can be used for a variety of activities, such as creating a vacuum or mixing and transferring materials in order to produce a desired product. While the list of auxiliary equipment is quite extensive, this chapter covers only a few of the items commonly used in the process industries.

Types of Auxiliary Equipment

There are many different types of auxiliary equipment. Some of the most common types, however, include agitators, mixers, eductors, centrifuges, and hydrocyclones.

AGITATORS

Agitators and mixers are used throughout the process industries for mixing and blending of feedstocks and products. Agitators and mixers come in many different shapes and sizes. An agitator is a device used to mechanically mix the contents of a tank or vessel. Process technicians operate agitators using motors located outside the vessels.

Agitators are used in a wide range of mixing duties (e.g., solids suspension, liquid blending, and de-stratification), and can be used in many different industries and operations (e.g., food and beverage manufacturing, wastewater treatment, and chemical production. Figure 15-1 shows an example of a tank with an agitator inside.

FIGURE 15-1 Tank with Agitator Inside
Courtesy of Sun Products Corporation

The components of an agitator are relatively simple. Agitators consist of an agitator shaft and agitator blades. Figure 15-2 shows an example of a simple agitator with the components labeled.

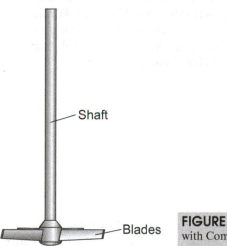

Shaft

Blades

FIGURE 15-2 Simple Agitator with Components Labeled

The **agitator shaft** is a cylindrical rod that holds the agitator blades. **Agitator blades** are devices that act on the fluid by creating turbulence, which promotes mixing. Agitator blades come in many different designs and provide various degrees of mixing within a specific tank or reactor. In some tanks or reactors, two sets of agitators, or a single agitator shaft with multiple blades, are used to increase the mixing process. Baffles (shown in Figure 15-3) are also used with agitators to facilitate mixing.

Did You Know?

Many home washing machines contain an agitator. As the agitator rotates back and forth, it moves the clothes through the water/detergent mixture.

Once the cleaning process is complete, the washer drum spins rapidly to produce the centrifugal motion necessary to force the excess water out of the clothes.

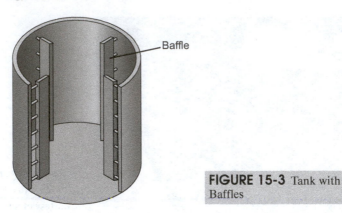

FIGURE 15-3 Tank with Baffles

Agitators function by stirring the contents of a vessel and creating turbulence. This keeps any suspended solids from settling to the floor of the vessel and forming a layer of solids.

An example of an agitator in the petrochemical industry is an agitator in a crude oil tank. Crude oil from an oil well contains sediment and other oil-wetted solids. If the crude oil were allowed to sit undisturbed in a tank, these solids would settle to the bottom. Over time, the solids layer (shown in Figure 15-4) would build up and require that the tank be removed from service for cleaning (a costly and time consuming process). By employing an agitator, it is possible to keep the solids in suspension, thereby resulting in less solids buildup and less tank maintenance.

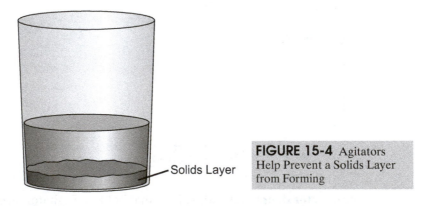

Solids Layer

FIGURE 15-4 Agitators Help Prevent a Solids Layer from Forming

MIXERS

A **mixer** is a piece of equipment that rotates to combine products. In the process industries, mixers can be designed to operate continuously or for a set period of time. Mixers can be used to mix wet or dry material during a process or to produce a final product. There are many different kinds of mixers and mixer blades, including dynamic mixers, inline mixers, agglomerators, and drum mixers. Figure 15-5 shows examples of various types of mixer blades.

FIGURE 15-5 Examples of Different Types of Agitator Mixer Blades

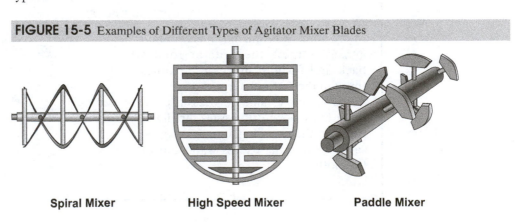

Spiral Mixer   **High Speed Mixer**   **Paddle Mixer**

Dynamic Mixers

A **dynamic mixer** (shown in Figure 15-6) is a piece of equipment that rotates to combine products. The components of a dynamic mixer include a shaft, blade, coupling, bearing housing, and driver.

Each type of mixer can introduce liquids through a series of nozzles or blend various powdered materials. For example, in the oil and gas industries, mixers are used to blend gasoline into different grades. In the pharmaceutical and cosmetics industries, mixers are used to make powders, creams, lipsticks, and ointments.

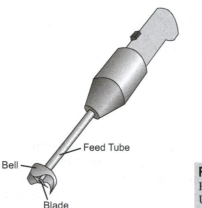

Feed Tube

Bell

Blade

FIGURE 15-6 Small, Hand-Held, Dynamic Mixer Used in Food Preparation

Inline Mixers

A **static** (inline mixer) operates by creating a complex flow path that results in intimate contact between the streams. The components of a static or inline mixer include a design element inside the pipe where the mixing occurs. Static or inline mixers are usually limited to the mixing of liquid streams. Figure 15-7 shows an example of an inline mixer.

FIGURE 15-7 Inline Mixer

Agglomerators

Some mixers are referred to as agglomerators. An **agglomerator** is a mixer that feeds dry and liquid products together to produce an agglomerated (clustered together) product like powdered laundry detergent. Figure 15-8 displays a diagram agglomerator; the perspective in the diagram is up from the bottom.

Some agglomerators are vertical mixers that rotate. During rotation, liquid material is introduced into the powder through a series of nozzles. Once the product has been hydrated (moisture added to it) and agglomerated (formed into clusters), it is then sent to a material dryer. The proper flow of powder and liquid, alignment of the nozzle air caps, and cleanliness of the nozzle assemblies is extremely important when trying to achieve correct density in agglomerated products.

FIGURE 15-8 Looking Up from the Bottom of an Agglomerator

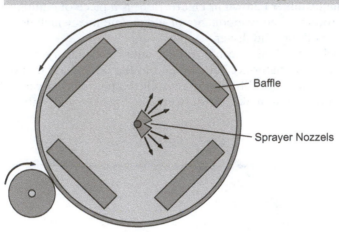

FIGURE 15-8 Looking Up from the Bottom of an Agglomerator

Baffle

Sprayer Nozzels

Drum Mixers

Drum mixers are screw-type devices used to blend two or more materials into a consistent composition. Drum mixers can be hand-held or stand-alone devices. Common components of a drum mixer are a motor driver and an agitator. The size and shape of a drum mixer can vary based on the design need. For example, hand-held drum mixers (shown in Figure 15-9) resemble a drill with a long, screw-like drill bit that is inserted into a mixing drum.

FIGURE 15-9 Hand-Held Drum Mixer

Rotating drum mixers are barrel-shaped devices that contain baffles or screws that facilitate mixing as the barrel rotates along its axis. Rotating drum mixers are used more frequently in industry than are hand-held devices because they produce less static electricity. Drum mixers can also be used for material that can be rotated slowly for a constant mixture (e.g. cement in a cement truck). Figure 15-10 shows the outside of a drum mixer, and Figure 15-11 shows the internal components.

EDUCTORS

An **eductor** is a device that transports fluid using a Venturi design. The internal components of an eductor are a Venturi section, nozzle, and sometimes a check valve.

A Venturi is a device that employs the **Bernoulli principle** (a principle stating that, as the speed of a fluid increases, the pressure of the fluid decreases) and consists of a converging section, a throat, and a diverging section. Within a Venturi, the flow speed

Did You Know?

A common kitchen mixer is a type of vertical (dynamic) mixer.

Like an industrial mixer, this vertical (dynamic) mixer can combine wet or dry materials to form a desired mixture.

FIGURE 15-10 Outside View of a Drum Mixer

Courtesy of Sun Products Corporation

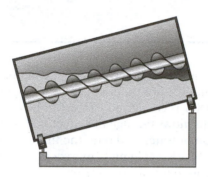

FIGURE 15-11 Simple Drum Mixer

Did You Know?

A cement truck is an example of a drum mixer.

When the drum rotates in one direction, the screw inside the drum mixes the concrete. When the drum turns in the other direction, the screw forces the cement out of the truck and onto a chute.

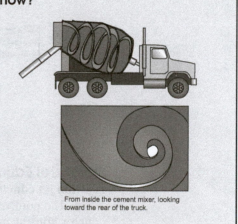

From inside the cement mixer, looking toward the rear of the truck.

of a gas or liquid increases as a result of the decreasing internal diameter or a constriction in the pipe or duct. Figure 15-12 shows an example of a Venturi tube.

Eductors are reliable pieces of equipment because they contain no moving parts. They are also very corrosion and erosion-resistant because they can be made from almost any material.

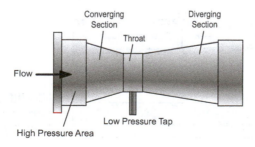

Converging Section

Diverging Section

Throat

Flow

Low Pressure Tap

High Pressure Area

FIGURE 15-12 Diagram of a Venturi Tube

Did You Know?

A garden hose chemical sprayer is a type of eductor.

As the water flows from the hose into the container, the chemicals and water are mixed and then discharged through the spray nozzle.

Eductors can be used in a variety of applications. For example, they can be used to proportion firefighting foam into the fire water stream when fighting oil spill and tank fires. In this situation, the eductor is connected to the fire hose and the stream of fire water is started. As the fire water flows through the eductor, it pulls the foam out of the bucket or drum, into the stream of water, and onto the fire. Eductors can also be used to pull a slight vacuum on a distillation column to extract unwanted gases. Figure 15-13 is a diagram of a foam eductor used for fire supression.

FIGURE 15-13 Diagram of a Foam Eductor

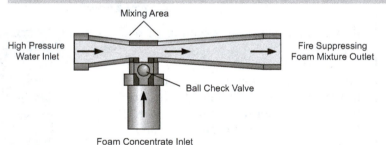

Steam Jet Eductor

A **steam jet eductor** (shown in Figure 15-14) uses a flow of steam to produce a Venturi effect. The components of a steam jet eductor system include the Venturi (diffuser), eductor housing, nozzle, and steam condenser.

In a steam jet eductor, the nozzle inside the eductor increases the velocity of the steam, thereby creating an area of low pressure at the low pressure suction inlet. This low-pressure area allows lower pressure gases to be drawn into the stream that is moving through the eductor.

The advantage of steam jet eductors is that they are easy to install and maintain. They can also be used in various types of industries because they can be made of a variety of

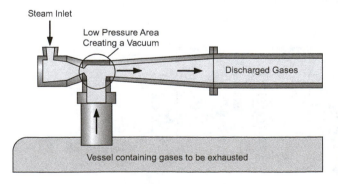

FIGURE 15-14 Steam Eductor Showing Flow of Vessel Gases to be Exhausted

materials (e.g., carbon steel, bronze, stainless steel, and Teflon) in order to meet the needs of the process.

The disadvantages of steam jet eductors are that a considerable amount of energy (e.g., high-pressure steam) is consumed, and the performance of the eductor is very sensitive to steam supply conditions (e.g., steam pressure and steam temperature).

Jet Pump Eductor

A **jet pump eductor** (also referred to as an aspirator) is a device that uses the movement of high-pressure fluid to create suction (Venturi effect). An example of a jet pump eductor is shown in Figure 15-15. The components of a jet pump eductor include a pump to provide a flow for the motive force and the eductor.

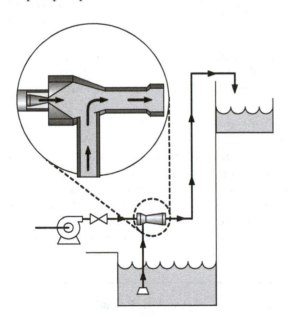

FIGURE 15-15 A Jet Pump Eductor Raising a Liquid to a Higher Elevation

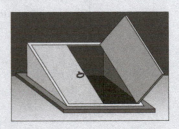

Jet pump eductors are effective at lifting fluids from wells, removing combustible fluids (which could ignite if exposed to a spark from an electric pump), and removing excessive amounts of debris (materials that would damage the screws or blades in a conventional pump). The advantages of jet pump eductors are that they have no moving parts, and they are simple and light weight.

CENTRIFUGES

A **centrifuge** (also called a centrifugal extractor or contactor) is a device that uses centrifugal force to separate solids from liquids or to extract a liquid from a liquid-liquid solvent. In a centrifuge, heavy liquids are spun from the center of an enclosed drum. As the liquid spins, centrifugal force (the force that causes something to move outward from the center of rotation) causes the fluid components to separate. The centrifugation process can be a single- or multistage operation depending on the nature of the material being separated. Figure 15-16 shows a diagram of a centrifuge with the components labeled.

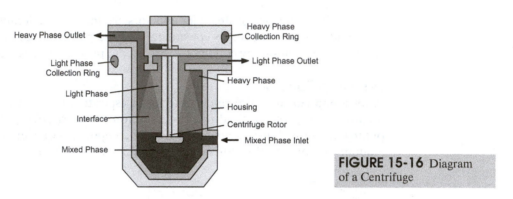

FIGURE 15-16 Diagram of a Centrifuge

The main benefit centrifugal extractors have over extraction columns or a mixer-settler system is they take up less space. The main disadvantage of a centrifugal extractor is that they have gearboxes and motors, which are more expensive and require additional maintenance.

HYDROCYCLONES

Hydrocyclones are mechanical devices that promote separation of heavy and light liquids by centrifugal (rotating) force. In a hydrocyclone, incoming liquid enters through the top of the vessel. The liquid then begins to rotate as a result of natural centrifugal force caused by the orientation of the inlet flow pipe. This forces the heavy liquid to the outer walls of the vessel, where it flows to the bottom of the cylinder and out the bottom. The lighter liquids remain in the center of the vessel, where they move up and out of the top of the vessel.

Did You Know?

Medical technologists use centrifuges spinning at 3000 revolutions per minute (rpm) to separate whole blood into its component parts (red blood cells, white blood cells, platelets, and plasma).

By separating and examining these parts, doctors can better diagnose and treat certain illnesses.

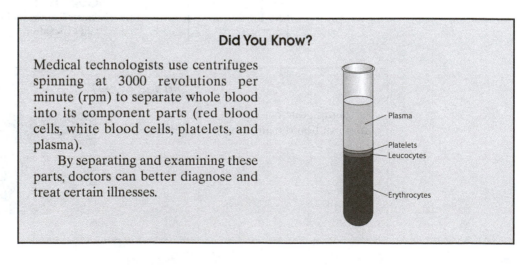

Hydrocyclones differ from centrifuges because they have no moving parts. One advantage of hydrocyclones is that they are less expensive than centrifuges. A disadvantage is that they are less effective than centrifuges.

In the process industries, hydrocyclones are used in a variety of applications. For example, hydrocyclones are used to remove solids and other impurities from seal flush systems on pumps.

DEMISTER

A demister is a mechanical device that promotes separation of liquids from gases. Demisters contain a porous, woven pad at the gas outlet. This pad allows gases to pass through while trapping liquids. These liquids coalesce (collect) and fall back into the vessel. Figure 15-17 shows a diagram of a demister.

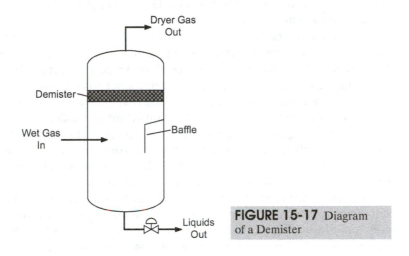

FIGURE 15-17 Diagram of a Demister

Potential Problems

When working around auxiliary equipment, process technicians should always be aware of potential problems such as leaks, wear and tear, higher revolutions per minutes (rpms), vibration on steam jets and nozzles, and steam conditions not being at design setting.

Typical Procedures

Process technicians should be familiar with the appropriate procedures for monitoring and maintaining auxiliary equipment components and conditions. Typical procedures, which include startup, shutdown, emergency, and lockout/tagout, vary at each individual site and for each type of auxiliary equipment. Because of this, process technicians should always follow the standard operating procedures (SOPs) for their assigned unit(s).

Process Technician's Role in Operation and Maintenance

The process technician's role in operation and maintenance includes following standard operating procedures for the facility and the unit. Process technicians may also be required to report readings, conduct field checks, and write lab reports. Process technicians should understand operational procedures related to each piece of equipment in order to maintain safe operating conditions, to report an impending violation to their supervisor, and to be able to respond appropriately in the event of an emergency.

Process technicians routinely monitor auxiliary equipment to ensure they are operating properly. When monitoring and maintaining auxiliary equipment, technicians must always remember to look, listen, and feel for the items indicated in Table 15-1.

TABLE 15-1	Process Technician's Role in Operation and Maintenance	
Look	*Listen*	*Feel*
• System operating at higher or lower speeds than normal • System pulling higher amps than normal • Gasses or solids leaking into the atmosphere	• Abnormal noises such as banging or hammering • Normal noises with abnormal pitches	• Excessive heat in areas • Surging through the lines (e.g., normal flow, then excessively high or low flow)

Summary

The auxiliary equipment described in this chapter is used in the process industries to mix or combine materials.

Agitators are used to combine contents in a tank or vessel and can be used to mix liquids by recirculating the content. The primary components of an agitator are the agitator shaft and agitator blade, which produce a mixing effect to obtain the final product.

A mixer rotates inside or outside a container to combine products. A drum mixer rotates using a drum attached to blades that rotate at a set speed. Mixers can be designed for continuous mixture or a set mixture as needed. Mixers are usually one of three types: a high-speed horizontal paddle mixer, a high-speed vertical mixer, and a low-speed horizontal screw mixer. Some mixers are referred to as agglomerators. Agglomerators feed dry and liquid products together to produce an agglomerated (clustered) product like powdered laundry detergent.

Eductors are devices that transport fluids using the Venturi effect. Eductors are reliable pieces of equipment due to their limited number of moving parts. Examples of eductors are used in firefighting to regulate the amount of foam distributed through the fire hose. Steam jet eductors use a flow of steam to draw air to its flow using a Venturi effect. Advantages of steam jet eductors are that they are easy to install and easy to maintain. A jet pump eductor is designed to lift liquids and/or mix materials inside a tank.

A centrifuge is a device that uses centrifugal force to separate liquids from liquids, or liquids from solids. Heavy liquids are forced to the outer walls of an enclosed drum while the lighter liquid remains in the center.

Hydrocyclones are used in industry to separate solids or liquids from liquid streams. They have no moving parts so they are less expensive than centrifuges.

A process technician's role in auxiliary equipment includes being aware of potential problems when an abnormal sound is heard or equipment is not operating properly. Process technicians should always follow standard operating procedures for their unit and/or facility.

Checking Your Knowledge

1. Define the following terms:
   a. Agitator
   b. Centrifuge
   c. Eductor
   d. Hydrocyclone
   e. Venturi effect
2. *(True or False)* The primary components of an agitator are the agitator shaft and agitator blades.
3. *(True or False)* A cement truck is an example of a drum mixer.
4. The internal components of an eductor are a Venturi section, nozzle, and sometimes a _____.
   a. relief valve
   b. ball valve
   c. check valve
   d. globe valve

5. *(True or False)* The decreased velocity at the throat creates a low-pressure area called the Bernoulli's principle.
6. *(True or False)* Steam jet eductors are often used in vacuum distillation columns.
7. A jet pump eductor is designed to be used in the following process(es) (select all that apply):
    a. areas that need to be drained
    b. areas that may contain high levels of debris
    c. areas that may contain combustible fluids
    d. areas that need more fluid added to the process
8. *(True or False)* In a hydrocyclone, incoming liquid enteres through the bottom of the vessel.
9. *(True or False)* A centrifuge and hydrocyclone both use centrifugal force.

Student Activities

1. Write a report on how agitators and mixers are used in the process industries. Discuss the similarities and the differences, and include a diagram of each.
2. Work with a team to prepare a presentation that discusses the theory of operation of eductors (including Venturi tube), steam jets, and jet pumps. Present the material to the class. Be sure to include drawings and a description of the process.
3. Write a report describing centrifuges and hydrocyclones. Discuss the similarities and differences. In addition to the written report, draw a flow diagram that illustrates how each device works.

16

Tools

Objectives

After completing this chapter, you will be able to:

■ List the types of tools used by technicians in the process industries.

■ Describe the appropriate uses of basic hand and power tools.

■ Describe basic hand and power tool safety.

■ Describe the appropriate care of hand and power tools.

Key Terms

Adjustable pliers—pliers that have a tongue and groove joint or slot that allows the jaws to widen and grip objects of different sizes. Also referred to as Channel Locks® or slip joint pliers.

Adjustable wrench—open-ended wrench with an adjustable jaw that contains no teeth or ridges (i.e., the jaw is smooth). Also referred to as a Crescent® wrench.

Crane—a mechanical lifting device equipped with a winder, wire ropes, and sheaves that can be used to lift and lower materials.

Dolly—a wheeled cart or hand truck used to transport heavy items such as barrels, drums, and boxes.

Drill—a device that contains a rotating bit designed to bore holes into various types of materials.

Electric tool—a tool operated by electrical means (either AC or DC).

Flaring tool—a device used to create a cone-shaped enlargement at the end of a piece of tubing so the tubing can accept a flare fitting.

Forklift—a motorized vehicle with pronged lifts. Forklifts are used to lift equipment or items placed on a pallet to higher elevations for use in a unit, for storage, or for maintenance.

Fuel-operated tools—tools powered by the combustion of a fuel (e.g., gasoline).

Gantry crane—crane (similar to an overhead crane) that runs on elevated rails attached to a frame or set of legs.

Grinder—a device that uses the rotating force of one material against another material (e.g., a stone grinding wheel against a metal blade) to reduce or adjust the shape and size of an object.

Hammer—a tool with a handle and a heavy head that is used for pounding or delivering blows to an object.

Hand tool—a tool that is operated manually instead of being powered by electricity, pneumatics, hydraulics, or other forms of power.

Hoist—a device composed of a pulley system with a cable or chain used to lift and move heavy objects.

Hydraulic tool—a tool that is powered using hydraulic (liquid) pressure.

Impact wrench—a pneumatically or electrically powered wrench that uses repeated blows from small internal hammers that generate torque to tighten or loosen fasteners.

Jib crane—crane that contains a vertical rotating member and an arm that extends to carry the hoist trolley.

Lineman's pliers—snub-nosed pliers with two flat gripping surfaces, two opposing cutting edges, and insulated handles that help reduce the risk of electrical shock.

Locking pliers—pliers that can be adjusted with an adjustment screw and then locked into place when the handgrips are squeezed together. Also referred to as Vise-Grips®.

Monorail crane—a type of crane that travels on a single runway beam.

Needle nose pliers—small pliers with long, thin jaws for fine work (e.g., holding small items in place, removing cotter pins, or bending wire).

Overhead traveling bridge cranes—a type of crane that runs on elevated rails along the length of a factory and provides three axes of hook motion (up and down, sideways, and back and forth).

Personnel lift—a lift that contains a pneumatic or electric arm with a personnel bucket attached at the end. Also referred to as a man lift or cherry picker.

Pipe wrench—an adjustable wrench that contains two serrated jaws that are designed to grip and turn pipes or other items with a rounded surface.

Pliers—a hand tool that contains two hinged arms and serrated jaws that are used for gripping, holding, or bending.

Pneumatic tool—a tool that is powered using pneumatic (air or gas) pressure.

Powder-actuated tool—a tool that uses a small explosive charge to drive fasteners into hard surfaces such as concrete, stone, and metal.

Power tool—a tool that is powered by electricity, pneumatics, hydraulics, or powder activation.

Reciprocating saw—a device that uses the back-and-forth motion of a saw blade to cut materials that range from wood to metal pipe.

Screwdriver—a device that engages with, and applies torque to, the head of a screw so the screw can be tightened or loosened.

Socket wrench—a wrench that contains interchangeable socket heads of varying sizes that can be attached to a ratcheting wrench handle or other driver.

Tool—a device designed to provide mechanical advantage and make a task easier.

Torque wrench—a manual wrench that uses a gauge to indicate the amount of torque (rotational force) being applied to the nut or bolt.

Valve wheel wrench—a hand tool used to provide mechanical advantage when opening and closing valves. These wrenches typically fit over the spoke of a valve wheel.

Wrench—a hand tool that uses gripping jaws to turn bolts, nuts, or other hard-to-turn items.

Introduction

Knowledge of hand and power tools is an integral part of the process technician's daily responsibilities. Knowing how to properly and safely use tools is critical. This chapter provides an overview of the types of tools most commonly used by process technicians, their applications, safety issues associated with the various tools, and proper tool care and maintenance.

Hand Tools

A **tool** is a device designed to provide mechanical advantage (the factor by which a device multiplies the applied force) and make a task easier. Process technicians use a variety of tools each and every day. The most common types of tools, however, are hand tools.

Hand tools are operated manually instead of being powered by electricity, pneumatics, hydraulics, or other forms of power. Many process technicians carry an assortment of hand tools with them as they make their daily rounds. Some of the most common hand tools include the following:

- Pliers
- Wrench
- Hammer
- Pry bar
- Bolt cutter
- Knife
- Level
- Square
- Tubing cutter
- Wire stripper

HAMMERS

A **hammer** is a tool with a handle and a heavy head that is used for pounding or delivering blows to an object. The most common applications of hammers include driving nails, fitting components together, or breaking objects apart. While hammer shapes and sizes can vary based on their applications, the most common features of a hammer are a long handle with a heavily weighted head. Figure 16-1 shows examples of some commonly used hammers.

PLIERS

Pliers are a hand tool that contains two hinged arms and serrated jaws that are used for gripping, holding, or bending. Pliers come in a variety of shapes and sizes.

FIGURE 16-1 Examples of Different Types of Hammers

Claw Hammer   Ball Peen Hammer   Rubber Mallet   Sledge Hammer

FIGURE 16-2 Examples of Different Types of Pliers

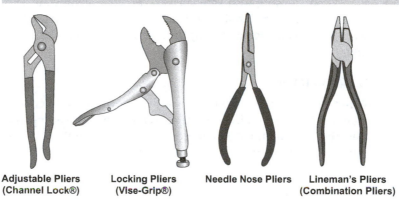

Adjustable Pliers   Locking Pliers   Needle Nose Pliers   Lineman's Pliers
(Channel Lock®)   (Vise-Grip®)   (Combination Pliers)

When using pliers, it is important to remember that the gripping surfaces are serrated. These serrated edges will damage finished surfaces, so they should not be used on the flat sides of packing nuts, fittings, bolts, or fasteners. Figure 16-2 shows examples of different types of pliers.

Adjustable pliers (sometimes referred to as Channel Locks® or slip joint pliers) are pliers that have a tongue and groove joint or slot that allows the jaws to widen and grip objects of different sizes. The jaw opening is typically adjusted by opening the pliers fully and selecting the desired groove or hinge pin position.

Locking pliers (sometimes referred to as Vise-Grips®) are pliers that can be locked into place when the handgrips are squeezed together. Locking pliers contain an adjustment screw that allows the distance between the jaws to be adjusted. Because of their locking mechanism, locking pliers provide a firm grip and prevent slippage.

Needle nose pliers are small pliers with long, thin jaws for fine work, for example, holding small items in place, removing cotter pins, or bending wire.

Lineman's pliers are snub-nosed pliers with two flat gripping surfaces, two opposing cutting edges, and insulated handles that help reduce the risk of electrical shock. Because of their versatility, lineman's pliers are most often used for cutting and bending wire, stripping wire insulation or cable jackets, and gripping pieces of wire so they can be twisted together.

SCREWDRIVERS

Screwdrivers are devices that engage with, and apply torque to, the head of a screw so the screw can be tightened or loosened. Screwdrivers, which come in a variety of shapes and sizes, consist of a handle and a shank. The end of a screwdriver shank (the head) is shaped according to the screwdriver type. For example, a Phillips screwdriver contains an end that looks like a cross; a regular screwdriver contains a flattened head. The most common types of screwdrivers used in the process industries are regular (also referred to as a flat head) and a Phillips head. Figure 16-3 shows examples of Phillips head and regular screwdrivers.

FIGURE 16-3 Examples of Different Types of Screwdrivers

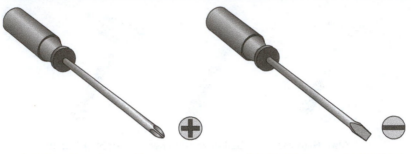

Phillips Head Screwdriver   **Regular (Flat head) Screwdriver**

When working with screwdrivers, process technicians should always ensure that the head is the proper size and shape for the screw head. Using an improperly sized screwdriver can damage (strip) the screw head or allow the screwdriver to slip, causing injury.

FLARING TOOL

A **flaring tool** (shown in Figure 16-4) is a device used to create a cone-shaped enlargement at the end of a piece of tubing so the tubing can accept a flare fitting. As a general rule, the maximum flare diameter can be no more than 1.4 times the diameter of the tube or the tubing material will begin to split.

FIGURE 16-4 Flaring Tool

WRENCHES

Wrenches are hand tools that use gripping jaws to turn bolts, nuts, or other hard-to-turn items. Wrenches come in many different shapes and sizes. Some examples of common wrenches include adjustable, impact, pipe, socket, torque, and valve wheel. Because the gripping surfaces on most wrenches are smooth, they can be used on finished surfaces and the flat sides of fittings, bolts, and fasteners. Figure 16-5 shows examples of different types of wrenches.

Adjustable Wrenches

Adjustable wrenches are also referred to as Crescent® wrenches. They are open-ended wrenches with an adjustable jaw that contains no teeth or ridges (i.e., the jaw is smooth).

Socket Wrenches

Socket wrenches are wrenches that contain interchangeable socket heads of varying sizes that can be attached to a ratcheting wrench handle or other driver. Most socket wrenches contain a ratcheting mechanism that allows a nut or bolt to be tightened or loosened in a continuous motion, as opposed to the socket or handle being removed and refitted after each turn.

FIGURE 16-5 Examples of Different Types of Wrenches

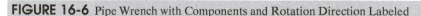

Adjustable Wrench   Impact Wrench   Non-adjustable Wrench
(Cresent® Wrench)

Socket Wrench and Socket   Torque Wrench   Valve Wrench

Pipe Wrenches

Pipe wrenches are adjustable wrenches that contain two serrated jaws designed to grip and turn pipes or other items with a rounded surface. In this type of wrench, pressure on the handle causes the jaws to move together more tightly. Pipe wrenches come in many different sizes and shapes. However, they all have the same purpose. Figure 16-6 shows an example of a pipe wrench with the components and direction of rotation labeled.

When using a pipe wrench, it is important to maintain space between the object being turned and the back of the jaws. It is also important to keep the pipe wrench in a suitable position to prevent injury or hand slippage.

FIGURE 16-6 Pipe Wrench with Components and Rotation Direction Labeled

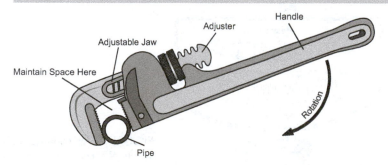

Handle

Adjuster

Adjustable Jaw

Maintain Space Here

Rotation

Pipe

Impact Wrenches

An **impact wrench** is a pneumatically or electrically powered wrench that uses repeated blows from small internal hammers that generate torque to tighten or loosen fasteners. Process technicians use impact wrenches to remove bolts from flanges or other pieces of equipment. Figure 16-7 shows an example of a pneumatic impact wrench.

Like socket wrenches, impact wrenches can accommodate removable sockets of many different shapes and sizes. Because they use repetitive blows instead of a single

FIGURE 16-7 Pneumatic Impact Wrench

stroke of brute force, impact wrenches are often more effective than traditional wrenches at removing bolts that are extremely tight.

Torque Wrenches

Torque wrenches are manual wrenches that use a gauge to indicate the amount of torque (rotational force) being applied to the nut or bolt. Torque wrenches are used when precise tightening is crucial. Figure 16-8 shows an example of a torque wrench scale.

FIGURE 16-8 Close Up of Torque Wrench Scale

Valve Wheel Wrenches

Valve wheel wrenches are tools used to provide mechanical advantage when opening and closing valves. Valve wheel wrenches usually fit over a spoke in the valve wheel to minimize the likelihood that the wrench will slip off the wheel when force is applied. Care must be taken when using this type of wrench to avoid applying so much force that it breaks the valve wheel or valve stem, or causes the wrench to slip, which can result in injury.

FIGURE 16-9 Examples of Power Tools

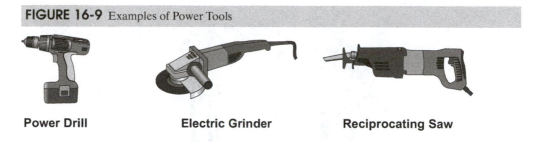

Power Drill   **Electric Grinder**   **Reciprocating Saw**

Power Tools

Power tools are tools powered by electricity, pneumatics, hydraulics, or powder charges (small charges filled with explosive materials such as gunpowder). Examples of common power tools (shown in Figure 16-9) include drills, grinders, and saws.

A **drill** is a device that contains a rotating bit designed to bore holes into various types of materials.

A **grinder** is a device that uses the rotating force of one material against another material (e.g., a stone grinding wheel against a metal blade) to reduce or adjust the shape and size of an object.

A **reciprocating saw** is a device that uses the back-and-forth motion of a saw blade to cut materials that range from wood to metal pipe.

Lifting Equipment

Many different kinds of lifting equipment are used in the process industries. These include hoists, cranes, forklifts, personnel lifts, and dollies.

HOISTS

A **hoist** is a device composed of a pulley or gear system with a cable or chain used to lift and move heavy objects. Basic safety requirements for hoists include the following:

* Operation must be within specified weight limits. This information is usually stamped or stenciled on the lift.
* Hoist systems should not be operated if the weight limit for the hoist is unknown or the weight of the object to be lifted is not known or accurately estimated by experienced personnel.
* Chains and ropes should be free from kinks.
* Prevent individuals from working beneath equipment suspended by a hoist.
* Chains should be inspected for bent or broken links and signs of corrosion.
* Ropes should be inspected for fraying and broken threads, and steel ropes should be inspected for corrosion.
* Loads should be properly secured and balanced before hoisting.

Figure 16-10 shows an example of manual lever and hand-chain (chain fall) operated hoists.

CRANES

A **crane** is a mechanical lifting device equipped with a winder, wire ropes, and sheaves that can be used to lift and lower materials. There are four main types of overhead crane systems: overhead traveling bridge, gantry, jib, and monorail.

Overhead traveling bridge cranes are cranes that run on an elevated rail system along the length of a warehouse and provide three axes of hook motion (up and down, sideways, and back and forth). Both single- and double-girder bridge designs are provided with great flexibility in allowing the hook to be positioned very precisely and for loads to be placed very gently. The bridge (which carries the hoist and trolley) is

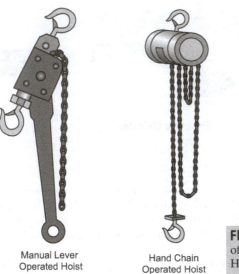

Manual Lever
Operated Hoist

Hand Chain
Operated Hoist

FIGURE 16-10 Examples of Manual Lever and Hand-Chain Hoists

supported by a pair of rails called tracks that, in turn, carry a pair of wheels. Rails can be installed at both the floor or ceiling level. Figure 16-11 shows an example of an overhead traveling bridge crane.

Gantry cranes (which are similar to overhead cranes) run on elevated rails attached to a frame or set of legs. Gantry cranes provide the same performance characteristics as overhead bridge cranes. Figure 16-12 shows an example of a gantry crane.

Jib cranes are cranes that contain a vertical rotating member and an arm that extends to carry the hoist trolley. Jib cranes consist of a pivoting head and boom assembly that carries a hoist and trolley unit. The pivoting head is supported either by a floor-mounted mast providing 360-degree boom rotation, or by an existing building column that provides 180 degrees of boom rotation. Figure 16-13 shows an example of a jib crane.

Monorail cranes are cranes that travel on a single runway beam. These types of cranes are quite specialized and very effective when properly incorporated into the factory layout. Only two directions of hook travel are afforded by the monorail: up and down, and along the axis of the monorail beam. It is not recommended (and quite dangerous) to push the load out from under the centerline of the monorail beam. Monorail systems are most often integrated into continuous production systems for material transport (e.g., hot metal operations in foundries, and material transport in paint booths). Figure 16-14 shows an example of a monorail crane.

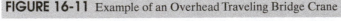

FIGURE 16-11 Example of an Overhead Traveling Bridge Crane

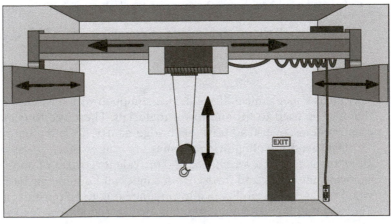

FIGURE 16-12 Example of a Gantry Crane

FIGURE 16-13 Example of a Jib Crane

FIGURE 16-14 Example of a Monorail Crane

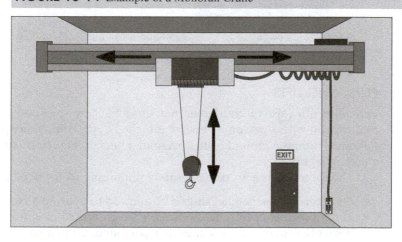

The following are basic safety requirements for all cranes:

- Only personnel who have been trained and certified should be permitted to operate a crane.
- Proper load lift angle and balance must be maintained.
- Crane load limits should always be known and respected.

- Operational weight limits should not be exceeded.
- Chains and ropes should be free from kinks.
- Chains should be inspected for bent or broken links and signs of corrosion.
- Ropes should be inspected for fraying and broken threads, and steel ropes should be inspected for corrosion.
- Loads should be properly secured and balanced before hoisting.
- Correct parts should always be used when repairing a crane.

FORKLIFTS

A **forklift** is a vehicle with prongs used for lifting and transporting pallets of product. Forklifts are most often used to lift items placed on a pallet to higher elevations for use in a unit or for storage. According to the Occupational Safety and Health Administration (OSHA), process technicians can operate a forklift only if they have been properly training and certified. Figure 16-15 shows an example of a forklift commonly used in the process industries.

The following is a list of basic safety requirements for forklifts:

- Respect load limits.
- Only trained and certified personnel should operate a forklift.
- Do not overextend, raise, tilt, or misalign the forks (this could cause load imbalance or slippage).
- Do not operate a forklift with leaky hydraulics.
- Use caution when operating a forklift on a ramp or loading dock edge to avoid falling off.
- The forklift tires and the surface where the forklift will be used must match use requirements, and these usage rules should be obeyed (e.g., many forklifts have small tires and can operate effectively only on smooth, solid surfaces).
- Operate fuel-powered forklifts only in properly ventilated spaces.

FIGURE 16-15 Example of a Forklift Commonly Used in the Process Industries

PERSONNEL LIFTS

Personnel lifts (also referred to as man lifts and cherry pickers) contain a pneumatic or electric arm with a personnel bucket attached at the end. Many personnel lifts may be operated from the ground or the personnel bucket. Figure 16-16 shows an example of a personnel lift.

The following is a list of basic safety requirements for personnel lifts:

- Only certified personnel should be allowed to operate a personnel lift.
- Have a second or third individual on the ground to monitor clearances and to warn other personnel when using a personnel lift in a busy or congested area.
- Avoid power lines and overhead obstructions.
- Follow proper safety procedures and wear proper personal protective equipment (PPE) (e.g., a secured harness) at all times.
- Secure tools so they do not drop onto personnel working below.
- Barricade the area when using a personnel lift.
- Properly position outrigger stabilizer feet (if available) onto a solid surface.

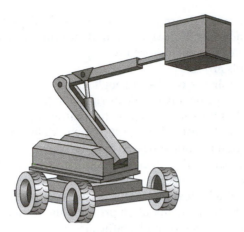

FIGURE 16-16 Example of a Personnel Lift Used in the Process Industries

DOLLIES

A **dolly** is a wheeled cart or hand truck that is used to transport heavy items such as barrels, drums, and boxes. Figure 16-17 shows examples of different types of dollies.

The following is a list of basic safety requirements for the operation of dollies:

- Operate dollies by pushing rather than pulling.
- Operate under the recommended weight requirements.
- Secure the load before moving.
- Make sure your path is clear and well lighted.
- Use slip-resistant footwear during operation.
- Maintain a safe angle to prevent tipping.

FIGURE 16-17 Examples of Dollies

Hand Truck   **Barrel Dolly**

Basic Hand and Power Tool Safety

OSHA states that employers shall be responsible for the safe condition of tools and equipment used by employees, but it is up to the employee to ensure that tools are used properly and safely.

GENERAL SAFETY TIPS

The following are some general tool usage safety tips:

- Never horseplay in the workplace or when working with tools.
- Always obtain required permits for dangerous or hazardous work (e.g., a hot work permit for grinding operations).
- Use the right tool for the job.
- Understand how to properly use and maintain the tools you are working with.

- Review safety procedures to learn the hazards associated with specific tools.
- Select the proper handgrip: left or right handed, or both (ambidextrous).
- Inspect the condition of the tool before use.
- Check that edged tools are sharp.
- Inspect metal tools for slivers, cracks, or rough spots.
- Check for wear or damage on any plastic or rubber parts or coatings.
- Inspect wooden handles for splinters, chips, or weathering.
- Do not use damaged tools. Instead, make sure they are labeled "Do Not Use" and report the tool condition to your supervisor.
- Wear the appropriate personal protective equipment (e.g., always wear eye protection when using hand tools.)
- Make sure the work area is clear and obstacles are removed.
- Keep a firm grip on the tool and maintain proper balance and footing.
- Do not leave tools in walkways or high traffic areas.
- Make sure the floor is clean and dry to prevent slips or falls when you are working with or around tools.
- Make sure the work area is properly lighted.
- Maintain a safe distance from other workers when using tools.
- Do not distract others while they are using a tool.
- Clean the tool when done and return it to the proper storage location.
- Be aware of other work that may affect the usage of the tool in that area.

GENERAL HAZARDS

General hazards that tools can cause include the following:

- Sparks (especially from grinding tools) in flammable or hazardous environments, which can result in a fire or explosion.
- Impact from flying parts or fragments (from the tool or material being worked on).
- Cuts, scrapes, bruises, broken bones, or similar injuries.
- Harmful dusts, fumes, mists, vapors, or gases released by the material being worked on.
- Impalement on sharp edges.
- Injury from a tool that is dropped from a height.
- Ergonomic hazards such as carpal tunnel syndrome or tendonitis from improper design or use, or repetitive motion.
- Noise hazards, such as a hand tool striking a surface or loud noises generated by power tools (e.g., a jackhammer).
- Falls while trying to use the tool at heights.
- Forceful ejections of projectiles (e.g., nails from a nail gun or pressurized water from a water blaster).

HAND TOOLS

Hand tools are tools that are operated manually instead of being powered by electricity, pneumatics, hydraulics, or other forms of power. Hand tools come in a wide range of types and designs.

Misuse and improper maintenance of hand tools pose the greatest hazards. The following are some general tips for proper use and safety when using hand tools:

- Check wooden-handled tools for splinters, cracks, chips, or weathering. Make sure the handle is securely attached to the hammer head.
- Never modify a tool to do a task for which it was not designed.
- Do not use screwdrivers as chisels or gouges.
- Check impact tools (e.g., chisels and wedges) for heads that have been flattened with repeated use (called mushroom heads) because these can shatter.
- Make sure the jaws of wrenches and pliers are not sprung to the point that slippage occurs.

- Use spark-resistant tools around flammable substances (e.g., tools made from nonsparking materials such as brass, plastic, aluminum, wood, titanium, bronze, and Monel™).
- Inspect spark-resistant tools for wear or damage because they are made of materials that are softer than other tools and can wear down quickly.
- Maintain a proper grip, holding the handle firmly across the fleshy part of your hand.
- Do not overexert yourself. If you feel your hand or arm strength weakening, take a break and resume the task later.
- In moist environments, make sure your hands remain dry and your vision is not blurred by sweat.
- Do not overtighten fasteners.

Wrenches

General safety suggestions for wrenches include the following:

- Valve wheel wrenches should fit over the valve wheel rim completely. Make sure that the wheel rim is in contact with the top of the wrench opening when applying force.
- Brace yourself firmly and pull rather than push. If it is necessary to push the wrench, do so with the flat of the hand rather than gripping around the wrench.
- Keep the wrench in the plane of the valve wheel rim when pulling because pulling at an angle increases the chance of slippage.
- Never apply excessive leverage to a wrench by means of a pipe extension or cheater bar.
- Never strike wrenches with hammers or other objects.
- Do not use a pipe wrench as a valve wheel wrench. Wrench teeth create sharp edges and burrs on a valve wheel that can cut an unprotected hand.

POWER TOOLS

There are various types of power tools; however, the main types are electric, pneumatic, fuel-operated, hydraulic, or powder-actuated. While each of these tools can enhance performance, they can also pose a variety of hazards. These hazards are listed below:

- Power-source related hazards such as electrocution (electrical) or line whip (pneumatic)
- High-speed impact from projectiles hurled by broken tools or materials
- Injuries from contact with moving parts or entanglement
- Fall or trip hazards from cords or hoses
- Vibration and noise hazards
- Slipping hazard from leaking fluids
- Hot surfaces caused by activity or prolonged operation (e.g., drill bits and motor housings).

In addition to the general safety procedures outlined previously, the following are some safety procedures for power tools:

- Follow all manufacturers' specifications for proper and safe use, and read all warning labels.
- Understand the capabilities, limitations, and hazards of the tool being used.
- Check that all safety features work properly.
- Never point the tool toward someone.
- Check that all guards and shields are in place.
- Do not use tools that are too heavy or too difficult to control.
- Inspect all cords or hoses for wear, fraying, or damage.

- Make sure the tool is kept clean and properly maintained, including lubrication, filter changes, and other manufacturer-recommended practices.
- Avoid using tools in dangerous environments (e.g., wet, flammable, extreme heat or cold).
- Use only intrinsically safe or explosion-proof tools in flammable environments.
- Do not carry the tool by its cord or hose.
- Do not yank the tool cord or hose to disconnect it.
- Keep cords and hoses away from heat, oil, and sharp edges.
- Do not let cords or hoses get knotted or tangled with other cords or hoses.
- Make sure cords or hoses are not lying across a walkway, thus presenting a trip hazard.
- Keep cords or hoses away from rotating or moving parts.
- Choose the correct accessories and use them properly.
- Check all handles to make sure they are secure and all grips are on tight.
- Keep your finger off the switch when carrying the tool or before it is in position to avoid accidental startup.
- Hold or brace the tool securely.
- Disconnect the tool before servicing it or when changing accessories (e.g., blades and bits).

Electric Tools

Electric tools are tools operated by electrical means (either AC or DC current). Some safety practices related to electric tools are listed below:

- Make sure all electric tools are properly grounded or double insulated.
- Always use a ground fault circuit interrupter (GFCI) when using electrical power tools.
- Know the hazards of electricity, including electrocution, shocks, and burns.
- Be aware of potential secondary hazards from electricity (e.g., a mild shock from an electric tool that startles a worker causing, him or her to fall off a ladder).
- Never remove the grounding plug from a cord.
- Understand the mechanical hazards of the tool.
- Remember to avoid wet or damp environments.
- Do not use a tool if it becomes wet.

Pneumatic Tools

Pneumatic tools are tools powered by pneumatic (air or gas) pressure. Safety practices related to pneumatic tools include the following:

- Do not use high pressure compressed air for cleaning purposes. Compressed air can blow dust or metal shavings into a process technician's eyes.
- Make sure attachments are properly secured using a tool retainer or safety clip.
- Check the hose and hose connection. Use only approved connectors.
- Do not exceed the manufacturer's safe operating pressure for hoses, pipes, valves, filters, and fittings.
- For hoses with a half inside diameter, a safety device must be used at the source of supply or branch line to reduce pressure in the event of hose failure to prevent line whip.
- Never kink the hose to cut off the air supply. Instead, turn off the air using the valve.
- Check for proper pressure before using the tool.
- Hose connections and overhead tools should be secured with a lanyard.

Hydraulic Tools

Hydraulic tools are tools powered by hydraulic (liquid) pressure. Many of the safety precautions for hydraulic tools are the same as those for pneumatic tools. However, one added hazard is exposure to hot and/or harmful hydraulic fluids.

Fuel-Operated Tools

Fuel-operated tools are tools powered by the combustion of a fuel (e.g., gasoline). Process technicians should always use caution and ensure that the device is properly grounded when filling fuel-operated tools because filling and tool operations produce static electricity, which can result in a fire or explosion. Tools should also be filled and operated in open air environments to avoid the buildup of harmful fuel vapors or lethal equipment exhaust.

Powder-Actuated Tools

Powder-actuated tools are tools that use a small explosive charge to drive fasteners into hard surfaces such as concrete, stone, or metal. Figure 16-18 shows an example of a powder-actuated tool designed to be struck by a hammer.

Because powder-actuated tools use powder charges that are explosive, they should always be handled with caution. Most of these types of tools will not activate unless the tip is pressed against a work surface. However, accidents do happen. Technicians should place only the tip of a powder-actuated tool toward the intended work surface, and never point it at another person.

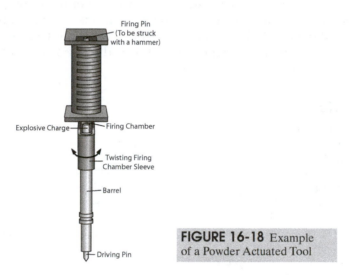

FIGURE 16-18 Example of a Powder Actuated Tool

Tool Care and Maintenance

Like pumps, compressors, or any other type of equipment, hand tools must be properly maintained for safe operation. Hand tools should be stored inside a properly sealed container or toolbox to prevent exposure to moisture and corrosion.

With power tools, routine checks and lubrication of various mechanical moving parts should always be performed before and after use. Electric tools should be inspected for frayed cords and bent or loose plugs.

With all hand tools, the handles should fit tightly. If the handle is cracked or otherwise damaged, replace or repair the tool before using. As a general rule, always inspect your tools for defects before using them.

These housekeeping chores may seem tedious and unnecessary at the time, but taking a few simple precautions can mean the difference between successful completion of a project and serious or fatal injury. Keep your equipment in top working condition to make your job both easier and safer.

Summary

There are many tool and construction hazards in the process industries. While process technicians are not likely to be involved in construction-related tasks, they may encounter construction areas around their facility. Thus, technicians must be aware of the safety hazards that construction areas and tools present.

Hand and power tools used on a regular basis include hammers, pliers, and drills. Wrenches are also important to the role of the process technician. Wrenches have jaws that are used to make adjustments and to tighten and remove fasteners as needed. There are various types of wrenches, including adjustable, impact, pipe, socket, torque, strap, and valve wheel.

Lifting equipment is used in industry to assist the process technician in lifting heavy objects to elevated positions. One type of lifting equipment that is used in the process industry is a hoist. A hoist is a fixed apparatus composed of a pulley system with a cable or chain. Cranes are used with ropes or chains and are connected to a boom or an arm. A forklift is a vehicle with pronged lifts to raise and lower items for use, maintenance, and storage.

Personnel lifts contain a pneumatic or electric arm with a bucket attached to the end to lift personnel to elevated locations as needed. A dolly is a cart or hand truck with casters used to transport barrels, drums, and boxes.

Proper care, use, and maintenance of tools are integral for a safe work environment. Before working with tools, process technicians should always select the proper tool for the job, and then inspect the tool for wear, damage, or other defects because these can make a tool unsafe. Never modify a tool or use it for a task that it was not designed to accommodate. In addition, technicians should always use proper techniques and safety practices when using tools. This includes wearing proper PPE, clearing obstacles or hazards from the work site, and ensuring that other individuals are a safe distance away before work begins.

Checking Your Knowledge

1. Define the following terms:
   a. Tool
   b. Dolly
   c. Crane
   d. Electric tool
   e. Hand tool
   f. Power tool
   g. Hoist
   h. Lift
   i. Hydraulic tool
   j. Pneumatic tool
   k. Powder-actuated tool

2. It is the _____ responsibility to ensure that tools are used properly.
   a. process technician's
   b. toolroom employee's
   c. supervisor's
   d. company's

3. Components of _____ lock into place when squeezed together.
   a. a flaring tool
   b. a power drill
   c. locking pliers
   d. a nonadjustable wrench

4. A _____ wrench is a device used to apply a precise amount of force to a fastening device such as a nut or bolt.
   a. pipe
   b. socket
   c. torque
   d. valve

5. A(n) _____ crane runs on an elevated runway system along the length of a warehouse and provides three axes of hook motion (up and down, sideways, and back and forth).
   a. overhead traveling bridge
   b. gantry
   c. jib
   d. monorail

6. *(True and False)* Process technicians do not have to be certified to operate a forklift.

7. Which of the following are good safety practices for tools (select all that apply)?
   a. Inspect the tool before use.
   b. It is permissible to distract someone while he or she is using a tool.
   c. Return the tool to its proper place when you are done.
   d. Don't worry about obstacles in the work area before using a tool.
   e. Keep a firm grip on the tool.

8. List five safety practices related to electric tools.
9. List four basic requirements for proper use of a hoist.
10. List five safety practices related to the use of a forklift.

Student Activities

1. Based on your personal experience, discuss some hazards you encountered working with tools, along with any injuries you or others suffered, at work or at home. How could these hazards have been prevented? Discuss this with your fellow students.

2. Research tool safety. Write a two-page paper for submission and prepare a presentation, focusing on a specific tool, its usage and maintenance, safe operation, and potential hazards.

3. Given a set of assorted tools, identify each tool type and describe or demonstrate proper and improper usage with a classmate. Be prepared to present this information to the class.

4. Consider the use of wrenches in the process industries. Work in a team and discuss the various types of wrenches, how they are used, and the safety-related effects of proper and improper usage.

5. Using sewing thread and weights, demonstrate the physics of properly and improperly loaded cables.

6. Demonstrate the proper procedures for tightening a bolt or a flange, and adjusting a pump or valve packing.

7. Research and present OSHA regulations for forklift, crane, and/or power tool use.

17

Separation Equipment

Objectives

After completing this chapter, you will be able to:

■ Explain the purpose of separation equipment in the process industries.

■ Identify the common types and applications of separation equipment and processes.

■ Identify the components of separation equipment and explain the purpose of each.

■ Explain the operating principles of separation equipment.

■ Identify typical procedures associated with separation equipment.

■ Describe the process technician's role in separation equipment operation and maintenance.

■ Identify potential problems associated with separation equipment.

Key Terms

Absorption—the process by which a gas or liquid is dissolved into a liquid or solid.

Adsorption—the adhesion of a gas or liquid to the surface area of a solid.

Azeotrope—a mixture of two or more substances that boil at a constant temperature and a fixed composition.

Azeotropic distillation—the process of adding a third substance (called an entrainer) to separate a mixture of two or more substances with close boiling points (e.g., adding benzene to a water/ethanol azeotrope).

Binary distillation—the process of separating two materials into an overhead product and a base product.

Bubble cap tray—a tray in a distillation column that uses slotted caps with fixed openings to disperse the vapor into the liquid.

Crystallization—the process of creating crystals from a supersaturated solution.

Distillation—the process of separating two or more liquids by their boiling points.

Extraction—a separation process that uses a solvent or other medium to separate components based on differences in solubility (the degree to which a solid dissolves in a solvent).

Extractive distillation—a method used for separating components with close boiling points by the addition of a solvent that alters the volatility of the components enough so they can be separated by normal distillation.

Flash distillation—the partial evaporation or vaporization that occurs when a saturated liquid stream undergoes a reduction in pressure.

Liquid-liquid extraction—a process that occurs when a feed solution is combined with a solvent that has an affinity for (attracts) one or more components of the feed.

Mother liquor—the portion of a solution remaining after crystallization.

Multicomponent distillation—the process of separating more than two liquids by the boiling point.

Nucleation—the growth of a new crystal from feed crystals.

Reflux—the overhead vapor (light components) that have been condensed to liquid and returned to the top of the tower to help cool and purify the overhead process or product.

Scrubber—a device used to remove contaminants from a gas stream by washing the stream with water or some other neutralizing agent.

Sieve tray—a tray in a distillation column that has equally spaced holes or openings.

Solubility—the degree to which a solid will dissolve.

Stripping—a process that uses a vapor stream to separate one or more components from a liquid stream.

Valve tray—a tray in a distillation column that has moveable devices over the tray openings.

Introduction

Distillation, which uses heat to separate substances by boiling point, is the most widely used process for separating liquids into their individual components. Through distillation, we can separate substances and create products that could not be produced through other separation methods. Depending on the substances to be produced, the distillation process can occur simply with heat, or it may require the addition of other substances such as solvents or pressure differentials.

Because of the considerable amount of monitoring that must occur during distillation, the process technician's role in distillation operations is critical. In-depth training is necessary to ensure knowledge and understanding of these processes.

Simple Separators

A separator is a device used to physically separate two or more components from a mixture. While separators have many different applications in the process industries (e.g., food and beverage manufacturing, oil and gas refining, pharmaceutical manufacturing, and water treatment), the basic operating principles are the same.

To illustrate how a separator works, consider the simple oil and water separator shown in Figure 17-1. In this type of separator, a large drum, a set of weirs, and different product densities are used to separate or stratify the components of an oil-water mixture.

FIGURE 17-1 Oil and Water Separator Used in Petroleum Production

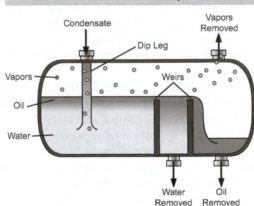

In the initial phase of this separation, liquid is discharged into the first section of a separator. As the liquid level rises, it eventually reaches a point that is slightly higher than the weir. When this occurs, the liquid begins to flow over the weir and into a second compartment. In the second compartment, the varying densities of the two substances cause them to separate (light on top, heavy on the bottom). The heavier substance (water) is then removed through the bottom of the drum and sent for further treatment, while the lighter substance (oil) is transferred to a third compartment. Once inside the third compartment, the oil is then removed and sent to storage.

If a third, gaseous substance had been included in the mixture, it would have floated to the top of the separator and been removed as a gas.

When working with simple separators, it is important to know that they are used only for bulk separation. Gravity may not force the smaller droplets of water to settle out of the oil stream, so there may be a small amount of water remaining in the oil. Because of this, separated oil is often routed to a coalescer, a device used to separate emulsions (mixtures of two or more liquids that don't easily combine) to further reduce the water content.

Distillation

One of the most popular separation methods used in the process industries today is distillation. Instead of using differing densities to separate materials, **distillation** uses heat to separate liquids by boiling point.

The distillation process has many uses. For example, it is used to remove particulates from water during water treatment, produce alcoholic beverages, separate gasoline from other crude oil products, and to purify chemicals and pharmaceutical products.

The process of distillation typically occurs in a device called a distillation column or tower, and it can be either a batch or continuous process. Distillation columns can range in diameter from 2 feet (65 centimeters) to 20 feet (6 meters) and can range in

Did You Know?

Many red wines contain sediment that must be removed before the wine is consumed. In order to remove this sediment, many wine connoisseurs use a decanter (a type of simple separator).

Once the wine has settled, the narrow neck of the decanter allows clarified wine to be poured out. The bowl-shaped base traps undesirable sediment and prevents it from leaving the decanter.

height from 20 feet (6 meters) to 200 feet (60 meters) or taller. They can also operate at a variety of pressures, ranging from less than atmospheric pressure (vacuum) to extremely high pressures, depending on the composition of the material being separated.

During the distillation process, feed comes into the process unit and is heated until it reaches its boiling point. Once boiling, the various components of the mixture begin to vaporize into fractions, a portion of the distillate that has a particular boiling point that is different from the boiling point of the other fractions in the column.

The effectiveness of a distillation tower depends on the type of trays, number of trays, and the spacing between the trays. Other factors that affect distillation include the height and diameter of the column.

TOWER SECTIONS

Distillation towers contain three sections: a flash zone, a rectifying (fractionating) section, and a stripping section. Figure 17-2 shows an example of each of these sections.

The flash zone is the entry point or feed tray for the process fluid. The portion of the tower below the feed point is the stripping section. The portion above the feed point is the rectifying section. The stripping section is where the bottom products (substances that boil at the highest temperature), or heavy ends, are located. Because they are the last products to volatize, they are located at the base of the column.

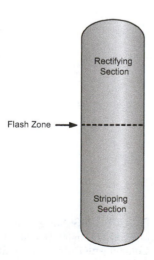

FIGURE 17-2 Rectifying and Stripping Sections in a Distillation Column

The rectifying (fractionating) section is where the overhead product is purified. Overhead products (also referred to as light ends) are the materials collected at the top of the distillation column. Light ends have the lowest boiling point of all the substances in the column.

During distillation, heat at the base of the column produces hot vapors that pass up the column into the rectifying section. As the hot vapors move up, cooler liquid flows down the column into the stripping section. Contact between the vapor and the liquid occurs on each tray or level as the lower boiling compounds convert to the vapor phase and the higher boiling compounds maintain the liquid phase.

TRAYS

To facilitate the distillation process, distillation columns are equipped with either trays or packing. These devices provide a contact point between the liquid and the vapor, and they help facilitate separation.

In a tray-type column, the liquid is introduced into the column through a feed tray located immediately below the feed line. The liquid then moves across the feed tray and down the tower, while vapors move up the tower through openings in the tray. Weirs (flat or notched dams or barriers) are used to help maintain a constant liquid level (also known as holdup) on the trays. The liquid is then routed from tray to tray through downcomers. Figure 17-3 shows the movement of fluid through a distillation column.

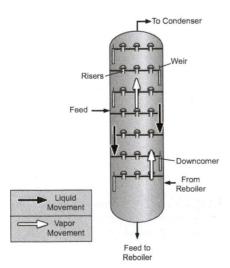

FIGURE 17-3 Movement of Fluid Through a Distillation Column with Bubble Cap Trays

FIGURE 17-4 Sieve Tray

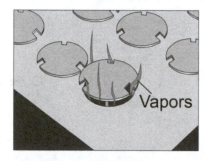

Vapors

FIGURE 17-5 Valve Tray

Slots allowing vapors to be released

FIGURE 17-6 Bubble Cap Tray

Distillation trays come in many different designs. The three most common designs, however, are sieve, valve, and bubble cap. Figures 17-4 through 17-6 show examples of these different types of trays.

Sieve trays contain small holes that are equally spaced over the surface of the tray deck. As the fluid passes over these trays, it flows down through these holes at the same time vapors flow up through the holes (this is known as countercurrent flow).

Valve trays contain valve lifters similar to those in a car engine. These lifters move up and down with the pressure of the vapor moving upward (i.e., the valve rises as the vapor rate increases).

Bubble cap trays contain risers with slotted caps on top of the risers. These fixed openings cause the vapor to move in a 180-degree pattern through the liquid on the tray.

While tray and packed towers both do an effective job of separating products, tray towers tend to be used for larger-scale operations, like separating gasoline from crude oil or refining and purifying chemical products.

PACKING

Like tray columns, packed columns use the same principles of fluid movement. However, instead of trays, packed columns use pieces of packing material to maximize contact area. Packing comes in many different shapes, sizes, and materials (e.g., plastic, metal, or ceramic), and can be installed randomly or in an organized pattern. The three most common types of packing used in distillation columns are raschig rings, berl saddles, and pall rings. Figures 17-7 and 17-8 show examples of different types of packing.

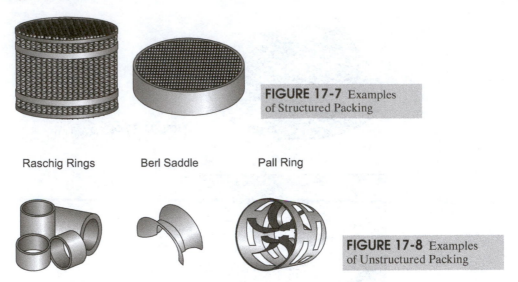

FIGURE 17-7 Examples of Structured Packing

Raschig Rings   Berl Saddle   Pall Ring

FIGURE 17-8 Examples of Unstructured Packing

In randomly packed columns, the packing material is basically poured into the column at random and is held in place with a screen. With structured packing, the packing material is installed in a specific, organized, and structured manner. Packed towers can be used for many tasks, including absorption of gases in liquid, stripping, evaporating liquids, condensing liquids, and drying operations.

HOW THE DISTILLATION PROCESS WORKS

Distillation is performed in large, vertical, cylindrical columns referred to as distillation towers, distillation columns, or fractionators. Distillation columns come in a variety of shapes and sizes with packing or trays. The most common components in a distillation column are the tower, packing or trays, a condenser, and a reboiler. Figure 17-9 shows examples of these components.

The tower is where the vaporization (fractionation) of the liquid occurs. In some processes, the temperature of the liquid coming into the tower is too low for vaporization to occur, so it must pass first through a preheater. A preheater is a heat exchanger that warms the liquid before it enters the tower. This preheating process helps the

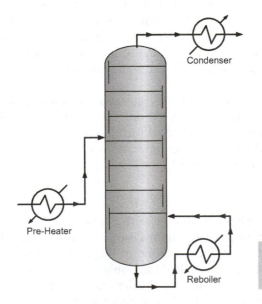

Condenser

Pre-Heater

Reboiler

FIGURE 17-9 Tray-Type Distillation Column with Associated Equipment

tower maintain a more constant temperature and reduces the amount of time required for the substance to reach its boiling point.

Inside the column, heat causes the lighter components (the low boilers) to vaporize and flow up the column to the overhead stream, while the heavier components (the high boilers) remain near the bottom of the column. Once the lighter vapors have reached the top of the column, they are cooled to their liquid phase and leave the column so they can be transferred to another part of the process, sent to a storage facility, or used as **reflux** (light, overhead vapors, condensed to a liquid form, that are used to cool and purify the overhead process or product).

A reboiler (a tubular heat exchanger placed at the bottom of a distillation column) is used to apply additional heat. This additional heat helps the column maintain a proper temperature and causes the substances at the bottom to vaporize. Once vaporized, these fluids then move to the top of the column, where some of the vapors are routed to a condenser (a heat exchanger that is used to condense a vapor to a liquid).

Figure 17-10 shows a diagram of a distillation system with reflux being returned to the column. Figure 17-11 displays a distillation column cross section and illustrates the movement of gas upward and liquid downward inside the column.

To prevent the entrainment of liquid droplets in the overhead vapors, mist eliminators or (demisters) are used in columns with high gas velocity rates. These devices are constructed of mesh pad or vane type (chevron) plate assemblies. They entrap

FIGURE 17-10 Distillation System with Reflux Being Returned to the Column

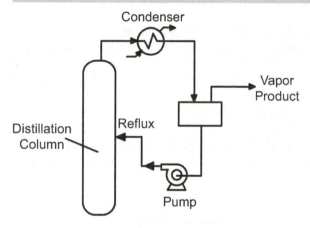

FIGURE 17-11 Distillation Column Cross Section Showing Valve Tray Flows

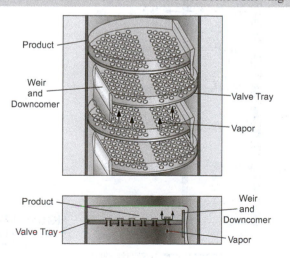

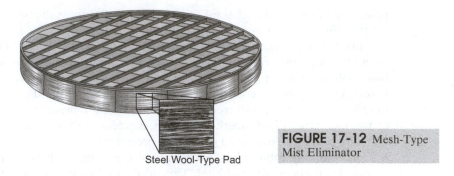

Steel Wool-Type Pad

FIGURE 17-12 Mesh-Type Mist Eliminator

liquid by forcing the vapor to change direction several times, thereby allowing the denser liquid droplets to coalesce and fall out of the vapor flow. Figure 17-12 shows an example of a mesh-type mist eliminator.

Pressure Conditions

As we mentioned earlier, distillation columns can operate at a variety of pressures, ranging from low to high. The pressure that is selected for a particular application is determined by the composition of the material being separated, the boiling range, and the degree of separation needed. For example, materials that are harder to separate require more heat and higher-than-atmospheric pressure settings. Materials that separate readily at lower pressures may decompose at high pressure and temperature settings.

Binary and Multicomponent Distillation

Binary distillation is the separation of two substances. During binary distillation, the lighter (lower boiling) component rises to the top of the column and is removed, while the heavier (higher boiling) component is removed from the base or bottom of the column.

   Multicomponent distillation is the separation of three or more substances. A good example of multicomponent distillation is crude oil refining (shown in Figure 17-13). In multicomponent distillation, there are usually several extraction points for removing materials of different boiling points (e.g., lubricating oil, kerosene, and gasoline).

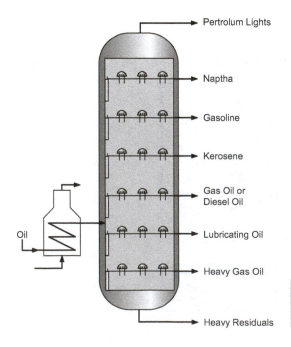

Pertrolum Lights

Naptha

Gasoline

Kerosene

Gas Oil or Diesel Oil

Lubricating Oil

Heavy Gas Oil

Oil

Heavy Residuals

FIGURE 17-13 Crude Oil Distillation Column Products and Boiling Points

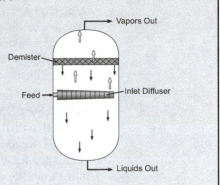
Flash Distillation

Flash distillation is the partial evaporation or vaporization that occurs when a saturated liquid stream undergoes a reduction in pressure by passing through a throttling valve or other throttling device. This type of distillation can be either single-stage or multistage.

Multistage flash distillation is used mainly in the process of desalinating sea or salt water. During this process, the water is heated and the pressure is reduced to allow the water to flash, thereby separating the steam from the salt deposits. This may be done in several stages, each one lowering the pressure to repeat the process. Once the process is complete, the salt is then returned to the ocean.

Azeotropic Distillation

An **azeotrope** is a mixture of two or more liquids that are combined in a way that prevents them from being separated by simple distillation. One characteristic of azeotropes is they boil at a constant temperature and a fixed composition. For example, water (which normally has a boiling point of 100 degrees C) and ethanol (which normally has a boiling point of 78.4 degrees C) can form a water/ethanol azeotrope that has a boiling point of 78.2 degrees C at atmospheric pressure. When this azeotropic mixture is heated, the entire volume, rather than the individual components, is vaporized.

To separate azeotropic mixtures or mixtures with very similar boiling points, the process industries use azeotropic distillation. During **azeotropic distillation**, an additional substance (called an entrainer) is added to the tower to facilitate separation. As this component bonds with or is attracted to the component, the boiling range changes, thereby allowing components to be separated.

One type of material added to help facilitate azeotropic distillation is a solvent. Solvents, which can be considered selective or nonselective, are substances that are

capable of dissolving other substances. Selective solvents (e.g., amines) have a narrow azeotropic range. Nonselective solvents (e.g., alcohols) have a wider azeotropic range. Which type of solvent is used depends on the mixture being separated. Because azeotropic distillation requires the use of solvents, it also requires additional equipment (e.g., a decanter) to then remove the solvent once the separation is complete.

Distillation equipment and solvent recovery systems are used extensively in recycling applications and to reduce manufacturing waste.

Extractive Distillation

Extractive distillation is another method for separating components with close boiling points. Like azeotropic distillation, extractive distillation also includes the use of a solvent, specifically a selective solvent. This solvent generally has a much higher boiling point than the components of the feed stream.

During the extractive process, this solvent attaches itself to the low boiling component and is removed from the bottom of the column. This differs from conventional distillation because the product with the lowest boiling point is not taken overhead. The solvent and desired products are then sent to another column, where the solvent is removed and recycled back to the first column to be used again. During this process, reflux is used to maintain the temperature at the top of the column and to prevent the loss of solvent into the overhead vapor line. Because the differential temperature between the feed stream and the solvent must be maintained precisely to have correct absorption of the desired product, extraction columns and mixer-settler systems must be constantly monitored.

Extraction

Extraction is a separation process that uses a solvent or other medium to separate components based on differences in **solubility** (the degree to which a solid dissolves in a solvent). During a **liquid-liquid extraction**, a feed solution is combined with a solvent that has an affinity for (attracts) one or more components of the feed. When these two substances are mixed, the attracted component combines with the solvent and is removed as extract, while the remaining substance is removed as raffinate (the part of a liquid that remains after the more soluble components have been extracted by a solvent). Figure 17-14 displays the liquid-liquid extraction process. Historically, the liquid-liquid extraction method has been used for separating components with wide boiling ranges.

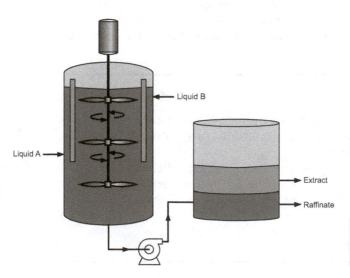

FIGURE 17-14 Diagram of Liquid-Liquid Extraction Process

In the mixer settler system, liquids are initially fed to a large tank that contains a motor-driven agitator. Extraction is achieved by agitating the liquids, which forces them into close contact with each other. After mixing, the liquids are allowed to settle, forming layers of extract.

In the extraction columns, the columns perform liquid-liquid extraction by using a countercurrent flow of liquids. Extraction takes place when liquids contact each other. Heavy liquid enters the column near the top and flows downward due to gravity, while light liquid enters at the bottom and moves upward due to the density difference of the liquids. The light and heavy liquid products are pumped out at opposite ends of the column.

There are two groups of extraction columns: stationary columns that use internal devices such as sieves or packing to perform extraction, while agitated columns use agitation to thoroughly mix liquids in the column. Agitation-type columns can utilize outside mixers, air pulses, and reciprocating plate columns to thoroughly mix the liquids.

Absorption and Stripping

Absorption is the process by which a gas or liquid is dissolved into a liquid or solid. Absorption is a common method for removing unwanted components from gases. The absorption process is completed in a column called a scrubber. A **scrubber** is a device used to remove contaminants from a gas stream by washing the stream with water or some other neutralizing agent.

In a scrubber column, gas enters from the bottom and flows upward and out the overhead line. Liquid enters the top of the column and flows downward, contacting the gas as it moves through packing, trays, or a spray chamber. The flows are countercurrent, giving maximum contact between the two components. When the liquid comes in contact with a gas, one or more unwanted components of the gas mixture are removed.

Water is the solvent that is most commonly used to remove specific components from gas mixtures exiting to the atmosphere. However, other solvents may also be used.

For absorption to be effective, the correct amount of mixing between gas and liquid solvents and the right amount of residence time are required. Without these qualifications, the absorption is less efficient, and the gas components may not completely dissolve. Industrial absorbers work by maximizing the mixing and providing enough time for the gas and liquid solvent to have an efficient absorption operation. Figure 17-15 shows a diagram of the absorption process.

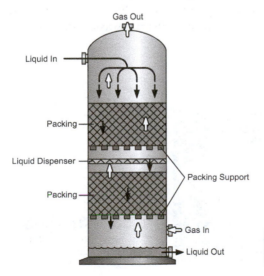

FIGURE 17-15 Diagram of Absorption Process

Did You Know?

Packages of silica gel can be found in electronics, pharmaceuticals, or other products that could be harmed by the presence of moisture.

Stripping is another process that uses a vapor stream to separate one or more components from a liquid stream. This process can be used because many volatile compounds do not have a strong ability to stay soluble in liquids.

Stripper columns are used to remove unwanted components from liquids such as wastewater. The liquid enters the column and a hot gas is injected into that phase to begin the process. The volatile compounds are stripped and leave the column through the overhead vapor line. This vapor can be recovered and the solvent recycled. Some examples of industrial materials removed by a stripper are alcohols, ketones, aromatics, acetaldehyde, isopropyl ether, sulfur dioxide, and carbon dioxide.

Adsorption

Adsorption is a process that occurs when a gas or liquid adheres or accumulates on the surface of a solid and is removable. Solids must be porous, containing a network of openings or pores that can trap liquids or gases. Adsorption is a result of surface energy, similar to surface tension. The amount of tension between specific adsorbents and liquids or gases is important when determining what kind of adsorbent to use in processing. Some common adsorbents are activated alumina, silica gel, and activated carbon.

Activated alumina is used to remove moisture from gases and liquids. This medium can sometimes be re-activated by the addition of dry heat.

Silica gel (shown in Figure 17-16) removes moisture from substances, typically gases, indicating adsorption by changing colors. When silica gel has adsorbed as much as possible, it must be replaced or regenerated by heating so that the adsorbed materials can be eliminated. Silica gel is commonly called the desiccant in industrial dryers.

Activated carbon (shown in Figure 17-17) is coal that is highly porous and is treated with steam in a process known as activation. It is used to remove foul-smelling and toxic vapors from vent gases before they are released into the atmosphere, to remove organic components from industrial wastewater, and to treat deionized water for use in steam boilers.

FIGURE 17-16 Silica Gel Used to Remove Moisture from Substances

FIGURE 17-17 Activated Carbon Tube with Two Types of Carbon to Remove Foul-Smelling and Toxic Vapors

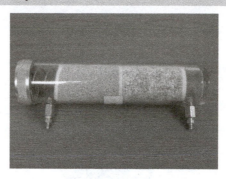

Evaporation

Evaporation is the process of changing a liquid into a vapor. The process of evaporation is carefully controlled with respect to the variables of temperature and pressure. Too much evaporation can cause the heavier component to solidify and plug operating equipment. Too little evaporation can create a vaporized component that does not meet customer specifications. Figure 17-18 displays a graphic of the evaporation process.

Evaporators separate components by vaporizing the lighter component of a feed stream and leaving a very concentrated heavy component in the liquid phase. (*Note:* Either or both of the components may be the desired product.) Evaporators can either be batch or continuous in their operation. Batch-type systems introduce specific amounts of raw material, complete the evaporation separation process, and then shut down so the concentrated material can be removed. Continuous-type systems operate on an ongoing basis, where the feed stream or raw material is continually fed to the system. A process shutdown is not required to remove the concentrate. Examples of the evaporation process include processing waste water sludge and creating powdered milk.

Several types of evaporators are in industrial use today, including falling film, rising (climbing) film, forced circulation, and plate evaporators. All of these units work in basically the same manner: Feed is introduced into the evaporator and heat is introduced either into a heating jacket or through heating tubes. As the heat is gradually increased, lighter components are vaporized and leave the top of the evaporator. Heavy material, typically solids in solution, are then removed from the bottom and sent on for further processing. This application is widely used in the dairy industry and in the production of monosodium glutamate (MSG).

FIGURE 17-18 Diagram of Evaporation Process Showing Raw Materials in and Phases of Condensate to the Vapor Out Process

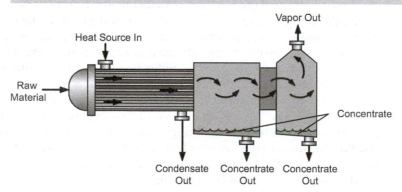

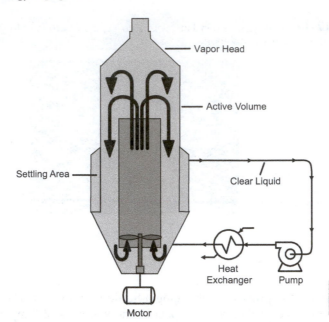

FIGURE 17-19 Draft Tube Baffle (DTB) Crystallizer

Crystallization

Crystallization is a process of creating crystal from a supersaturate solution. Crystallization is often used with evaporation units. The process of crystallization begins with the supersaturation of a liquid solution. This supersaturated solution is then cooled so that, when the **mother liquor** (the part of a solution that remains after crystallization) is drained, only crystals remain.

Nucleation is the growth of a new crystal or primary crystal from a liquid. Secondary nucleation is the growth of crystals using seeds, or primary crystals. Secondary nucleation is the heart of the crystallization process.

Some common types of crystallizers are tank crystallizers, scraped surface crystallizers, draft tube and baffle (DTB) crystallizers, and forced circulating liquid crystallizers. Figure 17-19 displays the parts of a draft tube and baffle (DTB) crystallizer. A few naturally occurring crystals include snowflakes, stalactites, stalagmites, and salt. Some common examples of industrially created crystals are ammonium nitrate, polyethylene glycols, and sucrose (table sugar).

Potential Problems

Because many of the materials separated in this equipment are volatile, proper process operation and control is critical to the safety of equipment and personnel. The nature of each process dictates its own set of problems. Plugging of equipment and piping is common during process upsets. One of the most misunderstood concepts in the operation of any column is the differential pressure variable and reading. This variable is a measure of the work taking place inside the column compared to the tower pressure.

Low differential pressure could mean that the column is not working hard enough. High differential pressure could mean overloading or flooding of the column. The temperature and pressure relationship are critical to the successful operation of these processes. Other problems include too much base heat on the column, thus causing the overhead product to be out of specification. Too much reflux on the column could mean that the product from the bottom does not conform to specification. Both of these problems are reflected as changes in the column's differential pressure reading.

Each process unit has a set of operating procedures that should be reviewed by the process technician before beginning work on that unit. The completion of appropriate training is critical to the success of any process unit or system operation. There are also startup and shutdown procedures that dictate changes to the process and time frames for the safe completion of that procedure.

Typical Procedures

Startup and shutdown procedures vary with the material of composition, whether the process is binary stage or multistage, and according to each company's specific procedures for these operations. These procedures also depend on whether the column is shut down for maintenance work or placed on total reflux while other process unit work is completed. Some common steps for these processes would be like those listed in the following sections.

STARTUP

1. Complete all lockout/tagout procedures.
2. Purge the column of unwanted materials.
3. Begin auxiliary services (i.e., cooling water, air flow, etc.).
4. Fill the column with the proper amount of material.
5. Start up heating and cooling processes.
6. Bring the column to normal operating temperatures and pressures with product in specification.

SHUTDOWN

1. Inform other units of process changes.
2. Reduce column rates and inventory.
3. Reduce and stop column heating and cooling processes.
4. Stop feed.
5. Depressurize the column.
6. Purge the column and prepare for entry.

EMERGENCY

1. Maintain safe operating parameters for pressure and temperature settings.
2. Stop feed to the column.
3. Depressurize the column using the process facility/department procedure for emergency operation.
4. Assess the equipment situation.

LOCKOUT/TAGOUT

Lockout/tagout procedures are mandated by the Occupational Safety and Health Administration (OSHA) for all operating equipment with an energy source that could cause injury or environmental or safety concerns. Lockout/tagout procedures should be followed in accordance with equipment-specific procedures designed by individual company requirements.

Process Technician's Role in Operation and Maintenance

Given the complexity of the equipment described in this chapter, it's important for process technicians to utilize the knowledge and skills learned on the job to ensure proper operation of distillation equipment. Process technicians are responsible for monitoring and controlling levels, temperatures, pressures, and flows for maintaining safe operation of the equipment and to ensure customer specifications are met. Preventive maintenance is also a must to prevent unscheduled downtime or other interruption of the process system.

Summary

The separation of process materials is a very complex and lengthy operation requiring multiple pieces of equipment and associated auxiliaries. Without the equipment and process capabilities discussed in this chapter, many chemicals and their derivatives could not be produced.

Distillation is the most widely used process to separate liquids into their individual components. This process (also called fractional distillation) is used to separate materials such as crude oil, ethane, propane, and many other process components. Materials with very close boiling ranges require other types of distillation such as azeotropic and extractive distillation processes. Materials that would decompose at high temperatures and pressures require a vacuum-type distillation process.

Other separation processes include adsorption, absorption, evaporation, stripping, and crystallization as a way of achieving the process or product composition. These and the distillation processes are often the last stage in a process facility or unit's operation to create a product for sale to the customer. Specifications must be maintained in order to continue the process and provide the needed quality and quantity to meet customer contracts. The process technician's role in these operations is critical, and in-depth training is provided to ensure knowledge and understanding of these processes.

Checking Your Knowledge

1. Define the following terms:
   a. Absorption
   b. Adsorption
   c. Distillation
   d. Evaporation
   e. Extraction
2. A common equipment component used to heat the base of a distillation column is:
   a. a base heater
   b. a condenser
   c. a reboiler
   d. both A and C
3. *(True or False)* A binary distillation system is mainly used in the process of desalinating sea or salt water.
4. *(True or False)* Extraction is the use of a solvent or other medium to separate components based on differences in solubility.
5. Which of the following processes is often used with evaporation units so that, when cooled and drained, only crystals remain?
   a. Nucleation
   b. Binary distillation
   c. Crystallization
   d. Adsorption
6. *(True or False)* An azeotrope is a combination of two or more liquids that prevents them from being separated by simple distillation.
7. *(True or False)* The process of adsorption and absorption are closely related in their process operation characteristics.
8. *(True or False)* A scrubber column is used to remove unwanted components from gases.
9. Which of the following processes is used to remove volatile compounds from a process steam?
   a. Stripping
   b. Multipurpose distillation
   c. Crystallization
   d. Absorption

Student Activities

1. Using a simulator or other process unit, review the components of a distillation system.
2. Take a process facility tour to see the operation of equipment discussed in this chapter.
3. Use the Internet to research a piece of equipment from this chapter and write a two-page paper on:
   - The equipment's role in the process
   - The auxiliary equipment involved
   - The operating parameters for pressure, temperature, and other controlled variables

18

Reactors

Objectives

After completing this chapter, you will be able to:

- Explain the purpose of reactors in the process industry.
- Identify the common types and applications of reactors.
- Identify the components of reactors and explain the purpose of each.
- Explain the operating principles of reactors.
- Identify typical procedures associated with reactors.
- Describe the process technician's role in reactor operation and maintenance.
- Identify potential problems associated with reactors.

Key Terms

Batch reaction—process in which reactants and catalysts or initiators are added as a single charge, rather than continuously, to the reactor to make a batch of final product.

Catalysts—substances that initiate or increase the rate of a chemical reaction at a given temperature and pressure but are not changed or consumed by the reaction.

Cold wall reactor—a reactor composed of an unheated or purposely cooled wall of some type of glass or refractory and an internal heating element.

Continuous reaction—a reaction in which products are continuously being formed and removed, while raw materials (reactants) are continuously fed into the reactor.

Endothermic—absorbing or requiring heat.

Exothermic—releasing energy in the form of heat.

Fixed bed reactor—a reactor in which the catalyst bed is stationary as the reactants pass through and around it.

Fluidized bed reactor—a reactor that uses high-velocity fluid to suspend or fluidize solid catalyst particles. The reactor feed is mixed with the suspended catalyst where the reaction takes place.

Hot wall reactor—a reactor in which heat is transferred to the reactants by conduction from the vessel wall.

Inhibitor—a substance used to slow or stop a chemical reaction. Inhibitors are generally consumed during the course of the reaction.

Initiator—a substance that starts or increases the rate of a chemical reaction. Initiators are consumed by the reaction.

Mixing system—a system consisting of an agitator, circulating pump, gas spargers, and a series of baffles to provide proper mixing of reactants and a catalyst. It is designed to promote adequate mixing of reactants and catalyst.

Relief system—a system designed to prevent too much pressure or a vacuum.

Residence time—the volume of material inside the reactor divided by the flow rate of the reactants through the reactor.

Runaway reaction—an uncontrolled reaction where the reaction rate increases, releasing heat that causes an increased reaction rate and results in the release of more heat.

Stirred tank reactor—a reactor that contains a mixer or agitator mounted inside the tank.

Temperature control system—a system that allows the temperature of a reaction to be adjusted and maintained.

Tubular reactor—a reactor composed of tubes used for continuous reactions with nozzles to introduce reactants.

Vessel head—the area on the top or bottom of a reactor shell that has access ports or nozzles.

Introduction

Reactors are vessels in which controlled chemical reactions are initiated and take place in either continuous or batch reactions. In these types of vessels, raw materials are reacted to form the desired chemical or chemical mixture. This final product is then separated or refined to recover the desired products. Some reactor products are final products, while others are intermediates that may become feedstock for downstream reactions or further processing.

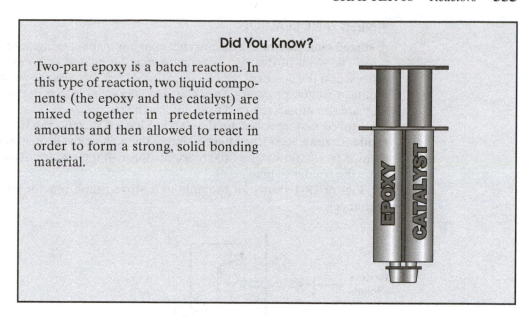

Did You Know?

Two-part epoxy is a batch reaction. In this type of reaction, two liquid components (the epoxy and the catalyst) are mixed together in predetermined amounts and then allowed to react in order to form a strong, solid bonding material.

Chemical Reactions

Reactors help facilitate chemical reactions. The rate of chemical reactions can be altered (sped up or slowed down) through the application of heat, a catalyst, an inhibitor, or an initiator. **Catalysts** are substances that initiate or increase the rate of a chemical reaction at a given temperature and pressure but are not changed or consumed by the reaction. **Initiators** are substances that start or increase the rate of a chemical reaction and are consumed by the reaction. **Inhibitors** are substances that slow or stop a chemical reaction.

The catalyst phase and reaction kinetics (the speed of the reaction) determine the design of the reactor. The catalyst may be in the same phase as the reactants (homogeneous) or different phase (heterogeneous). The product stream from a reactor consists of the desired product, and it may also contain by-products, such as unreacted raw materials, carrier solvents, and catalyst.

Types of Reactions and Reactors

In the process industry, chemical reactions are utilized to generate new products. These reactions occur in a reactor and can be categorized as batch or continuous. Reactors come in many shapes and sizes. The most common types of reactors are stirred tank, tubular, fixed bed, fluidized bed, and hot wall.

BATCH VERSUS CONTINUOUS

A **batch reaction** is a process in which reactants and catalysts or initiators are carefully metered and added as a single charge to a reactor. These substances are then mixed and allowed to react. After a predetermined time, the reaction is stopped and the desired product is removed from the reactor. The next batch can then be started. In a batch reactor, the composition of the materials inside the reactor is constantly changing as the reaction proceeds to the desired composition.

A **continuous reaction** occurs when products are continuously being formed and removed, while raw materials (reactants) are continuously fed into the reactor.

STIRRED TANK REACTOR

A **stirred tank reactor** is a reactor that contains a mixer or agitator mounted inside the tank. The shell of this type of reactor may be heated or cooled depending on the process and the design of the reactor. Stirred tank reactors can run both batch and continuous reaction processes. They can also operate at higher pressures. Because of this, they are sometimes referred to as autoclave reactors.

Stirred tank reactors are designed to have a specific reaction time, referred to as **residence time**. Residence time is simply the volume of the material inside the reactor divided by the flow rate of the reactants through the reactor. Reaction kinetics determine the residence time.

Figure 18-1 shows an example of a stirred tank reactor with an external heat exchanger.

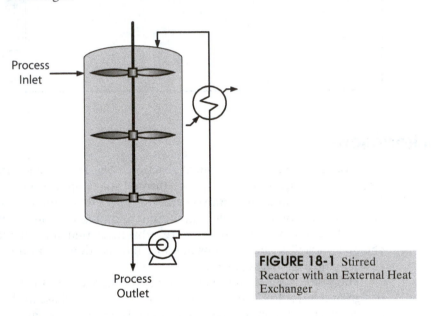

Process Inlet

Process Outlet

FIGURE 18-1 Stirred Reactor with an External Heat Exchanger

TUBULAR REACTOR

A **tubular reactor** is composed of tubes used for continuous reactions with nozzles to introduce reactants. Tubular reactors may be required if the reaction kinetics must occur quickly. Figure 18-2 shows an example of a tubular reactor. Based on process requirements (e.g., volume, surface area, and pressure), the design of a tubular reactor can range from a simple jacketed tube to a multipass shell and tube arrangement. Tubular reactions are always continuous in nature.

One type of tubular reactor is a section of pipe designed with a shell, baffles, jackets, and inlet and outlet ports to mix together two or more fluids (and catalyst if

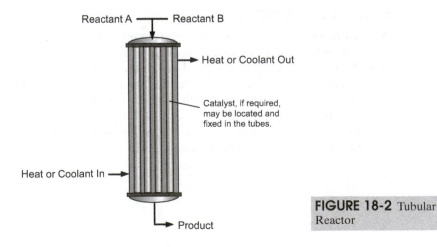

Reactant A — Reactant B

Heat or Coolant Out

Catalyst, if required, may be located and fixed in the tubes.

Heat or Coolant In

Product

FIGURE 18-2 Tubular Reactor

required) to form a chemical reaction. The shell is used to control the temperature inside the reactor because the shell can be heated or cooled as needed. Due to the design of the tubular reactor, a larger amount of surface area per unit volume is available for heat transfer than is available in a jacketed reactor.

Another type of tubular reactor is simply a tube inside a furnace. The tube may or may not contain a catalyst. Examples of these reactors are gas reformers and cracking furnaces.

FIXED BED REACTOR

A **fixed bed reactor** is a reactor in which the catalyst bed is stationary and the reactants are passed through and around it. In this type of reactor, the catalyst occupies a fixed position and is not designed to leave the reactor with the process. Figure 18-3 shows an example of a fixed bed reactor.

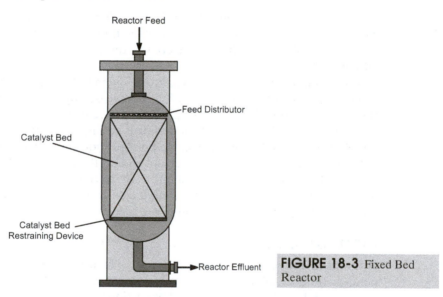

FIGURE 18-3 Fixed Bed Reactor

Fixed bed reactors contain a catalyst bed, catalyst restraining device, and inlet and outlet ports. The catalyst restraining device keeps the catalyst in place during operation. This device contains a screen or plates with holes that allow liquid or gas to flow through the reactor.

In a fixed bed reactor, products are formed when the reactant flows into the reactor and over the catalyst. The product then flows out of the reactor as finished product or as a product that requires further processing. Proper contact between the reactants and the catalyst is extremely critical. Uniform distribution of the reactants across the catalyst bed is accomplished by a distribution device such as a tray, spray nozzle, or gas diffuser.

Figure 18-4 shows a tubular distribution from an aerial view, and Figure 18-5 shows distribution from the nozzles.

In a fixed bed reactor, the composition of material inside the reactor is constantly changing as the material descends through the catalyst bed. This is sometimes referred to as plug flow.

FIGURE 18-4 Tubular Distributor with Outlet Holes on the Underside

FIGURE 18-5 Tubular Distributor Nozzles for the Distribution of the Liquid

The amount of reaction is normally controlled by the reaction temperature and the contact time. The contact time is the average time it takes for the material to enter and leave the catalyst bed. The surface area of the catalyst available for reaction directly influences the reaction rate.

A common type of fixed bed reactor found in both refineries and chemical process facilities is a hydrotreating reactor. An example of a hydrotreating reactor is a naphtha (heavy gasoline) hydrotreater. Naphtha hydrotreaters contain a fixed bed of catalyst that contacts the naphtha feed and removes sulfur, nitrogen, and oxygen and converts double-bonded molecules into single-bonded molecules. Any metal containing compounds found in the naphtha are removed and deposited on the catalyst. Over time, these metals can foul the catalyst and require the replacement of the catalyst bed. Sulfur and iron are examples of common elements that can cause fouling. Figure 18-6 is an example of a naphtha hydrotreating reactor.

FLUIDIZED BED REACTOR

A **fluidized bed reactor** is a reactor that uses high-velocity fluid (gas or liquid) to suspend or fluidize solid catalyst particles. The reactor feed is then mixed with the suspended catalyst at the point where the reaction takes place. Once the catalyst is used, it is sent to a regenerator so it can be regenerated (cleaned) and returned to the reactor. Equipment abrasion, particle agglomeration, particle fracturing, particle size distribution, particle density, gas velocity, gas density, and pressure drop are all issues when operating a fluidized bed.

An example of a fluid bed reactor is the fluid catalytic cracking unit (FCCU), which is used to "crack" (break down) heavy gas oils into gasoline and diesel fuel in a refinery. Figure 18-7 shows an example of a fluidized catalytic cracking unit.

HOT WALL VERSUS COLD WALL

In a **hot wall reactor**, heat is transferred to the reactions by conduction with the vessel wall. In a **cold wall reactor**, the vessel is composed of an unheated or purposely cooled

Did You Know?

The catalytic converter on your car is a fixed bed reactor. Catalytic converters are used to remove harmful pollutants from engine exhaust.

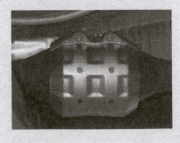

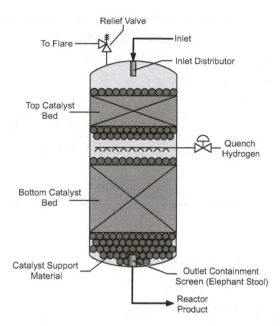

FIGURE 18-6 Naphtha Hydrotreating Reactor

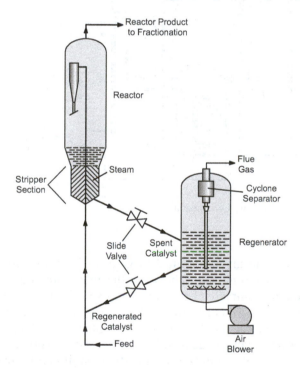

FIGURE 18-7 Fluid Catalytic Cracking Unit (FCCU)

wall of some type of glass or refractory and an internal heating element. Cold wall reactors are vessels that have external cooling systems or internal insulation so the reactor walls are cooler than the process. This allows the use of less expensive metals in the reactor walls.

NUCLEAR REACTOR

Nuclear reactors are high-temperature and high-pressure reactors used to generate the steam necessary to produce electricity. Rather than using a chemical reaction (e.g., burning coal) a nuclear reaction (splitting uranium atoms through fission) is used to produce heat. The reactor, its components, and the water pumped through the reactor control the nuclear chain reaction and carry away the heat generated. Figure 18-8 is an example of a nuclear reactor.

Did You Know?

In 2008, the United States had 104 commercial nuclear reactors in thirty-one states.

FIGURE 18-8 Example of a Nuclear Reactor

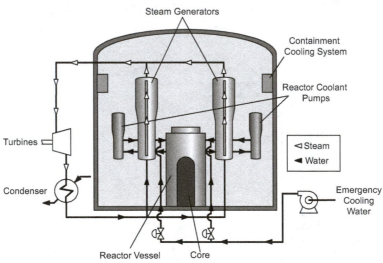

Courtesy of Nuclear Regulatory Commission

When uranium atoms are split, they give off immense heat. Water is pumped around the fuel rods to remove this heat. For this reason, the water is also referred to as the reactor coolant. Commercial nuclear reactors in the United States can be divided into two groups: boiling water reactors and pressurized water reactors. The classification of the reactor depends on whether the reactor coolant is used directly or indirectly.

In boiling water reactors, the heat from the nuclear fission reaction is used to heat the reactor coolant water and boil it into steam so it can be used to drive a steam

Did You Know?

Nuclear reactor employees typically receive less radiation exposure than what others receive from everyday sources, like cosmic rays from the sun, medical X-rays, and radon.

turbine. Pressurized water reactors are slightly different. In this type of reactor the water is kept at a higher pressure so it remains a liquid and does not boil. The water is then pumped through heat exchangers, where it is used to boil water. Because they produce steam, these heat exchangers are often referred to as steam generators. This secondary system fluid (the water boiled into steam) is then used to drive a steam turbine.

The fuel core in a nuclear reactor typically lasts from 18 to 24 months. After that, about half of the fuel is replaced with fresh fuel and the remaining fuel is rearranged so that it is consumed more evenly. The spent nuclear fuel is very radioactive, so it must be shielded at all times.

Components of Reactors

Reactors consist of various components, depending on the type of reactor. However, some of the main components include a shell, vessel head, agitator, baffles, tank, mixing system, heating and cooling system, and a relief system. Figure 18-9 shows an example of a reactor and its components.

The shell is the outer surface, casing, or external covering of a vessel. This surface may consist of various types of materials that protect it from chemical effects, depending on the chemical reaction taking place inside the equipment. A shell consists of a side wall, top and bottom heads, and the discharge area. The **vessel head** is the area on top of the shell that has access ports or nozzles. These ports and nozzles provide multiple uses, including introduction of chemicals; removal of gaseous products or solvents; location for instrumentation ports; and connection to relief devices, agitation, and vacuum systems.

The agitation system of a stirred tank reactor typically consists of a motor, shaft, and blades. The agitator shaft is a cylindrical rod that holds the agitator blades, which stir the chemicals in the tank or reactor. Agitator blades are similar to those in a kitchen blender. They are devices that act on the fluid by generating turbulence, which promotes mixing. In some reactors, two or more sets of agitator blades are used to increase the mixing process. Agitator blades come in different designs.

Because turbulence is required for proper mixing, baffles are installed. A baffle is a stationary plate, mounted on the wall of the tank or vessel, which is used to alter the flow of chemicals and facilitate mixing. Baffles are typically vertical or spiral strips, or horizontal rings, fixed to the reactor walls.

The reactor vessel is the portion of the reactor that has nozzles for inlet feeds, product discharge, circulation, instrumentation, agitation systems, vacuum, and pressure relief. The vessel must be the proper size, rated at the recommended pressure rating, and constructed of the proper materials for the process. The vessel must also be

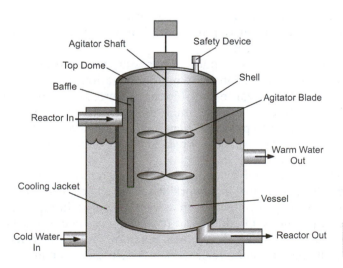

FIGURE 18-9 Reactor Components

equipped with the proper instrumentation to control the reaction. This instrumentation typically includes level, temperature, reactant composition, pressure, and flow transmitters.

The **mixing system** consists of an agitator, circulating pump, gas spargers, and a series of baffles to provide proper mixing of reactants and catalyst. The mixing system is critical in many reactions. Because of this, instrumentation (e.g., motion indicators and process interlocks) may be incorporated to ensure proper operation.

The **temperature control system** allows the temperature of a reaction to be adjusted and maintained. Simple systems may contain a cooling or heating jacket. More complex systems may contain a circulation system with heat exchangers. Depending on the requirements, some systems may contain both jackets and circulation systems. The temperature control system is critical in all exothermic reactions. Because of this, instrumentation (e.g., process interlocks) may be incorporated to ensure proper operation.

The **relief system** is designed to prevent overpressurization or a vacuum. It includes a piping system that collects the discharge of the pressure relief valve and routes it to the atmosphere, to a flare, or to another containment system.

Many reactor systems also require filtration or decanting systems downstream of the reactor to recover catalyst or product, or to remove unwanted reaction byproducts. Table 18-1 contains an overview of common reactor types and their components.

TABLE 18-1   Reactor Types and Their Common Components

Reactor Type	*Example*	*Common Components*
Stirred tank reactor	Sulfuric acid alkylation reactor	Vessel, agitator, internal baffles, heater/cooler, catalyst mix tank, and relief systems
Tubular reactor	Steam methane reformer	Relief systems, gas quench system, reformer reactor
Fixed bed reactor	Hydrotreating reactor	Inlet distributor, catalyst support material, gas quench system, catalyst support tray, multiple beds, inner screen and outer screen, and relief systems
Fluidized bed reactor	Fluid catalytic cracking reactor	Relief systems, baffles, cyclone separators, risers, standpipes, reaction chamber, plenum chamber, slide valves, and stripping section

Theory of Operation

Chemical reactions can either be endothermic or exothermic. **Endothermic** describes reactions that absorb or require heat. **Exothermic** describes reactions that release energy in the form of heat.

Reactors provide the conditions necessary for controlled reactions to occur efficiently and safely. Reactions may require agitation for mixing or solids suspension, heat addition or removal to control temperature, gas addition or removal to control pressure, or analyzers to control composition. The reaction rate (the rate at which reactants are converted to products) generally increases as the temperature increases. To achieve the desired conversion of reactants to products, sufficient residence time must also be provided in the reactor.

In a reactor, excessive heating or failure to remove heat may produce off-specification products that must be reworked, sold at a loss, or sent to waste. Excessive cooling may also cause problems such as crystallization in liquids, higher viscosities, or freezing of polymers. In addition to producing off-spec products, loss of temperature control can also result in a **runaway reaction**, an uncontrolled reaction where the reaction rate increases and releases heat, which causes an increased reaction rate and

Did You Know?

A simple way to remember the difference between exothermic and endothermic reactions is to think of **ex**othermic reactions as having heat **ex**iting the reactor. **En**dothermic reactions must have heat **en**tering the reactor.

- Endo = Enter
- Exo = Exit

results in the release of more heat. Runaway reactions are extremely dangerous and can lead to a catastrophic failure of the system.

Endothermic and exothermic reactions both require energy (heat) for the reaction to occur. This heat energy allows the reaction to overcome a so-called energy hill, also called the energy of activation. As the energy of activation increases, so does the amount of energy required to initiate the reaction.

There are two main ways to overcome the energy of activation: increase the temperature or use a catalyst. Increasing the temperature increases the amount of energy to the system, which in turn increases the reaction rate. Some reaction rates are very sensitive to temperature. As little as a 10-degree C change in reaction temperature can double the reaction rate. Increasing the reaction temperature to increase the reaction rate to overcome the energy hill has some disadvantages. For example, the temperature could become excessively high and pose a safety risk: as the temperature increases, the system pressure increases. The reaction rate of unwanted side reactions (an often undesirable chemical reaction that occurs in addition to the main reaction) could also increase, resulting in increased production of waste products. Higher temperatures can also accelerate equipment corrosion and sometimes require the use of more expensive construction materials.

Using a catalyst lowers the energy of activation and allows reactions to occur more easily, It increases the reaction rate without increasing the temperature. Catalysts can also be designed to lower the energy of activation for the desired reaction and thus prevent unwanted side reactions.

The rate of reaction can also be increased by increasing the number of collisions between molecules. Molecules must collide with each other for a reaction to occur. The more often these molecules collide, the faster the reaction. In a gas phase reaction, the collision rate can be increased by raising the pressure. In a liquid phase reaction, mixing the reactor contents increases the collision rate.

In summary, the rate of reaction is a key variable in process operations. Process technicians must understand the variables (e.g., temperature, using a catalyst, pressure, concentrations, residence time, and mixing) that are critical for controlling the rate of reaction for the process they are operating.

Auxiliary Equipment Associated with Reactors

Auxiliary equipment associated with reactors may include internal heating and cooling coils, agitators and/or jackets, ultraviolet light or gamma radiation, various types of internal lining, static mixer elements, continuous circulating water systems, and temperature control systems.

Table 18-2 lists examples of various types of auxiliary equipment and their benefits.

TABLE 18-2	Auxiliary Equipment Associated with Reactors and Their Related Benefits	

Auxiliary Equipment	*Benefits*
Internal heating and cooling coils or external circulating reaction mass heating and cooling systems	Used to heat and cool the material in a reactor
Reactor shell jacket	Provides a shell through which heating or cooling media may pass
Initiating sources like ultraviolet light or gamma radiation	Increased reaction kinetics
Corrosion-resistant liners and construction materials (e.g., glass, eramics, polymers, precious metals, or alloys)	Prevent the corrosive properties of the reactants and products from damaging the reactor shell
Static mixer elements in tubular or in-line reactors	Increase mixing and turbulence. Without adequate mixing, complete reaction may not occur. Also, without adequate mixing, "hot spots" may occur leading to product deterioration and/or poor yields, and they could initiate a runaway reaction
Continuously circulating water, oil, or other fluid systems on jackets and internal and external heat exchangers	Allows for heating and cooling of the reaction materials
Temperature control systems	Controls the temperature in the reactor by regulating the flow rate of the heating/cooling fluid to the reactor heating/cooling system
Pressure or vacuum control and relief systems (including rupture discs)	Prevents overpressurization or a vacuum that could result in the catastrophic failure of the reactor
Fire and explosion suppression systems (e.g., sprinklers, explosion barriers, reaction quench systems, inert gas, or foam blanketing)	Protects personnel and equipment and controls runaway reactions or other abnormal conditions such as solvent or reactant fires
Programmable logic and digital control systems	Manage reaction steps and ensure correct conditions for continuous reactions

Potential Problems

When working with reactors, process technicians should always be aware of potential problems. For example, reactor vessels require close attention because rapid changes in conditions can cause process upsets or other hazards. When other factors, such as reaction temperatures, pressures, catalyst concentration, and the ratio of reactants change from desired control points, by-product or contaminant formation may increase to levels that cause process upsets or off-specification product. Process temperature extremes (both high and low) should also be avoided to ensure safe, uninterrupted operation.

Table 18-3 lists potential problems associated with reactors, as well as the causes and solutions for each. It is important to note that with each potential scenario, process technicians need to have knowledge of the standard operating procedures for their process unit.

Typical Procedures

Process technicians should be familiar with the appropriate procedures for monitoring and maintaining reactor components and conditions. The typical procedures (including startup, shutdown, emergency, and lockout/tagout) vary at each individual site and for each type of reactor. Because of this, process technicians need to follow the standard operating procedures (SOPs) for their assigned unit(s).

TABLE 18-3   Potential Problems Associated with Reactor Operation (Not an Inclusive List)

Problem	Cause	Solution
Runaway reaction	• Inadequate heat removal, inadequate agitation, or an excessive amount of catalyst • Excessive gas generation resulting from above • Contaminated or wrong reactants	• Maximize heat removal • Reduce pressure to a minimum • Stop all feed streams • Add inhibitors if possible
Agitator or agitation failure	• Motor trip out • Coupling failure • Excessive reaction mass viscosity	• Reset breaker on motor starter • Repair broken coupling • Address reaction problem
Leaks	• Corrosion • Overpressure	• Reduce pressure • Repair leak
Lining failure	• Corrosion • Poor installation • Glass liner breakage	• Repair or replace liner
Incomplete reactions, product deterioration, or undesirable by-products	• Thermocouple or temperature sensor failure • Improper reaction conditions	• Repair or replace thermocouple or sensor • Return conditions to standard operating procedure • Confirm proper instrument operation and repair or recalibrate if required
Product yield low/off-spec product	• Inadequate mixing • Catalyst addition rate high	• Correct agitation • Reduce catalyst addition
Slow reaction/low product yield	• Inadequate mixing • Catalyst addition rate low • Deterioration of catalyst	• Correct agitation • Increase catalyst addition • Replace catalyst
Temperature increasing or decreasing abnormally	• Inadequate heat removal • Inadequate agitation • Excessive amount of catalyst	• Correct heat removal • Correct agitation • Decrease catalyst addition
Competing reactions	• Temperature, pressure, and reactant concentrations are incorrect • Jacket or coil heating/cooling temperatures are incorrect • Feed, catalyst, or solvent (reaction media) pump failure • Feed ratio controller system failure	• Bring temperature and pressure back into control • Return temperature to normal • Restart pump • Restart ratio control system • Reestablish proper quantities
High pressure drop across catalyst bed	• Catalyst has deteriorated • Fouled catalyst	• Replace or regenerate catalyst
Low reaction and production rates	• Poor mixing in stirred tank • Poor distribution in fixed bed • Low catalyst activity	• Check agitator • Replace or redistribute catalyst during turnaround

Process Technician's Role in Operation and Maintenance

Process technicians play an important role in the safe operation and maintenance of reactors. Through frequent monitoring, process technicians can ensure that the equipment is operating properly. During this monitoring, technicians must be alert to unusual sounds that can signal reactor problems.

Process technicians must also monitor outside instrumentation on the reactor on a regular basis and as requested by the process board operator, and collect samples and monitor items such as reactor feed rate, composition, and catalyst concentration. Process board technicians must also monitor all controllers to ensure that the process is being held to the desired set point and that all alarm conditions have been properly addressed.

Summary

Reactors are specialized vessels that are used to contain a controlled chemical reaction and change raw materials into finished products. Reaction variables include temperature, pressure, time, material concentration, flow rates, pressure, surface area, and other factors. Like vessels, reactor designs vary widely based on the chemical reaction that must occur in the process.

In a reactor, the rate of chemical reactions can be altered through the application of heat, a catalyst, or an inhibitor. Catalysts are substances that affect the chemical reaction but are not consumed by the reaction. Initiators are substances that affect the chemical reaction and are consumed by the reaction. An inhibitor slows or stops a chemical reaction.

Reactors allow for batch or continuous reactions inside the reactor. In a batch reaction, product is made in a single operation. In a continuous reaction, raw material is continuously added to the reactor while product is continuously removed.

A stirred tank reactor contains a mixer or agitator that is mounted inside the tank and is used to mix products. This type of reactor is also referred to as an autoclave reactor if it operates at high temperatures and pressures.

A tubular reactor has a shape similar to a tubular heat exchanger. Many tubular reactors contain a section of pipe designed with a shell, baffles, jackets, and inlet and outlet ports to mix two fluids together to form a chemical reaction.

A fixed bed reactor contains a stationary bed of catalyst over which liquid or gas flows to form the desired product. A fluidized bed reactor is a reactor in which a fine catalyst is fluidized and contacted with the reactants.

Nuclear reactors are used within the power industry and can be divided into boiling water reactors and pressurized water reactors. In this type of reactor, water is pumped around fuel rods to remove the heat of the nuclear reaction and ultimately form steam. In both cases, steam turbines drive large electrical generators that allow low-energy steam to reenter the water.

Chemical reactions are either endothermic or exothermic. Endothermic reactions require heat to react and exothermic reactions generate heat. Reactors are designed to accommodate the conditions needed for either of these reaction types.

Process technicians are responsible for knowing standard operating procedures associated with the reactors in their units. Potential problems associated with reactors include operation, design, and human error.

Checking Your Knowledge

1. Define the following terms:
   a. Catalysts
   b. Endothermic
   c. Exothermic
   d. Inhibitors
   e. Runaway reaction
2. *(True or False)* In a batch reaction, products are continuously being formed as raw materials are fed into the reactor.
3. What reactor type is also called an autoclave reactor?
   a. Stirred tank reactor
   b. Fixed bed reactor
   c. Tubular reactor
   d. Fluidized bed reactor
4. This type of reactor contains a stationary catalyst bed, and product flows over the catalyst to form the desired product.
   a. Stirred tank reactor
   b. Fixed bed reactor
   c. Tubular reactor
   d. Fluidized bed reactor

5. What is the most common type of reactor?
    a. Stirred tank reactor
    b. Fixed bed reactor
    c. Tubular reactor
    d. Fluidized bed reactor
6. *(True or False)* Measures of control in fixed bed reactors can be differential pressure and temperature across a catalyst bed.
7. The two types of nuclear reactors generating electricity in the United States are.
    a. boiling water reactors
    b. fixed bed reactors
    c. pressurized water reactors
    d. cold wall reactors
8. *(True or False)* The benefit of temperature control systems is that they prevent the corrosive properties of the reactants and products from damaging the reactor.
9. In a reactor operation, which of the following problems is caused by an incomplete reaction, product deterioration, undesirable by-products, or runaway reactions?
    a. Slow reaction in a low product yield
    b. Abnormal increase or decrease in temperature
    c. Loss of raw material feed
    d. Thermo-well temperature sensor failure

Student Activities

1. Write a paper describing stirred tank reactors, tubular reactors, fixed bed reactors, and fluidized bed reactors. In the paper, discuss how each reactor works, its components, the type of process, and examples of when it is used.
2. As a group, research how catalytic cracker systems ("cat crackers") work to produce gasoline.
3. Draw a schematic of a reactor system with associated auxiliary systems attached, and include a description of the process. Present your work to the class.
4. Research additional uses for reactors. Summarize and report your findings to the class.

Filters and Dryers

Objectives

After completing this chapter, you will be able to:

- Explain the purpose of filters and dryers in the process industry.
- Identify the common types and applications of filters and dryers.
- Identify the components of filters and dryers, and explain the purpose of each.
- Explain the operating principles of filters and dryers.
- Describe safety and environmental hazards associated with filters and dryers.
- Identify typical procedures associated with filters and dryers.
- Describe the process technician's role in filter and dryer operation and maintenance.
- Identify potential problems associated with filters and dryers.

Key Terms

Bag filter—a tube-shaped filtration device that uses a porous bag to capture and retain solid particles.

Breakthrough—the condition of any adsorbent bed when the component being adsorbed starts to appear in the outlet stream.

Cartridge filter—a filtration unit that uses a fine mesh that is tightly folded or pleated to remove suspended contaminants from a liquid.

Cyclone—a mechanical device that uses a swirling (cyclonic) action to separate heavier and lighter components of a gas or liquid stream.

Desiccant—a substance that attracts and absorbs moisture.

Fixed bed dryer—a device that uses a desiccant to remove moisture from a process stream.

Flash dryer—a device that forces hot gas (usually air or nitrogen) through a vertical or horizontal flash tube causing the moist solids to become suspended.

Fluid bed dryer—a device that uses hot gases to fluidize solids and promote the contact of dry gases and solids.

Leaf filter—a type of filter that uses a precoated screen, located inside the filter vessel, to hold the filtered material.

Micron—a unit of measure equal to .000001 meters. Filters are rated based on their ability to filter a minimum micron-size particle.

Plate and frame filter—a type of filter used to remove liquids from a slurry feed by using a deep filtration process that includes clarification and prefiltration.

Pleated cartridge filter—a type of filter used as a filter medium in a pleated form to provide more surface area.

Rotary drum filter—a type of filter that contains a cloth screen and is located on a drum to hold solids being filtered.

Rotary dryer—a device composed of large, rotating, cylindrical tubes; it reduces or minimizes the moisture content of a material by bringing it into direct contact with a heated gas.

Slurry—a liquid solution that contains a high concentration of suspended solids.

Slurry dryer—a device designed to convert wet process slurry into dry powder.

Spray dryer—a device that sprays a moist feed stream into a vertical drying chamber using a nozzle or rotary wheel.

Introduction

Process operations often depend on the use of filters and dryers. The purpose of filters and dryers in the process industries is to remove undesirable components from process streams. Filters are used to remove particles, and dryers are used to remove moisture.

Types of Filters

A filter is a device that removes particles from a process and allows the clean product to pass through the filter. The main component of a filter system is the filter unit itself, which collects particles and allows clean product to pass through. All filter systems have a process flow inlet and outlet, and an opening to clean or replace the filter. The most common types of filters are cartridge, bag, screen, leaf, rotary drum, pleated, and plate and frame.

CARTRIDGE FILTER

Cartridge filters, the most common types of filters used in the process industries, are filtration units that use a fine mesh that is tightly folded or pleated to remove suspended contaminants from a liquid. These types of filters consist of a housing and a filter tube that fits inside the housing. Within the filter tube, there is a central core that is wrapped or surrounded with a filter medium (the material that performs the filtering action). This filter medium, which can be made of various materials, including stone, metal mesh, and nonmetallic fabrics, works like a strainer. As the fluid moves through the fine mesh, solid particles become trapped inside the cartridge. Figure 19-1 shows an example of a cartridge filter.

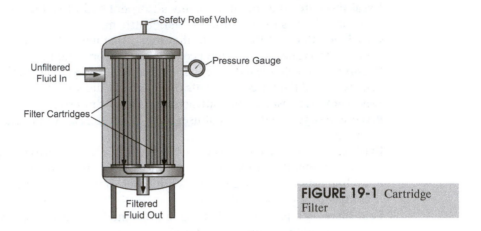

FIGURE 19-1 Cartridge Filter

Within cartridge filters, the process fluid flows from the outside of the cartridge to the inside. Over time, the cartridge filter fills with particles, requiring the filter to be replaced or cleaned. One indication that the cartridge filter is full is when the process flow becomes restricted and the differential pressure (ΔP) across the cartridge increases.

BAG FILTER

A **bag filter** (shown in Figure 19-2) is a tube-shaped filtration device that uses a porous bag to capture and retain solid particles. Depending on the filter design, process fluids can flow from the inside of the bag to the outside or from the outside to the inside. In some industrial applications, bags are shaken in order to release the particles into a chamber so they can be returned to the process.

Common examples of bag filters include the bags used in household vacuum cleaners and those attached to automatic swimming pool cleaners. An industrial example

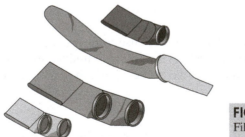

FIGURE 19-2 Examples of Filter Bags for Bag House

Did You Know?

The oil filter in your car is a cartridge filter.

FIGURE 19-3 Examples of a Baghouse System

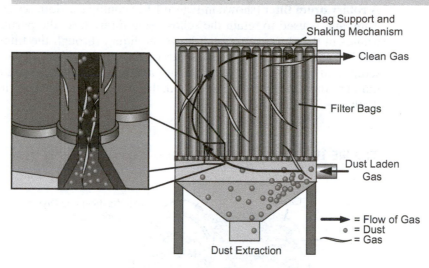

of a bag filter is a baghouse system. Baghouse systems are pollution control devices used to remove particles from an air stream prior to discharge into the atmosphere. Figure 19-3 is a diagram of a baghouse system.

LEAF FILTER

A **leaf filter** uses a precoated screen, located inside the filter vessel, to hold the filtered material. The granular filtered material is caught on the screen and coats it. When the filtered material is expended (clogged), it is backwashed and discarded, and a new precoat is applied. Filter precoat systems consist of a mix tank, an agitator, and a pump. An example of a leaf filter is a water filter that uses retaining screens, such as a pool filter that is found in many backyard pools. Figure 19-4 shows an example of a leaf filter.

Precoated slurry is prepared in the mix tank, circulated by the pump, and deposited on the filter. During operation, the filter precoat becomes plugged with the materials being separated from the process fluid. If left unattended, this plugging would eventually restrict the flow to the point of zero flow. When this occurs, the filter must be disassembled and recoated with precoat material.

FIGURE 19-4 Leaf Filter

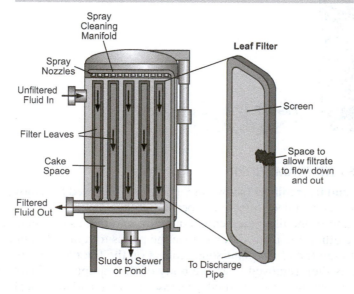

ROTARY DRUM FILTER

A **rotary drum filter** (shown in Figure 19-5) contains a cloth screen that is located on a drum and is used to retain the solids being filtered. As the partially submerged drum rotates in the slurry, a vacuum draws the liquid through the filter medium (cloth) on the drum surface and leaves the solids on the outside of the drum (cake) so they can be scraped off. The vacuum pulls air (or gas) through the cake and continues to remove liquid as the drum rotates. If required, the cake can be washed prior to final drying and discharge.

FIGURE 19-5 Rotary Drum Filter Showing Rotation Through the Drum

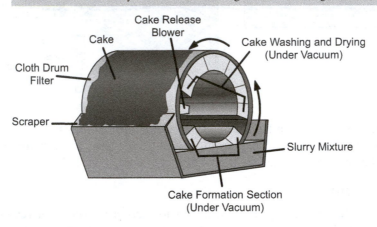

PLEATED CARTRIDGE FILTER

In a **pleated cartridge filter** (shown in Figure 19-6), the actual filter medium is in a pleated form to provide more surface area for filtration. This results in more pounds of filtered material per cubic foot of available space (volume). The pleated material is supported by a perforated core that is coarse to give minimum pressure drop and good strength to the cartridge. Examples of filter media include nylon, polypropylene, fluropolymers, and glass fiber. A common example of a pleated cartridge filter is a swimming pool pump filter or the filter on a wet/dry shop vacuum.

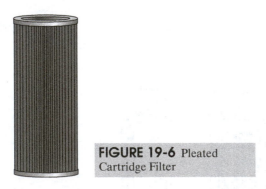

FIGURE 19-6 Pleated Cartridge Filter

PLATE AND FRAME FILTER

Plate and frame filters (shown in Figure 19-7) are used to remove liquids from a slurry feed by using a deep filtration process that includes clarification and prefiltration. In a plate and frame filter, feed enters the filter and travels between the plates through the filter cloth as the plate shifters move the plates to retain the solids. As the flow travels toward the top of the plate, solids collect on the filtered cloth material. The remaining liquid is then removed from the body of the equipment. Plate and frame filters are used in coatings, resins, and ink processes in small and large-scale applications.

FIGURE 19-7 Plate and Frame Filter

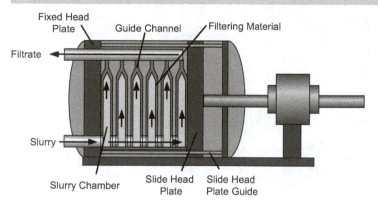

Filter Ratings

Filters are rated on their ability to remove particles of various sizes from a fluid. Filters have a size rating that refers to the specific size of particle that can be prevented from flowing through the filter. For example, a filter with a 10-micron rating can stop a particle that is 10 microns or larger in size. A **micron** is a unit of measure (1 micron = .000001 meters).

Filters are usually given an absolute rating or a nominal rating. With an absolute rating, the filter is capable of removing 99.9 percent of the particles larger than a specified micron rating. With a nominal rating, a majority (85 to 95 percent) of an approximate particle size does not pass through the filter.

Types of Dryers

Dryers are devices used to remove moisture from a process stream. There are several different types of dryers used in the process industries. However, the most common are fixed bed, fluid bed, flash, and rotary dryers, and cyclone separators.

FIXED BED DRYER

Fixed bed dryers are devices that use **desiccant** (substances that attract and absorb moisture) to remove moisture from a process stream. In this type of dryer, the desiccant is evenly distributed in a bed and remains there throughout the life of the bed. Desiccant selection is based on its strong affinity for moisture or other contaminant material, and on its capacity to contain the material being removed.

Over time, the desiccant loses its capacity for retaining the contaminant material and must be replaced or regenerated. **Breakthrough** is the condition of any adsorbent bed when the component being adsorbed starts to appear in the outlet stream. Analyzers are often used to detect breakthrough so that the dryer can be serviced. Analyzers continuously monitor the moisture content of the outlet product in dryers.

DUAL-BED AIR DRYER

The dual-bed air dryer (shown in Figure 19-8) is an example of a fixed bed dryer. It is used on instrument air systems. This type of dryer uses a desiccant such as a silica gel installed in two beds. One bed removes water from the air while the other bed is being regenerated. As the drying agent becomes saturated with water, the beds are switched so that the air flows through the fresh bed. The saturated bed is heated and purged with hot gas (usually air) to strip the water from the drying agent so that it can be reused. Air dryers are used to dry the air for instrumentation and for process use where moisture in the air is undesirable (e.g., unloading a liquid chlorine car).

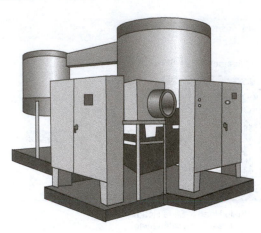

FIGURE 19-8 Dual-Bed Air Dryer

FLUID BED DRYERS

A **fluid bed dryer** (shown in Figure 19-9) uses hot gases to fluidize solids and promote the contact of dry gases and solids. This type of dryer is best suited for the drying of crystals, granules, short fibers, and some powders.

In a fluid bed system, a solid desiccant is fed continuously into the liquid or gas process stream. The gas and liquid phases are separated. The desiccant is typically regenerated and recycled to the dryer. Solids, slurries, and liquids with high moisture content must use either inert beds or dry solids recirculation.

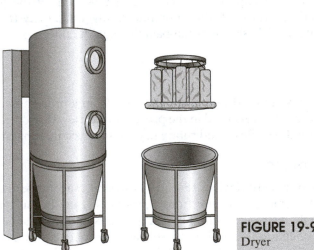

FIGURE 19-9 Fluid Bed Dryer

Desiccant selection is based on its strong affinity for the water or contaminant in the process stream. Some desiccants trap the contaminant by means of a chemical reaction. In this type of system, the undesired contaminant attaches to the desiccant particles and is physically removed from the process. This removal may be performed in a downstream device such as a cyclone separator. The desiccant is then disposed of or regenerated for reuse, depending on the desiccant and contaminant types.

Flash dryers are devices that force hot gas (usually air or nitrogen) through a vertical or horizontal flash tube, thus causing the moist solids to become suspended. Because the feed enters the hot gas stream near the gas inlet, drying takes place within 1 to 3 seconds. The dry product is then carried to the collection system, which is a cyclone or bag collector in most cases. Figure 19-10 shows an example of a flash dryer system.

In this type of system, the dryer must achieve good gas dispersion in order to increase the surface area of the wet solid exposed to the hot gases. To accomplish this, the flash tube incorporates a Venturi (a tube with a constriction used to control fluid

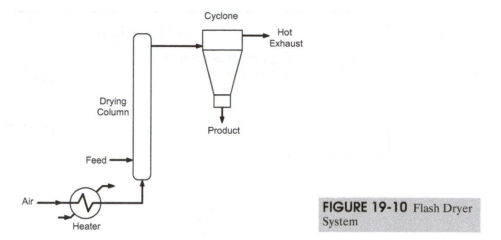

FIGURE 19-10 Flash Dryer System

flow), which allows high-velocity gas to aid in product dispersion. Flash dryers are the most economical choice for drying nonsticky, moist, powdery, granular, and crystallized materials that are generally less than 500 microns. They are simple and take up less space than other drying systems such as a fluid bed dryer. Due to the relatively short residence time, they are also good for heat-sensitive materials. They are not good, however, for materials that require greater dryer residence time.

SPRAY DRYER

A **spray dryer** is a device that sprays a moist feed stream into a vertical drying chamber using a nozzle or rotary wheel. The nozzle or rotary wheel atomizes (makes into a fine spray) the feed stream where it comes into contact with hot gas. The moisture is quickly evaporated, leaving small, dry particles. The wet gas and dry particles exit from the bottom of the drying chamber and flow to a cyclone, where the dry particles are separated from the wet gas. A bag filter is used downstream of the cyclone to capture very fine particles (fines). The wet gas then exits the bag house and is vented to the atmosphere, recycled, or sent to a recovery system. Figure 19-11 shows an example of a typical spray dryer.

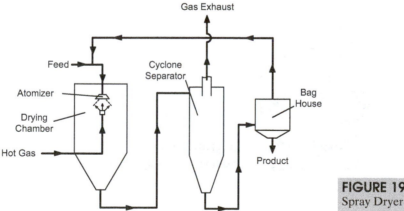

FIGURE 19-11 Typical Spray Dryer

ROTARY DRYER

Rotary dryers are devices composed of large, rotating, cylindrical tubes that reduce or minimize the moisture content of a material by bringing it into direct contact with a heated gas. Rotary dryers employ a slightly inclined rotating cylinder (see Figure 19-12) in which the product discharge end is lower than the inlet end where moist feed enters

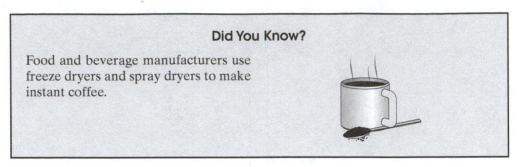

Did You Know?

Food and beverage manufacturers use freeze dryers and spray dryers to make instant coffee.

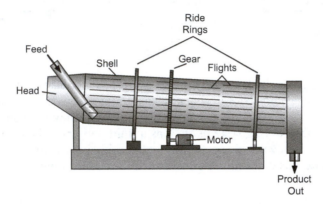

FIGURE 19-12 Outside View of a Rotary Dryer

the dryer. Depending on the application, hot gases may be introduced into either end of the cylinder.

In this type of dryer, the feed is lifted by a series of fins known as flights, which are mounted to the cylinder wall in a helical or spiral configuration (see Figure 19-13). The moist feed collects on the flights until the cylinder rotates to the top of the dryer and the feed falls by gravity from the top to the bottom of the cylinder. The product continues to cascade down the length of the dryer. This process is known as rolling forward or down hill. Finally, the dry product and wet exhaust exit the cylinder at the lowest end of the dryer.

FIGURE 19-13 Cross Section of a Rotary Dryer

SLURRY DRYER

A **slurry dryer** is a device designed to convert wet process **slurry** (a liquid solution that contains a high concentration of suspended solids) into dry powder. In one type of slurry system, wet slurry is pumped to a set of rings in the tower and is atomized by a set of nozzles. This atomized slurry then passes through heated air in the tower and is dried. The nozzles in the tower are designed to atomize the slurry in a consistent pattern so the tower can readily dry the particles into powder form.

Did You Know?

The gas or electric clothes dryer in your home is a type of rotary dryer.

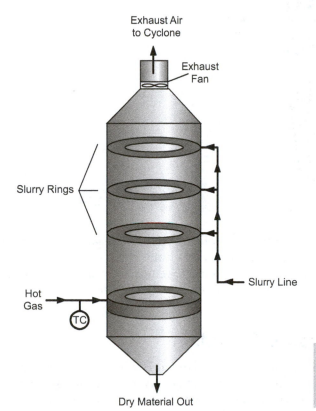

Exhaust Air
to Cyclone

Exhaust
Fan

Slurry Rings

Slurry Line

Hot
Gas

TC

Dry Material Out

FIGURE 19-14 Diagram of a Standard Counter-Current Tower Dryer Used to Dry Slurry

Figure 19-14 is a diagram of a standard countercurrent tower dryer where slurry enters the tower at the top through three nozzle rings and heated air enters at the bottom. The furnace heats the air before it enters the tower. This heated air flow is where the slurry dries to a powder. The dried solid material falls to the bottom of the tower, while the moist air exits the top of the tower into a cyclone.

A **cyclone** is a mechanical device that uses a swirling (cyclonic) action to separate heavier and lighter components of a gas or liquid stream. Because of their design, cyclones are essentially free of moving parts.

Inside a cyclone, air and entrained solids enter the side of the cylinder tangentially through a slot, which creates a rotation of the gas and solids from the top to the bottom of the cylinder. As the gas and solids rotate toward the bottom of the cylinder, the heavier material moves by centrifugal force to the cyclone wall and drops to the conical bottom, where it collects. The lighter particles and gas move to the center of the cyclone and exhaust out of the top (see Figure 19-15).

The majority of cyclones have a conical bottom section. This section spins the solids even faster toward the bottom outlet, thus allowing for disengagement (separation) from the gas stream. At the top, there is a small outlet pipe protruding from the cyclone through which the cleaned gas stream exits. This protrusion into the center of

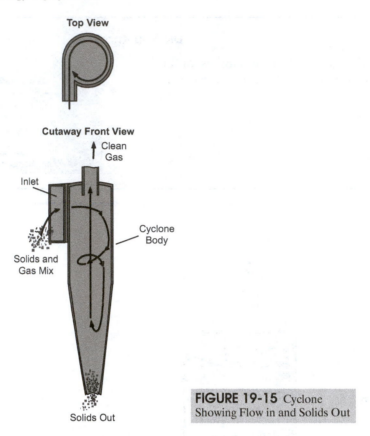

FIGURE 19-15 Cyclone Showing Flow in and Solids Out

the cyclone ensures that few solids leave with the gas stream. This process is very similar to a tornado or to the vortex that forms when water goes down the drain.

Cyclones are used in saw mills to remove sawdust from extracted air. In the petrochemical field, cyclones are used in catalytic crackers to remove the catalyst from the vapor.

Consequences of Improper Operation

Improper operation of filters and dryers can have a serious negative impact on downstream processes and final product quality. The severity of the impact is relative to the consequences of contamination. For example, the presence of contaminant material in the final product may result in costly product rejection and rework. Even worse, if contaminated product reaches the customer, it may be rejected or shipped back to the manufacturer, which may result in monetary penalties, diminished company reputation, and possible loss of business. The following are some additional examples of negative consequences associated with improper filter and dryer operations:

- In a polymer process facility, moisture breakthrough in a feedstock guard bed dryer can cause major loss of polymer catalyst activity and the production of off-spec product.
- A dryer failure in moisture-containing process streams containing highly corrosive compounds such as hydrochloric acid (HCl) can lead to massive mechanical failure of the downstream equipment due to corrosion.
- A filter releasing solids may damage pumps and plug downstream equipment, requiring operations to be shut down for equipment repair and limiting process facility production rates.
- Traces of water can poison a catalyst bed, leading to costly downtime and catalyst replacement.
- Moisture in the instrument air system can eventually lead to broad failure of control system components. For example, water in an instrumentation air line could freeze during cold weather, thus rendering instrumentation inoperable.

- Failure of a mud filter to properly remove sulfurous compounds in a lime kiln can lead to elevated environmental emissions that are above permissible limits.
- The manufacture of chlorine dioxide requires filtered water and air. Failure of the filtering systems can lead to chlorine dioxide decompositions and process shutdown.
- Failure to filter solids from a fluid process stream can plug or damage downstream equipment such as pumps. This reduces facility production rates and may lead to environmental violations because of releases to the atmosphere and/or liquid streams.

Potential Problems

This section describes some of the typical problems associated with filter and dryer system operations and how they can be prevented or corrected.

MOISTURE BREAKTHROUGH

In the case of an air dryer, moisture breakthrough occurs when the dryer output contains excessive moisture. Causes of breakthrough include poor bed regeneration due to insufficient heating and poor contact of the material to be dried with the hot drying gases. Analyzers are used to monitor for moisture breakthrough and ensure that dryer beds are switching and regenerating properly.

Typical air dryer systems use a moisture knockout drum or other type of vessel equipped with a condensate trap to remove most of the liquid collected in the bottom of the knockout drum. Air, now free of liquid, may enter the air dryer.

Compressor systems use filters to capture oil entrained in the compressed air prior to entry into the air dryer. Filter failure allows oil to coat the desiccant, rendering it useless and leading to premature moisture breakthrough. Moisture breakthrough may also result from an instrument malfunction, resulting in incomplete bed regeneration. To prevent moisture breakthrough, the oil filter element and the prefilter system should be inspected and/or replaced on a schedule.

Certain products may contain trace amounts of impurities that are unacceptable to the downstream process and can be economically removed only by guard bed adsorbers. For example, certain polymer streams contain gels that must be filtered out to maintain suitable product quality.

FOULING

Fouling occurs in filters when there is a buildup of solids or if the filter medium becomes coated with sticky materials. Excessive fouling leads to increasing buildup in differential pressure across the filter. Because of this, differential pressure should be continuously monitored to determine when the filter needs to be taken off line for cleaning or replacement.

CHANNELING

An excessive pressure drop across a portion of the dryer bed can cause a condition known as channeling. During channeling, a path is formed in the dryer that allows the material to flow through the dryer without proper contact. If channeling is suspected, the bed should be regenerated to try and disrupt the channel. Failure of an orifice or valve setting may allow too much flow through the unit, causing channeling. Ultimately, physical inspection of the dryer's internal components may be required to ensure that internal baffling, supports, and other components are in place and in order.

To prevent channeling, follow these steps:

- Do not overpressurize dryers, and do not exceed the flow limits.
- Operate within the design limits (e.g., temperatures and flows).
- Ensure that the correct drying agent is installed.

Safety and Environmental Hazards

This section identifies some of the hazards usually associated with normal and abnormal operation of dryers and filters, and describes appropriate safety precautions to avoid injury.

PHYSICAL CONTACT

There is always the potential for tripping or head bumping on equipment. During process upsets and emergencies, there are additional hazards such as chemical contact with the skin or eyes. A potential burn hazard can occur from the hot, high-pressure steam (e.g., when lines are steamed to clear plugs) and from the sudden, loud bursts during compressed air system blow downs. Also, compressed air systems have automatic blow downs that, when engaged, create sudden, loud bursts.

PERSONAL PROTECTIVE EQUIPMENT

In addition to the standard personal protective equipment (PPE) worn at all times in process facilities, goggles must be worn when sampling or operating valves around filters and dryers. If there is the possibility of toxic gases being released, the proper breathing protection apparatus is required. If the line contains either acids or caustics, additional PPE must be worn (e.g., rubber gloves, face shield, slicker suit, and, in some cases, full body acid suits). Face shields must be worn when using steam or high-pressure water to clean equipment or unplug lines.

HOUSEKEEPING

Housekeeping is a routine part of daily operations. Spills and leaks should be cleaned up promptly. During routine or preventive maintenance, lockout/tagout and isolation/clearing procedures must be followed.

ENVIRONMENTAL IMPACT

Depending on the service that the filter or dryer was in and the material that was filtered out (spent material, residue, or sludge), the filter medium itself may be considered hazardous waste. Hazardous wastes have one or more of the following characteristics: ignitability, corrosiveness, reactivity, or toxicity. It is the responsibility of the facility generating the material to make an initial determination. This can be done by testing according to facility procedures or by applying knowledge of the hazardous characteristics of the waste in light of the materials used. After the initial determination, a recharacterization must be performed at least every twelve months or whenever there is a process change.

Some filter systems are used as a treatment method to control environmental emissions on regulated systems. Poor performance or failure of these filtering systems can lead to equipment shutdown, fines from federal or state environmental agencies, and poor working conditions.

Spent filter cartridges, desiccants, precoat materials, bags, and other filtering media should be disposed of according to process facility policies. Depending on the nature of the contaminant, the treatment of the media may simply be to throw the material in the trash or to go through a more elaborate process of removing the contaminant from the media prior to disposal. Each contaminant and medium has a different set of protocols and process facility procedures for disposal.

Typical Procedures

Process technicians are responsible for monitoring filter and dryer system equipment for proper operation and for detecting small problems before they have a chance to become large problems. Process technicians also perform routine and preventive maintenance while complying with all clearing, isolation, and lockout/tagout requirements. Process technicians may also routinely capture samples for lab testing.

All process technicians must be familiar with normal and emergency shutdown procedures. They must know how to respond to emergencies and any regulatory issues concerning the process involved.

MONITORING

Process control systems and log sheets are used to document equipment and process operations. Information found on log sheets and in control system history databases provide trends and operating conditions that allow process technicians to troubleshoot and problem-solve issues as they arise. In many regulated systems, these log sheets and databases are required as proof of proper operation for the systems.

LOCKOUT/TAGOUT

Lockout/tagout procedures ensure that the equipment is in a zero-energy state while maintenance or inspection is performed. Process technicians are required to follow the lockout/tagout procedure for their process unit. A generic summary of lockout/tagout procedures is as follows:

1. The equipment is isolated from the process; that is, pump breakers are de-energized, equipment is depressurized and drained, and the equipment is cleaned (purged) of process fluids or gases.
2. All forms of possible energy sources (for example, pneumatic, hydraulic, electrical, mechanical, and nuclear [radioactive]) are eliminated and prevented from recurring,

ROUTINE AND PREVENTIVE MAINTENANCE

Maintenance is generally scheduled and may include cleaning basket strainers, changing filter elements, or other duties as needed.

EMERGENCY

Most companies have emergency procedures in place to respond to sudden conditions that could result in catastrophic releases or personnel injuries, such as a line rupture or fire, power failure, or instrument air failure. At times, this involves an emergency shutdown of the unit.

SAMPLING

Process technicians are responsible for keeping a process facility operating safely and efficiently to ensure the production of a high-quality product. More specifically, process technicians are responsible for controlling and monitoring process systems, inspecting equipment, and conducting routine system operations. Process technicians also routinely sample and test process fluids and solids at various stages of the production process. These tasks are necessary to help ensure that a high level of product quality is maintained.

Samples are taken in process systems to detect problems early so that equipment is not damaged and products are not wasted. If the testing of a sample indicates that a material is unacceptable, the final product could be wasted unless actions are taken to correct the problem. Samples are also taken to verify that waste products discharged into the environment are in compliance with company and government regulations.

Depending on the materials and work areas involved, there may be certain hazards associated with taking samples. Process technicians should be familiar with the materials used in their process systems to determine the hazards, if any, that exist. For instance, several hazards are associated with acidic and caustic materials. If these types of materials come in contact with skin or eyes, they could cause burns or blindness.

Toxic fumes are another hazard that may be present when some materials are sampled. If these fumes are inhaled, they could cause serious injury or even death.

Some materials may be under pressure or at very high temperatures. When such materials are sampled, process technicians should exercise care to avoid being sprayed or burned. Other materials are flammable or explosive, and they could be ignited by the smallest spark or flame, or even static electricity in the atmosphere. To avoid serious injuries, no smoking or open flames are allowed around these materials.

In addition, special care should be taken to prevent spills when materials are sampled. A spill could cause someone to fall and be injured. If a spill occurs, report it immediately to first-line management and clean it up promptly and properly to prevent an accident.

When sampling or handling materials, process technicians must take all applicable precautions to prevent exposure. For example, appropriate protective clothing must be worn. Any excess or unused portions of samples should be disposed of properly according to company procedures. Depending on the sample to be retrieved, the container that the sample is placed in may be a glass jar, a vial, or a bottle. In the case of pressurized gas, metal pressure cylinders may also be used.

An important part of taking samples in a process system is knowledge of where the sample points are located. The locations of sampling points can be found on process facility diagrams or by consulting facility procedures. In addition, the central lab should have a master printout of all the sample points in the process facility. A sample point generally consists of a small diameter pipe extending off the main process piping system. A valve in the sample line can be operated to start and stop the flow of the sample to the container.

Another important point to consider when taking samples is to make sure that the samples are not contaminated with other materials. The basic purpose of sampling is to take a small portion of the process fluid so that it can be tested to see if it meets the desired specifications. If other materials are allowed to contaminate the sample, the test results could be inaccurate. To safeguard against this, the process technician taking the sample must ensure that the sample container is clean and free from contaminants. Once the sample container is filled, the process technician should immediately seal it off from the atmosphere by tightly closing the cover or cap to prevent contamination while in route to the laboratory.

When obtaining samples, it is important to ensure that the amount of material drawn is representative of the actual material in the process system. It may be necessary to allow the fluid to drain into a drainpipe or other draining system for a minute or more in order to retrieve a "pure" sample that has not degraded while sitting static in the line over a long period of time.

Some samples are difficult to collect. For example, the process to be sampled may be under negative pressure, or vacuum. In this case, a vacuum pump is connected to the sample line and the sample is drawn into the sample container.

Process Technician's Role in Operation and Maintenance

The role of the process technician is to ensure the proper operation of the filters and dryers and to be aware of problems that can affect the system. Table 19-1 lists some of the details that a process technician should monitor when working with filters and dryers.

TABLE 19-1 Process Technician's Role in Operation and Maintenance

Look	Listen	Feel
• Check for desiccant on the surface floor or slab under dryers. • Look for pools of liquid on the floor or slab. • Observe instrumentation regularly. • Check for excessive pressure drops across the filter. • Examine the filtered product for evidence of contamination. • Check for leaks around the filter closure. • Insure that a proper micron rating filter is used.	• Listen for unusual noises.	• Check for high temperatures. • Feel for excessive vibrations.

Summary

Filters are used to remove particles, and dryers are used to remove moisture. Proper operation of these systems prevents potentially serious problems from developing in downstream equipment and processes.

Cartridge filters force process fluids through pores and remove solid particles. Bag filters contain a bag through which process fluids flow and particles are collected. In a leaf filter, a precoated screen is mounted inside the filter vessel to hold filtered material. Rotary drum filters use a cloth mounted to the inner wall of the filter to retain solids. Pleated cartridge filters employ a filter medium in a pleated form to provide more surface area for filtration. Plate and frame filters are used to remove liquids from slurry and are used in coatings, resins, and ink processes. All cartridges must be replaced when they become filled with particles.

A fixed bed dryer uses a desiccant to absorb moisture. Distributors may be used to promote even distribution of the feed across the dryer bed and prevent channeling. Many fixed bed dryers have two beds: one in service, and the other on standby. When the in-service bed is saturated, the dryer is switched to the fresh, standby bed while the saturated bed is regenerated. Some fixed bed dryers use two beds in series so that any contaminant material that escapes the first bed is caught in the second bed.

Fluid bed dryers use hot gases to fluidize a solid desiccant or process product to promote contact between the dry gases and the solids. Gas and liquid phases are separated, and the desiccant is usually rejuvenated and recycled to the dryer. Product dried in a fluid bed dryer is typically metered out of the dryer at the same rate that the flow enters the dryer to ensure a stable, continuous bed depth.

Many filter and dryer systems use screens to prevent particles from carrying over with the process stream. Leaf filters and drum filters use a precoated screen to catch and hold the filter material.

Flash dryers are used to remove moisture from hot process fluids. A sudden pressure drop across an orifice causes the moisture to flash, converting it from a liquid to a vapor so it can be routed away from the process and removed.

Improper operation of filters and dryers can cause serious problems in downstream equipment and affect the quality of the finished product. For example, moisture breakthrough can diminish drying and filtering performance, causing the product to be off-spec. Solids that are not trapped can damage or plug downstream equipment, creating a need for shutdown and limiting production rates.

Process technicians are responsible for the routine monitoring of filter and dryer equipment and for the early detection of problems. Process technicians may also be responsible for performing routine housekeeping tasks, catching samples for testing, and performing routine and preventive maintenance according to established procedures while complying with all lockout and tagout requirements. Technicians must also know how to perform normal and emergency shutdowns, and how to respond to emergencies.

Problems associated with filter and dryer systems include moisture breakthrough, fouling, and channeling. Close monitoring can prevent these problems from occurring or minimize their impact.

Checking Your Knowledge

1. Define the following terms:
   a. Breakthrough
   b. Desiccant
   c. Dryer
   d. Filter
   e. Micron
2. One purpose of filters is to:
   a. add air or other gases to a process stream
   b. remove moisture from a process stream
   c. remove particles from a process stream
   d. none of the above

3. A bag filter is used to:
    a. control process variables in process streams
    b. direct liquids or slurries into the filter or dryer
    c. circulate precoat through the filter
    d. remove solids from vent gases received from dryers and air-conveying systems
4. Where is a leaf filter located?
    a. Outside the filter vessel
    b. Inside the filter vessel
    c. None of the above
5. One purpose of dryers is to:
    a. remove trace chemical components of a stream
    b. remove moisture
    c. clean the water in process streams and products
    d. none of the above
6. A _____ is a piece of equipment that uses swirling (cyclonic) action to separate heavier and lighter components.
    a. cyclone
    b. conveyor
    c. dryer
    d. decanter
7. This _____ dryer uses hot gases to fluidize solids and promote the contact of dry gases and solids.
    a. fluid bed
    b. flash
    c. fixed bed
    d. rotary
8. A dryer used to dry sugar in a sugar mill is a:
    a. fluid bed dryer
    b. flash dryer
    c. rotary dryer
    d. fixed bed dryer
9. *(True or False)* Filters are rated based on their ability to remove particles of various sizes from a fluid.
10. A _____ occurs when the fixed bed desiccant is so saturated it allows significant amounts of contaminant material to pass through the dryer.
    a. breakthrough
    b. saturation limit
    c. contamination replacement
    d. desiccant trap

Student Activities

1. Write a two-page paper on the types of filters and dryers used in the process industries. Be sure to explain the purpose of each type.
2. With a classmate, prepare a presentation on the differences among fluid bed dryers, flash dryers, and fixed bed dryers.
3. List three safety and environmental hazards associated with filters and dryers.
4. Research moisture breakthrough with a classmate and list possible corrective actions and steps for prevention.

CHAPTER

20

Solids Handling Equipment

Objectives

After completing this chapter, you will be able to:

■ Explain the purpose of solids handling equipment in the process industry.

■ Identify the common types and applications of solids handling equipment.

■ Identify the components of solids handling equipment and explain the purpose of each.

■ Describe the operating principles of solids handling equipment.

■ Describe safety and environmental hazards associated with solids handling equipment.

■ Identify typical procedures associated with solids handling equipment.

■ Describe the process technician's role in solids handling equipment operation and maintenance.

■ Identify potential problems associated solids handling equipment.

Key Terms

Bin—a container used to store solid products or materials; typically designed to hold a finished product destined for distribution, but may also store off-specification product earmarked for re-work or blending.

Bucket elevator—a continuous line of buckets attached by pins to two endless chains running over tracks and driven by sprockets.

Conveyor—a mechanical device used to move solid material from one place to another in a continuous stream.

Die—a metal plate with a specifically shaped perforation designed to give materials a specific form as they pass through an extruder.

Extruder—a device that forces materials through the opening in a die so it can be formed into a particular shape.

Feeder—a piece of equipment that conveys material to a processing device (e.g., an extruder) at a controllable and changeable rate.

Gravimetric feeder—a device designed to convey material at a controlled rate by measuring the weight lost over time.

Hopper—a funnel-shaped, temporary storage container that is typically filled from the top and emptied from the bottom.

Live bottom—a vibratory device, attached to the bottom of a bin or silo, that is designed to facilitate the smooth discharge of material.

Pneumatic conveyor—a device used to pneumatically move material from one point to another in a continuous stream.

Rotary valve—a valve that uses a set of rotating pockets to meter a fixed volume of solids at a fixed rate. Also referred to as an airlock.

Screener—a piece of equipment that uses sieves or screens to separate dry materials according to particle size.

Silo—a tower-shaped container for storage of materials; typically used to store product destined for distribution.

Solids handling equipment—equipment used to process and transfer solid materials from one location to another in a process facility; may also provide storage for those materials.

Trickle valve—a valve used to continuously transfer a fixed weight of solids between two different pressure zones at a constant rate.

Volumetric feeder—a device designed to convey materials at a controlled rate by running at a set motor speed.

Weighing system—a system used to weigh material before shipping.

Introduction

The process industries use many types of equipment to process, transfer, store, and ship solids. In some processes, solid material is required as a feed stock. In these processes, process technicians must often unload solids from their containers, such as railcars, into larger, long-term storage equipment. This solids movement occurs through both mechanical and pneumatic means.

Solids handling equipment is designed so process solids can behave similarly to liquids. For example, a conveyor moves solids from one place to another in a continuous stream, much like liquid flowing in a pipe. Bins hold a large volume of solid material much like a chemical storage tank holds a large volume of liquids. Compressed air is often used to push material from a bin to a hopper car or truck like a pump is used to push liquids from a tank to a tank car or truck.

Types of Solids Handling Equipment

Solids handling equipment is used to process and transfer solid materials from one location to another in a process facility; it may also provide storage for those materials. Some of the most common types of solids handling equipment include feeders, extruders, screening systems, conveyors, elevators, storage containers, valves, and weighing systems.

FEEDERS

A **feeder** is a piece of equipment that conveys material to a processing device (e.g., an extruder) at a controllable and changeable rate. This ability to change feed rate is the distinguishing difference between feeders and conveyors.

Feeders can operate gravimetrically (based on weight) or volumetrically (based on volume). The rate of the feeder can be set based on material rate (e.g., pounds per hour) for gravimetric feeders or motor speed (e.g., revolutions per minute) for volumetric feeders.

Gravimetric feeders (shown in Figure 20-1) are devices designed to convey material at a controlled rate by measuring the weight lost over time. While in operation, gravimetric feeders take an internal weight measurement for the material loaded, record that weight, feed material for a specific amount of time, and then take another internal weight measurement. The feeder then compares the weight lost over time (actual rate) with the target rate and makes automatic adjustments if needed. This type of feeder is usually referred to as a loss weight feeder.

FIGURE 20-1 Example of a Gravimetric Feeder

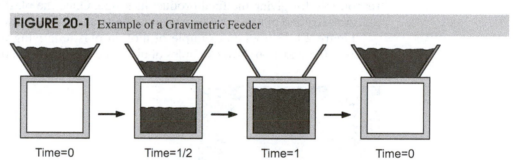

| Time=0 | Time=1/2 | Time=1 | Time=0 |

Volumetric feeders (shown in Figure 20-2) are devices designed to convey material at a controlled rate by running at a set motor speed. Volumetric feeders must have a predetermined maximum feed rate that is calculated by the feeder's maximum motor speed and material density. For example, if a feeder is capable of 400 pounds per hour (pph) at its maximum motor speed (100 percent), then the controller should be set to 50 percent output to run the feeder at 200 pph.

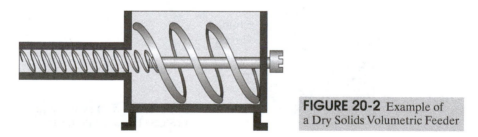

FIGURE 20-2 Example of a Dry Solids Volumetric Feeder

EXTRUDERS

An **extruder** is a device that forces materials through the opening in a die so it can be formed into a particular shape. Extrusion is a continuous process for making many products, including pasta, chewing gum, certain candies, plastic films, sheets, tubes, plastic pipes, and many other items.

In the plastics extruding process, plastic material (e.g., granules, pellets, or powder) from a hopper is fed into a long, cylindrical, heated chamber where the material moves by the action of a continuously rotating screw. This cylindrical chamber is known as an extruder and is composed of three zones: the feed zone, the compression zone, and the metering zone (see Figure 20-3).

Material enters the extruder through the feed zone (no melting takes place in this zone) and then moves to the compression zone. At the start of the compression zone,

FIGURE 20-3 Example of a Continuous Extrusion Feeder Showing the Feed Zone, Compression, Zone, Metering Zone, and Extrusion

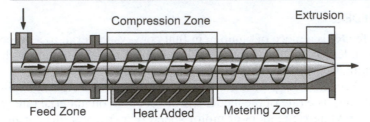

friction and compression-induced heat begin the melting process. Once the machine is running, most of the energy required for melting is mechanical energy resulting from the compression and shearing of feed material by the screw. However, additional heat is added to the extruder wall.

At the end of the compression zone is the metering zone, where the molten plastic is forced out through a small opening called a **die** (a metal plate with a specifically shaped perforation designed to give extruded material a specific form). The die is responsible for giving the final product its shape. Once the plastic passes through the die, the extruded plastic is cut by a rotating knife and fed into a water bath for cooling.

Figure 20-4 shows an example of molten plastic moving through an extruder. Figures 20-5 and 20-6 show examples of dies used to shape final products.

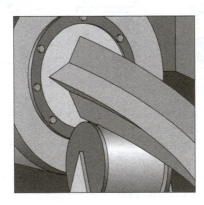

FIGURE 20-4 Molten Plastic Passing Through an Extruder Die

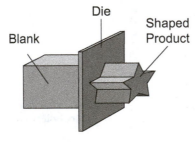

FIGURE 20-5 Dies are Used to Shape Extruded Products

FIGURE 20-6 Examples of Extrusion Dies for Fibers and Plastics

SCREENING SYSTEMS

Screening systems (e.g., sieves, sifters, and screens) are used to separate particles by size. Particles that are too small (called fines) are reprocessed or recycled, and particles that are too large (called overs) are re-milled or recycled.

Screeners are pieces of equipment that use sieves or screens to separate dry materials according to particle size. In a screening system, materials enter the top of the screener and are distributed across the surface of the screen. Vibration is then used to further distribute the material across the screen and convey it toward the discharge end.

Some screening systems contain multiple levels; the top tier contains the largest screen openings and the bottom tier contains the smallest (see Figure 20-7). Some screening systems also contain balls confined in pocket areas beneath the screen surface. The bouncing action of these balls dislodges particles by direct contact, thereby cleaning the screens. These resilient balls also keep the screen surface alive, providing agitation to separate particles that may stick together.

FIGURE 20-7 Separating Screens Go from Largest to Smallest

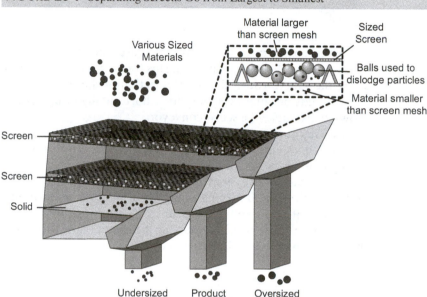

CONVEYORS

A **conveyor** is a mechanical device used to move material from one place to another in a continuous stream. Types of conveyors include belt, screw, moving floor, drag chain, vibratory, and powered roller.

A belt conveyor consists of a belt supported by pulleys that run in a continuous loop. Belt conveyors can be used to move items such as coal, petroleum coke, or grain to a storage or packaging area. Conveyor belts are not always horizontal. In fact, they often move material on an incline, either up or down. An example of commonly used belt conveyors are the ones found in grocery store checkout lines. These conveyors help move the groceries toward the cashier and speed up the checkout process. Figure 20-8 shows an example of a belt conveyor.

FIGURE 20-8 Belt Conveyor

Courtesy of Sun Products
Corporation

FIGURE 20-9 Inside View of a Screw (Helical) Conveyor

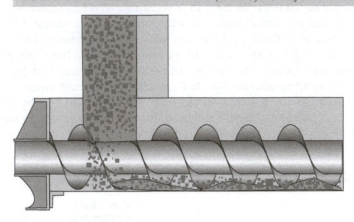

Screw conveyors are conveyors that use a helical screw, rotating within a stationary tube or trough, to convey material. This rotating screw propels liquids or fluidized solids along the tube through the pushing action of the screw blades. Screw conveyors can move liquids or fluidized solids either horizontally or upward at an incline. Figure 20-9 shows an example of a screw conveyor.

Moving floor conveyor systems are ground-based conveyer systems designed to transport both light and heavy loads such as pallets into a trailer. Figure 20-10 shows an example of a moving floor conveyor.

Drag chain conveyors consist of two or more parallel chains moving in the same direction along a conveying duct. A gripping mechanism made from a rubber-like material is attached to the chain links. These gripping mechanisms grab and push the conveyed material along the duct in the direction of the moving chains. Figure 20-11 shows an example of a drag chain conveyor.

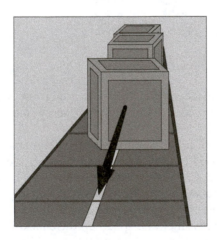

FIGURE 20-10 Example of a Moving Floor Conveyor

Did You Know?

An escalator in an office building or department store is considered a moving floor conveyor.

FIGURE 20-11 Example of a Drag Chain Conveyor

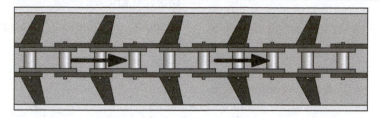

Vibratory conveyors typically include a conveyor trough or deck that is shaken by a vibratory driver to produce motion in the material carried in the trough. Mechanical vibratory drivers typically include a number of eccentrically mounted weights that, when rotated, create vibration in the trough or deck, thus conveying the materials along the deck. Figure 20-12 shows an example of a vibratory conveyor.

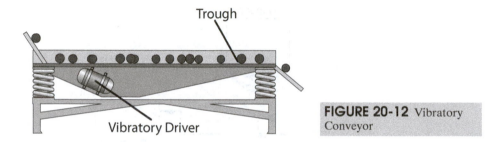

Trough

Vibratory Driver

FIGURE 20-12 Vibratory Conveyor

Roller systems are conveyor systems that use a series of rotating cylinders (rollers) to move materials. Roller movement in these types of systems can be either gravity-driven or powered. In a gravity-driven system, the cylinders are mounted in parallel and suspended in a frame by bearings so they turn freely as materials (e.g., bags or boxes) move across the rollers. Powered roller systems use chains or belts to turn the rollers and move the load. Because they are mechanically driven, powered roller systems can be used to move materials horizontally, at an upward incline, or in a downward direction. Gravity-driven rollers, on the other hand, are designed to move only horizontally and in a downward direction (by the force of gravity). Figure 20-13 shows an example of a powered roller system.

FIGURE 20-13 Example of Roller System

Courtesy of Sun Products Corporation

Pneumatic Conveyors

Pneumatic conveyors are devices used to move material pneumatically from one point to another in a continuous stream. There are two types of pneumatic conveying systems: dilute phase and dense phase. Both systems use gas (usually air) to transfer suspended solids both horizontally and vertically through pipe lines. These systems can convey a wide range of solids, from fine powders to pellets.

FIGURE 20-14 Dilute Phase Pressure Operation

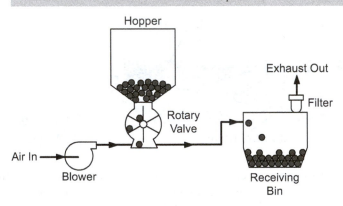

Dilute phase systems use a conveying gas volume and velocity sufficient to keep the solids that are being transferred in suspension. These systems are used to convey solids continuously so that the material does not accumulate in the bottom of the conveying line at any point. There are three types of dilute phase conveying systems: a pressure driven operation, a vacuum driven operation, and a pressure/vacuum operation. A pressure-driven dilute phase system (shown in Figure 20-14) uses a blower to blow (push) gas (air) through a line, where it picks up solids dropped into the line from a hopper/silo rotary valve. The solids entrained in the flowing gas are transferred to a bin where they drop and are then stored. The solids-free gas is then vented from the bin.

Dense phase systems operate with product velocities below the saltation velocity (the critical velocity where particles fall from suspension in the pipe); thus, the flow of material is not continuous but moves in waves or plugs. Typical dense phase conveying systems use a small volume of high-pressure gas to push plugs or waves of solids through a conveying line. Wave movement occurs as the air continuously flows over the material that has collected on the bottom of the pipe, whereas plugs are separated by zones of gas as they move through the transfer line. Figure 20-15 shows a comparison of dilute and dense phase conveying systems.

The advantage of a dense phase conveying systems is lower gas requirements. This translates to lower energy requirements and lower solids velocities. Lower velocities mean that abrasive and friable materials may be conveyed without significant transfer line erosion or product degradation.

PROCESS ELEVATORS

A **bucket elevator** (shown in Figure 20-16) is a continuous line of buckets that are attached with pins to two endless chains running over tracks and driven by sprockets. Bucket elevators are designed to move bulk material vertically. The materials carried by bucket elevators can be light or heavy, and they can range in size from large to fine (e.g., large pieces of gravel to small pieces of grain).

FIGURE 20-15 Comparison of Dilute and Dense Phase Conveying Systems

Dilute Phase Conveying System
(constant suspension)

Dense Phase Conveying System
(plugs or waves)

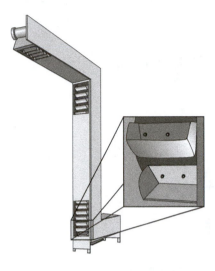

FIGURE 20-16 Example of a Bucket Elevator

Most modern bucket elevators use rounded rubber buckets mounted on a rubber belt that is rotated by very large pulleys (several feet in diameter), both at the top and bottom of the elevator. The top pulley (the drive pulley) is driven by an electric motor. Once the load is picked up from a pit or a pile of material, it is then carried to an alternate location and discharged.

True vertical conveyors run at high speeds, so they discharge the material at the top of the elevator by centrifugal force (see Figure 20-17). Incline conveyors run at a slower speed and discharge under the head pulley, and they often use a trip or wiper device to empty the bucket.

FIGURE 20-17 Bucket Elevator Emptying by Centrifugal Force

SILOS, BINS, AND HOPPERS

Silos, bins, and hoppers are used in the process industries to store materials (usually dry) that are to be used later in a process or sold to a client. Bins, silos, and hoppers can be made of many materials (e.g., concrete, wood, and steel) and normally have a valve (slide or rotary) fitted at the bottom.

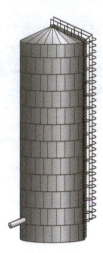

FIGURE 20-18 Example of a Silo

FIGURE 20-19 Example of a Bin

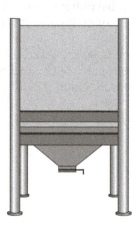

FIGURE 20-20 Example of a Hopper

A **silo** (shown in Figure 20-18) is a large, tower-shaped container designed to store materials. Silos are typically used to store products destined for distribution.

Bins (shown in Figure 20-19) are containers used to store solid products or materials. They are typically designed to hold a finished product destined for distribution.

Hoppers (shown in Figure 20-20) are funnel-shaped, temporary storage containers typically filled from the top and emptied from the bottom. Hoppers are smaller than bins or silos and are normally mounted to other types of equipment that are fed directly by them. An example of a hopper would be the storage vessel that directly feeds an extruder.

Bins, silos, and hoppers are designed to receive and store feed from the process until it is eventually transferred to bagging operations, hopper cars, or hopper trucks. To facilitate the transfer of materials, many of these containers are fitted with a gyrating bin discharger called a live bottom.

Live Bottom

A **live bottom** is a vibratory device, attached to the bottom of a bin or silo, that is designed to facilitate the smooth discharge of material. Live bottoms contain a flow-inducing, vibrating device that does the following:

- Keeps the weight of the product inside from compacting onto the equipment below and jamming the exit area of the cone
- Changes the powder draw from rat holing (center feed) to a more uniform feed.

Figure 20-21 shows a detached live bottom, and Figure 20-22 shows an attached live bottom. Figures 20-23 and 20-24 show how products flow with and without a live bottom, respectively.

FIGURE 20-21 Live Bottom Attachment

Courtesy of Sun Products Corporation

FIGURE 20-22 Live Bottom Attached to the Base of a Bin

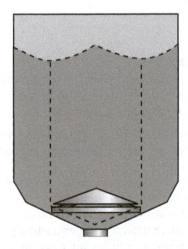

FIGURE 20-23 How Product Flows with a Live Bottom

FIGURE 20-24 How Product Flows without a Live Bottom ("Rat Holing")

Flow Inducers, Bin Vibrators, and Bin Activators

Because of the characteristics of some solid materials, flow must sometimes be assisted. Flow inducers, bin vibrators, and bin activators are used to assist flow.

Flow inducers and bin vibrators are connected to chutes, hoppers, and other storage devices to release accumulated product from the interior walls of the equipment. A bin vibrator refers to a piece of equipment that is mounted to a bin or silo. A flow inducer refers to one that is mounted to chutes or other types of equipment. Figure 20-25 shows an example of a piston vibrator attached to a bin. Figure 20-26 shows an example of the vibrator mechanism from a vibratory hopper.

FIGURE 20-25 Piston Vibrator Attached to a Bin

Courtesy of Sun Products Corporation

FIGURE 20-26 Rotary Vibrator Attached to a Bin

Courtesy of Sun Products Corporation

Hopper Cars and Hopper Trucks

Hopper cars and hopper trucks are designed to carry solid materials by rail or over roads. The descriptions included in this text are specific to hopper cars. However, the very same technology is also employed by hopper trucks.

Hopper Cars Hopper cars are used in the process industries to transport material in large quantities from one facility to another. Some hopper cars are a single bin, while others have multiple bins (chambers).

Figure 20-27 shows a hopper car for transporting solid particles. Figure 20-28 illustrates the inside configuration of a multi-bin hopper car.

There are two types of access covers on the top of most solids handling hopper cars. The first is a standard access cover used to gain entry to a product bin. The other is a suspiration (breather) vent designed to allow air to come in and out of the bins while preventing rain or other contaminants from entering.

FIGURE 20-27 Outside Picture of a Hopper Car Used to Transport Solids

Courtesy of Sun Products Corporation

FIGURE 20-28 Illustration of a Multi-Bin Hopper Car

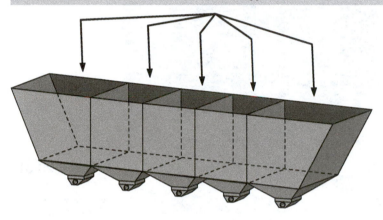

Figure 20-29 shows a series of access covers on the top of a railcar, and Figure 20-30 shows the design of a standard access cover. Figure 20-31 shows a regular suspiration cover (breather vent). Figure 20-32 shows an old-style breather vent.

FIGURE 20-29 Top Loading Area of a Hopper Car

Courtesy of Sun Products Corporation

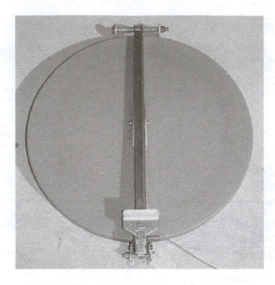

FIGURE 20-30 Standard Access Cover

Courtesy of Sun Products Corporation

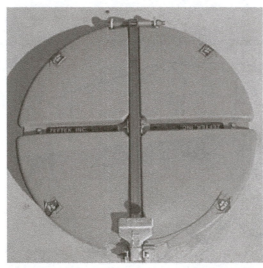

FIGURE 20-31 Regular Air Suspiration Cover

Courtesy of Sun Products Corporation

FIGURE 20-32 Old Style Air Suspiration Cover

Courtesy of Sun Products Corporation

Figure 20-33 shows the opening bin at the bottom of a railcar. Figure 20-34 shows a transfer hose from the bottom of a hopper car pipeline used for loading or unloading. In this case, either the material will be transferred pneumatically to storage, or the hopper car will be pressurized in order to push the product from the hopper car to an atmospheric bin for storage or further processing.

FIGURE 20-33 Rail Car Bottom Outlet and Valve

Courtesy of Sun Products Corporation

FIGURE 20-34 Hose Used to Load and Unload Product from a Railcar

Courtesy of Sun Products Corporation

Some railcars are designed to be unloaded by dropping the material from the railcar into a pit below the tracks. The interior of this type of bin is very simple. There are no working parts other than the slide gate, which operates from the outside. Figure 20-35 shows the inside of a railcar with a small amount of material (commonly known as a heel) remaining.

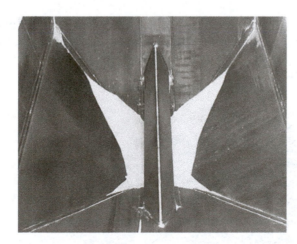

FIGURE 20-35 Top View of a Railcar Containing a Small Amount of Material

Courtesy of Sun Products Corporation

Figure 20-36 shows the boot extended at the bottom of a railcar. Individual railcar bins for this type must be lined up directly over the top of the pit to begin offloading. The operator activates the boot to seal off the railcar bin. The solid particles drop into the pneumatic conveyor system.

FIGURE 20-36 Slide Gate at Bottom of Railcar with Boot Extended

Courtesy of Sun Products Corporation

At times there may be more than just the standard one-gate opening linkage. In Figure 20-37, we can see two gate opening linkages and a pneumatic conveyor port, both shown here at the bottom of the railcar. There are two standard linkages to assist in the alignment of the railcar bin and the drop area. The pneumatic port gives the option of using an aboveground pneumatic conveyor system.

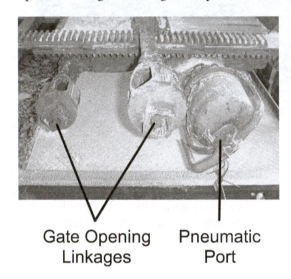

Gate Opening Linkages   Pneumatic Port

FIGURE 20-37 Gate Opening Linkages and Pneumatic Ports at the Bottom of a Railcar

Courtesy of Sun Products Corporation

Some powder railcars are designed as a single bin with multiple unloading ports. Although this type of railcar still has multiple unloading ports, there are no divider walls between the areas. Figure 20-38 shows an interior picture of a single bin railcar with multiple unloading ports.

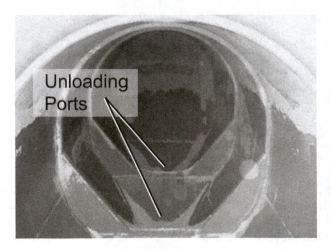

Unloading Ports

FIGURE 20-38 Interior Picture of a Single Bin Rail Car with Multiple Unloading Ports

Courtesy of Sun Products Corporation

Figure 20-39 shows the outside of a railcar. Figures 20-40 and 20-41 show the inside of a container railcar with the piping systems displayed.

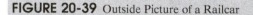

FIGURE 20-39 Outside Picture of a Railcar

Courtesy of Sun Products Corporation

FIGURE 20-40 Drawing of a Containerized Railcar with Piping System Displayed

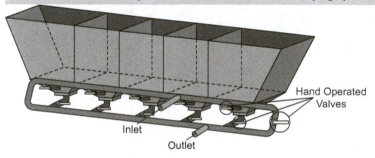

Hand Operated Valves

Inlet

Outlet

FIGURE 20-41 Bin Area Being Offloaded with Flow Displayed

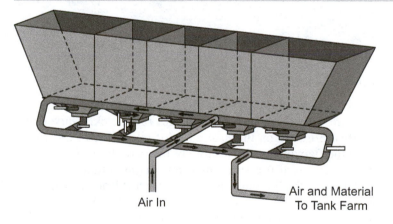

Air In

Air and Material To Tank Farm

The access doors at the top of the railcars are all the same as regular solid access doors (i.e., simple open/close doors). However, they have different latching mechanisms and sizes, and they are designed to maintain pressure environment adjustments inside the car. Notice that the access doors in Figure 20-42 appear thicker and have multiple latches. This is because the railcars are pressurized during offloading. The access doors are set up to be safer under pressure. This type of railcar has one open product bin, and each bin section is usually offloaded individually.

Figure 20-42 shows the top side of a railcar with latching systems, and Figure 20-43 is a close-up picture of the access door.

FIGURE 20-42 Railcar Showing Latching System to Maintain Pressure Adjustments Inside the Car

Courtesy of Sun Products Corporation

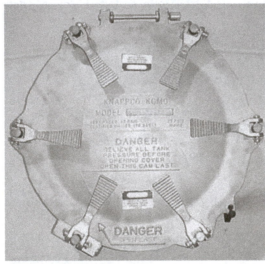

FIGURE 20-43 Close Up Picture of an Access Door

Courtesy of Sun Products Corporation

VALVES

A **trickle valve** is used to transfer a fixed weight of solids continuously between two different pressure zones at a constant rate. Trickle valves operate with a vacuum or negative pressure on the upstream (or inlet) side of the valve flap, and that pressure holds the flap against the bottom of the valve housing. This free-hanging flap between the two different pressure environments maintains an air seal. This type of valve is often used at the bottom of a cyclone. Product from the cyclone collects on the upstream or negative air pressurized side of the channel. When the accumulated weight of the product overcomes the negative air pressure, the product will begin to trickle through the flap. Figure 20-44 shows how the trickle valve flap works in a negative air pressure environment within a shoot.

A **rotary valve** (also referred to as an airlock) uses a set of rotating pockets to meter a fixed volume of solids at a fixed rate. Rotary valves are designed to operate in the same environment as the trickle valve. Generally, the low-pressure environment is located above, or upstream from, the valve, and the higher-pressure environment is located downstream from, or below, the valve. This type of valve uses a chambered paddle wheel that rotates inside a cylindrical housing. The valve is open to the process

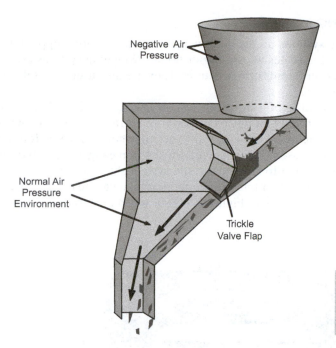

Negative Air
Pressure

Normal Air
Pressure
Environment

Trickle
Valve Flap

on the upstream side or top of the valve, where material flows into a chamber. The paddle wheel rotates through the sealed chamber, and the material drops from the valve on the downstream side or bottom of the valve, thus maintaining the differential pressure between the upstream and downstream pressure environments. Figure 20-45 is a diagram of an airlock, and Figure 20-46 is an actual picture of a rotary valve.

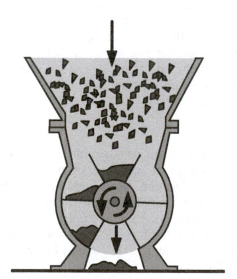

WEIGHING SYSTEMS

Weighing systems are used to weigh material before shipping. Weighing systems are similar to the scales we use in our homes. Weighing systems are important because of weight requirements or restrictions. They are also used to establish the weight shipped to a customer.

An example of a weighing system is a series of load cells used on surge bins to show the current amount of material in storage. Load cells work by measuring the changes in electrical resistance that occurs when a resistive foil mounted on a backing material is stressed (strain gauge). Most strain gauges are of a double-switch type. This allows one switch to fail without the system shutting down. Because strain gauges are often mounted under bins and hoppers, replacing one usually necessitates considerable downtime and expense. Figure 20-47 shows an example of a strain gauge.

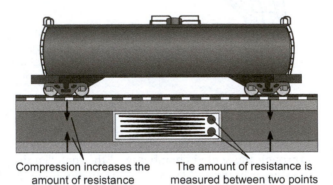

Compression increases the amount of resistance

The amount of resistance is measured between two points

FIGURE 20-47 Train Scale Using a Strain Gauge System

Multiple strain gages are typically used to measure one bin. The use of three gauges is preferred because it allows a more stable system than a two-point system, where overloading on one side or the other may occur.

Weighing systems for hopper cars and hopper trucks use a system of load cells mounted to a steel frame under the rails to provide the weight measurement. Modern railroad weigh scales allow the automated weighing of entire trains as they cross the scale at speeds up to 10 km/hour (6 mph).

BULK BAG STATION

A bulk bag (or super sack) station is a piece of equipment designed to fill bulk bags to a predetermined weight. Bag filling can be manual or automated. Solid material flows into the bulk bag from a storage location. Bag weight is constantly measured using a load cell, and material flow is terminated when the desired weight is achieved. Automated systems are designed to have an automatic feed system that fills, weighs, and seals the bag, and removes and transfers it to a storage location for shipment. Bulk bags typically weigh between 1,000 and 2,000 pounds. Figure 20-48 shows a finished bag after being loaded in a bulk bag station.

Large bags can be desirable because they contain a large quantity of product in one container. Bulk bag handling equipment can be costly, however, so smaller bags (typically 50 pounds) are often used instead.

BAGGING OPERATIONS

Some customers require finished products to be packaged in smaller quantities. In these instances, bags are often the preferred container. Bags come in all sizes, for example, 50-pound bags of corn, 25-pound bags of dog food, and 30-kilogram bags of concrete. In general, these bags use the same basic packing equipment. Packaging operations range from fully automated to fully manual.

FIGURE 20-48 Bulk Bag
Finished and Ready to Ship
from Bulk Bag System

Courtesy of Sun Products
Corporation

After the finished product is classified (e.g., run through to screening equipment so it can be sorted to the desired size), the material is pneumatically transferred to a hopper that is part of the bagging equipment. Depending on the type of bag to be filled, modern bagging equipment will grab an empty bag, place it on a filling nozzle, and meter the product into a bag, which sits on weigh cells until the bag reaches the target weight. At that point, the flow of product stops. In an open mouth bag (shown in Figure 20-49), the bag is then transferred immediately to a sealing mechanism that glues or sews the bag closed. In the case of a valve bag (shown in Figure 20-50), the bag automatically seals or a sealer is used to seal the bag. After being filled and sealed, the bag is then conveyed to a palletizer.

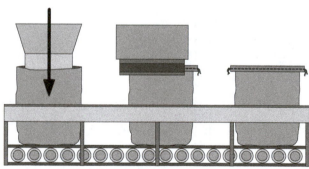

Product Fills Bag   Sewing Mechanism   Completed Bag
                     Seals Bag

FIGURE 20-49 Open Mouth
Bag Filling and Sealing

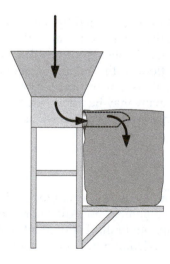

FIGURE 20-50 Valve Bag
Filling

A palletizer (shown in Figure 20-51) is a piece of equipment designed to stack the filled bags in a specific pattern on a pallet. The pattern (shown in Figure 20-52) is determined by the size of the bag and the number of layers or tiers required to achieve the maximum allowed height or weight and still maintain a stable stack for shipping. The last step in the bagging process is automated stretch wrapping. A stretch wrapper (shown in Figure 20-53) wraps the pallet with a wide film of polyethylene wrap (36 to 48 inches wide) to prevent the bags from shifting. The number of revolutions around the pallet depends on the thickness of the stretch wrap needed to provide a secure and stable bag stack on the pallet. The pallet is now ready for shipping to customers.

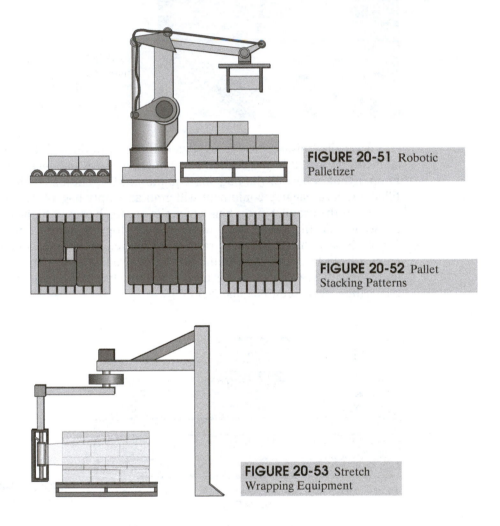

FIGURE 20-51 Robotic Palletizer

FIGURE 20-52 Pallet Stacking Patterns

FIGURE 20-53 Stretch Wrapping Equipment

The final pallet height and weight is determined by the shipping container. Pallet design must take into consideration the available room in the truck trailer, sea container, or railcar. It also must consider the availability of the equipment needed to move the pallet from the packaging floor to the shipping container and the ultimate weight of the container filled with pallets. For instance, if the tractor trailer is over a legal weight limit, the trucking company may be fined when the truck is weighed at an interstate weigh station. All this must be considered when designing the packaging system for a specific product.

Potential Problems

When working with solids handling equipment, process technicians should always be aware of potential problems such as equipment breakdown, equipment plugging, electrical power surges, pinch points, slips, and falls. Equipment malfunction can create additional burdens on the process environment. If equipment breaks down, it can cause production interruptions throughout the process facility. Plugged equipment can cause facility downtime and require process technicians to have to enter equipment to clear the plug, which can result in exposure to dangerous materials and result in a product spill. Spill hazards, which can cause slipping hazards and lost production, are also potential problems that face process facilities.

Human error can result in equipment downtime for clearing plugs or unplanned cleaning. It can also result in equipment failure. But the most important reason to eliminate human error is that it may result in injury or worse. Because of this, it is very important that process technicians follow all standard operating procedures to ensure that equipment operates properly and safely.

Safety and Environmental Hazards

Hazards associated with normal and abnormal solids handling equipment can affect personal safety, equipment, production, and the environment. Table 20-1 lists several potential hazards and their effects.

TABLE 20-1 Hazards Associated with Solids Handling Equipment

Improper Operation	*Possible Effects*			
	Individual	*Equipment*	*Production*	*Environment*
Product spills	Injury from slipping or chemical exposure	Damage and repair Shutdown and cleaning before re-start	Loss of production caused by downtime	Ground or water contamination
Excessive dust accumulation	Explosion Dust exposure	Damage and repair Shutdown and cleaning before re-start	Loss of production caused by downtime	Atmospheric dust pollution
Solids plugging	Time to unplug the system Chemical exposure	Shutdown and cleaning before re-start	Loss of production due to equipment shutdown	Ground or water contamination
Equipment failure	Potential injury Chemical exposure	Repair or replacement	Loss of production caused by downtime	

Typical Procedures

Process technicians should be familiar with the appropriate procedures for monitoring and controlling solids handling equipment components and conditions. The typical procedures, which include startup, shutdown, emergency, and lockout/tagout, vary at each individual site and for each type of solids handing equipment. Process technicians must follow the standard operating procedures (SOPs) for their assigned unit(s).

Process Technician's Role in Operation and Maintenance

A process technician's role in the operation and maintenance of solids handling equipment is the same as for all processing equipment (i.e., following standard operating procedures for the facility and the unit). Process technicians are required to report all abnormal events, such as spills and injuries, to their supervisor. Process technicians should understand operational procedures relating to each piece of equipment in order to maintain safe operating conditions, report an impending violation to their supervisor, and be able to respond in the event of an emergency.

Process technicians should routinely monitor solids handling equipment to ensure that it is operating properly. When monitoring and controlling this equipment, process technicians should always remember to look, listen, and feel for the items indicated in Table 20-2.

TABLE 20-2	Process Technician's Role in Operation and Maintenance	
Look	*Listen*	*Feel*
• Leaks or the accumulation of solids or dust in new areas	• Abnormal noises such as banging or hammering	• Excessive heat in areas
• Systems operating at higher or lower than normal speeds	• Normal noises with abnormal pitches	• Surging through the lines (e.g., normal flow, then excessively high or low flow)
• System pulling higher amps than normal		
• Gas or solids leaking into the atmosphere		

Summary

Solids handling equipment is used to process, transfer, store, package, and ship solids. Solids can be moved or transferred in a variety of ways.

A conveyor is a mechanical device used to move solid material from one place to another in a continuous stream. Types of conveyors include belt, screw, moving floor, vibratory, powered rollers, and both dilute phase and dense phase pneumatic. Although each conveyor is designed differently, the basic use is the same: to move material from one place to another.

A feeder is a piece of equipment that conveys material from one place to another. Feeders operate gravimetrically or volumetrically. Gravimetric feeders convey material at a controlled rate, measuring the weight lost over time. Volumetric feeders convey material at a controlled rate by running a set motor speed.

Extruders create shaped products (e.g., pellets, strings, or sheets) by forcing malleable solids through a perforated plate called a die.

Bucket elevators use a series of rotating buckets attached to a belt to move material from a pit or pile to an elevation above the pit or pile.

Screening systems are used for particle-size distribution. Screeners classify (separate) dry materials according to particle size. Material enters at one end and is conveyed across the screen. Smaller particles fall through the screen, while larger particles remain above the screen surface and exit the screener at the lower end.

Silos, bins, and hoppers are storage containers used to handle material (usually dry). A silo is a large container used for on-site storage. Silos are typically vertical storage containers that can reach up to 275 feet high. Bins are smaller containers that are designed to hold a finished product destined for distribution. Hoppers are the smallest of the three and are normally mounted directly to solids processing equipment. Bins and silos may have mechanisms at the bottom (e.g., a live bottom) that allow material to be removed easily.

Flow inducers, bin vibrators, and bin activators are used to shake loose accumulated product from the interior walls of bins, chutes, hoppers, and other storage devices.

A trickle valve uses negative air pressure to contain a quantity of product until that product reaches a specific weight. A free-hanging flap is used between two air environments to maintain an air seal. Once the weight of the product reaches a set limit, the flap opens and the product passes through.

A rotary valve uses a set of rotating pockets to meter a fixed volume of solids at a fixed rate from a low-pressure environment to a higher-pressure environment, or vice versa.

Hopper cars and hopper trucks are designed to carry solid materials by rail or over the road. There are single-bin hoppers and multiple-bin hoppers. Hoppers discharge their load by either dropping directly from the bottom of the hopper into a pit, or by pneumatic transfers of the material through hoses and piping systems. Whether the solid material is stored in bins or hopper cars or trucks, strain gages or load cells are used to determine the weight of the stored material.

Packaging systems transfer product into containers like bags, supersacks, or boxes. Bulk bags contain quantities up to 1,000 pounds. Bagging equipment packages smaller bags manually or automatically. The bagging equipment also transfers the bags or boxes to pallets and secures them with plastic wrap called stretch wrapping.

Process technicians must be aware of potential problems surrounding solids handling equipment. Technicians must also monitor safety and environmental hazards and follow proper operating procedures at all times.

Checking Your Knowledge

1. Define the following terms:
   a. Extruder
   b. Screeners
   c. Bin
   d. Conveyor
   e. Feeder
   f. Hopper
2. *(True or False)* A conveyor is a piece of equipment used to mix raw product to form a new product (usually dry material).
3. *(True or False)* A gravimetric feeder conveys material at a controlled rate by running at a set motor speed.
4. *(True or False)* A volumetric feeder must have a predetermined maximum feed rate that is calculated by the feeder's maximum motor speed and material density.
5. *(True or False)* A die is a predesigned piece of equipment used to shape a piece of metal or other objects to the desired shape.
6. Which of the following type of storage equipment is medium-size and is more commonly used in day-to-day operation?
   a. Silo
   b. Bin
   c. Hopper
   d. Cyclone
7. Flow inducers and bin vibrators can be attached to _____ to free accumulated products and facilitate flow.
   a. mixers
   b. extruders
   c. chutes
   d. railcars
8. *(True or False)* Access doors at the top of a railcar are the same as regular solid access doors.
9. *(True or False)* A trickle valve is a free-hanging flap between two different pressure environments and maintains an air seal inside a flow area.
10. Which of the following is *not* a piece of bag packaging equipment?
    a. Stretch wrapper
    b. Fluid bed dryer
    c. Valve bag filler
    d. Palletizer

Student Activities

1. Write a one-page paper on how a bucket elevator system works. Describe the pieces of equipment, give an example of when a bucket elevator system is or can be used in industry, and draw a diagram of the system.
2. Describe, from beginning to end, how a pneumatic conveyor in a grain process facility works. Explain how you would know how much material is moving.

21

Environmental Control Equipment

Objectives

After completing this chapter, you will be able to:

- Explain the purpose of environmental control equipment in the process industry.
- Identify the common types and applications of environmental control equipment.
- Identify the components of environmental control equipment and explain the purpose of each.
- Describe the operating principles of environmental control equipment.
- Identify rules and regulations associated with environmental control equipment.
- Identify typical procedures associated with environmental control equipment.
- Describe the process technician's role in environmental control equipment operation, maintenance, and compliance.
- Identify potential problems associated with environmental control equipment.

Key Terms

Baghouse—air pollution control equipment that contains fabric or bag filter tubes, envelopes, or cartridges designed to capture, separate, or filter particulate matter.

Dike—a wall (earthen, shell, or concrete) built around a piece of equipment to contain any liquids should the equipment rupture or leak.

Electrostatic precipitator—a device that contains positively charged collecting plates and a series of negatively charged wires.

Flare system—a device used to burn unwanted process gases before they are released into the atmosphere.

Incinerator—a device that uses high temperatures to destroy solid, liquid, or gaseous wastes.

Landfill—a human-made or natural pit, typically lined with an impermeable, flexible substance (e.g. rubber) or clay and is used to store residential or industrial waste materials.

Introduction

A combination of the ever increasing need to protect Earth's natural resources and the advent of emerging technologies has spurred the process industries to maximize the use of environmental controls in processes and operating equipment. Environmental disasters such as the release of chemicals into the drinking water aquifers of Love Canal, a neighborhood in Niagra Falls, New York; the release of deadly chemicals into the air in Bhopal, India; and the Exxon Valdez oil spill off the Alaskan coast have reinforced the need for stricter environmental controls and procedures when dealing with hazardous and nonhazardous chemicals.

If the emission of impurities, by-products, or other materials into the air, land, or water becomes excessive, the environment suffers. Laws are in place at the federal, state, and local levels to ensure that the environment is protected at all times. Industrial companies nationwide have stepped forward to enhance the operations of their facilities in order to maintain, and in some cases, exceed those standards.

The purpose of designing and implementing environmental controls is to maintain the operation of process facility equipment within the prescribed standards allowable by law. The process of environmental control is accomplished through monitoring, sampling, and using control devices. Control devices can be categorized by the natural resource being preserved and the type of operating equipment that is regulated by each control mechanism.

Types of Environmental Control Equipment

AIR POLLUTION CONTROL

Many industrial components (e.g., furnaces, boilers, incinerators, separation or distillation towers, gas turbines, and other types of equipment) vent to the atmosphere and require the combustion of fossil fuels in order to operate. As the hydrocarbon-based fossil fuels are burned, nitrogen oxides (NO_x), carbon monoxide (CO), and sulfur dioxide (SO_2) are produced. NO_x, CO, and SO_2 emissions are extremely harmful to living organisms, contribute to acid rain, and join with hydrocarbons to create ground-level ozone. Because of this, these emissions must be carefully monitored and maintained according to the terms of the Clean Air Act.

The Clean Air Act, which was enacted by Congress in 1977 and amended in 1990, requires industry to reduce and maintain hazardous emissions that affect ambient air (any part of the atmosphere that is accessible and breathable by the public). Furthermore, Title V, one of several programs in the 1990 amendment to the Clean Air

Did You Know?

Forty-two percent of the world's total NO_x emissions are generated from the burning of hydrocarbon and fossil fuels.

These fuels are commonly used in furnaces, boilers, and incinerators at petrochemical, power-generating, and other process facilities.

Act, requires local and state air quality agencies to issue operating permits for facilities that produce significant amounts of air pollution.

In light of these new rules and regulations, new processes such as selective catalytic reduction and selective noncatalytic reduction are being developed and implemented in order to reduce NOx emissions. These modifications to the combustion process utilize a catalyst or free radical to convert the harmful NO_x emissions into harmless nitrogen and water. These types of emission control processes maintain high NO_x conversion rates while minimally affecting the performance of the process equipment.

Baghouses and Precipitators (Electrostatic)

Boilers and power generation facilities that burn fossil fuels like coal use electrostatic precipitators and baghouses as pollution control devices. These devices prevent residue, in the form of fly ash, from escaping the stacks and passing into the atmosphere.

Baghouses consist of fabric or bay filter tubes, envelopes, or cartridges designed to capture, separate, or filter particulates as the effluent gas stream passes through them. **Electrostatic precipitators** contain positively charged collecting plates and a series of negatively charged wires. As particulates in the gas pass over the wires, they become negatively charged and are attracted to the positively charged collecting plates. Once these particles have been captured, the clean effluent gas is then discharged to the atmosphere and the particulates are collected for disposal or incorporation into products like cinder blocks (blocks made of ash and cement that are used in construction). Figure 21-1 shows a baghouse; Figure 21-2 shows how an electrostatic precipitator processes particles.

Coal Gasification

Another environmental control technology being implemented today is coal gasification. Coal is converted from a solid to a gaseous state before being combusted at the burners. This gasification helps minimize carbon monoxide (CO), nitrogen oxide (NO_x), and sulfuric oxide (SO_x), and reduces the amount of carbon dioxide (CO_2) emitted to the atmosphere. Many scientists believe that CO_2 contributes to the greenhouse effect, which is commonly thought to promote global warming.

Vapor and Gas Emission Controls

Several other types of vapor and gas emission equipment are utilized to capture and/or recycle process materials rather than emit them to the atmosphere. For example, tanks and other storage vessels incorporate vapor recovery systems that take the vapor from

FIGURE 21-1 Baghouse

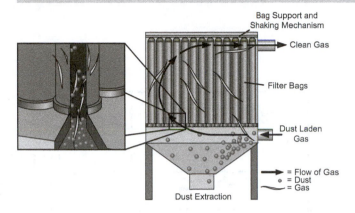

FIGURE 21-2 Electrostatic Precipitator

Courtesy of John Zinc LLP

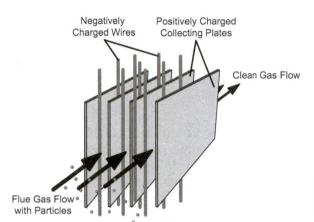

these vessels, cool it to a liquid phase, and then return the liquid back to the storage vessel. During emergencies, when vapor cannot be recovered, an attempt must be made to clean it before venting it to the flare. Figure 21-3 shows a flare gas recovery system. Figure 21-4 is a schematic of a flare gas recovery system.

FIGURE 21-3 Photo of Flare Gas Recovery System

Courtesy of John Zinc LLP

FIGURE 21-4 Schematic of Flare Gas Recovery System

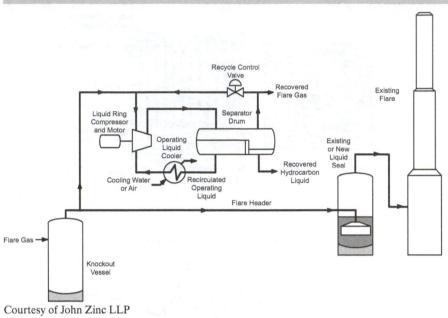

Courtesy of John Zinc LLP

Gas Treatment (Catalytic versus Noncatalytic)

Catalytic treatment is a process that uses a catalyst to remove SO (sulfur monoxide), H_2S (hydrogen sulfide), and NO_x (nitrogen oxide) from the effluent gas streams of chemical manufacturing, natural gas production, and oil and gas refining facilities. These gases are also commonly referred to as flue gas, tail gas, or off gas. In a gas treatment facility, the gas flows over a catalyst bed that converts the SO_2 (sulfur dioxide) and H_2S (hydrogen sulfide) to elemental sulfur, and the NO_x to nitrogen gas (N_2)— elements that are more compatible with the environment.

A common method of noncatalytic treatment of gas streams is to scrub or wash the gas with NH_3 (ammonia). This scrubbing converts the NO_x to water and nitrogen, elements that are more environmentally friendly. This process is called selective noncatalytic reduction. Another material that can be used for this process is urea (an organic compound of carbon, nitrogen, oxygen, and hydrogen that is produced from the reaction of synthetic ammonia and carbon dioxide). When injected into the flue gas, urea converts the NO_x to nitrogen, carbon dioxide (CO_2), and water (H_2O).

Scrubbers

Scrubbers are devices that remove unwanted gases or particles from a gas stream by spraying the stream with liquid (usually water). In a scrubber, the gas is passed through a liquid that is moving in the opposite direction (contra-flow). The liquid absorbs the contaminants from the gas before it is released to the atmosphere or sent for further treatment elsewhere. Figure 21-5 shows an example of a scrubber system.

Another type of contaminant removal system is a carbon adsorber. In a carbon adsorber, the gas passes through a bed of activated carbon. As the gas flows through the system, the contaminants, which are attracted to the pores in the carbon granules, attach to the carbon and are removed while the clean gas is released to the atmosphere or sent for further treatment.

Incinerators (Thermal Oxidizers)

Incinerators are devices that use high temperatures to destroy solid, liquid, or gaseous wastes. The result of inefficient or incomplete incineration can be hazardous and may produce extremely hot vapors. Because of this, these systems are strictly regulated by the Environmental Protection Agency (EPA) and must meet the quality mandates for waste disposal before these vapors can be introduced into the atmosphere.

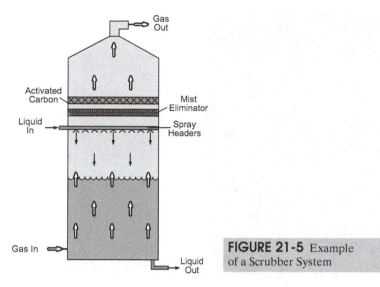

FIGURE 21-5 Example of a Scrubber System

Hazardous waste incinerators utilize extremely high temperatures to thermally destroy waste material. The resulting flue gas stream must be cooled before allowing it to return to the atmosphere in order to conserve energy and to prevent thermal pollution.

During incineration there is usually enough heat energy left in the flue gas stream to generate steam in a waste heat boiler. This secondary steam production helps save the company money and reduces the costs associated with incinerator operation. Figure 21-6 displays a schematic for a thermal oxidizer with a recuperator and a waste heat boiler.

FIGURE 21-6 Schematic of Thermal Oxidizer with Recuperator and Waste Heat Boiler

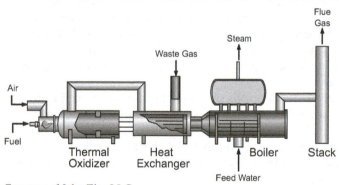

Courtesy of John Zinc LLC

Flare System

Flare systems are devices that are used to burn unwanted process gases before they are released into the atmosphere. These units burn a small amount of pilot gas at all times to ensure there is a continual flame present. Flare systems, which are most often used during emergency situations to rapidly dispose of gases during a process facility upset, can be vertical or ground mounted. Figure 21-7 shows an example of a vertically mounted flare stack.

Most flares are equipped with steam systems to ensure that the gas is completely combusted and to protect the tip of the flare from damage by heat and flame. The gases disposed of through the flaring method are reduced to carbon dioxide and water vapor.

Flare systems can be noisy and may emit smoke if they are not properly controlled. During an emergency, when the flare is receiving maximum load, it is a fine balancing act to ensure that enough steam is applied to the flare without creating undue noise pollution for those living and working in the surrounding area. Flare systems are highly regulated by the Environmental Protection Agency as well as state agencies in the states where these facilities are built.

FIGURE 21-7 Vertical Flare Stack

WATER AND SOIL POLLUTION CONTROL

Other areas of concern in relation to environmental control are the Earth's waterways and soil. Dumping into the soil or pollution of waterways can lead to the destruction of ecosystems and the leaching of contaminants into the drinking water supply. To prevent these types of contamination, the Water Pollution Act was enacted by the U.S. government in 1972. This act was then revised in 1977 and renamed the Clean Water Act. The Clean Water Act regulates and prohibits the discharge of pollutants into waterways in the United States.

Water in and around process facilities must be constantly monitored for waterborne contaminants. Some of these contaminants come from process leaks, overburdened process sewer flows, and surface water or rain that has been contaminated by its association or proximity to a piece of operating equipment. Other pollutants are produced as a direct result of a company's process facility operation. For example, condensate used for heating and cooling a process, water used for cleaning, and wastewater all pose disposal problems.

Wastewater is one of the main disposal concerns in process facilities. Most of this water comes from process sewers or from storm drain overflow during periods of heavy rainfall. Before it can be introduced into the environment, this water must be cleaned of all hydrocarbon material; thermally adjusted (heated or cooled); made pH-compatible; and properly oxygenated to meet all federal, state, and local environmental regulations.

Activated Sludge Process

The 1980s brought a way to treat wastewater with the aid of microorganisms. These microorganisms (bacteria) are attracted to and like to ingest (consume) hydrocarbons and other pollutants.

Did You Know?

Did you know that the United States spends more money on environmental control than any other country in the world?

In 1990, the United States spent over $100.00 per person!

This wastewater cleanup process is accomplished in several steps. First, the water must be cooled and the pH adjusted to make it as attractive as possible to the microorganisms. The water is then aerated at the point where the bacteria are introduced because the bacteria (often referred to as bugs) respond better when the environment is oxygen rich.

Once the bacteria has had a chance to do its job, the water is clarified in a basin, where the bacteria settle to the bottom and are removed as waste sludge or are recycled for further use in the aeration basins. The bacterial sludge is then dried into a cake and burned to ash in a fluidized bed incinerator.

During the incineration process, exhaust gases are passed over water tubes in a waste heat boiler to remove excess heat. This heat is then used to make the steam for facility use and the gases become thermally cooler. The effluent from this incinerator is scrubbed or washed with water and a caustic spray mixture to ensure that the ash or ash combustion products do not reach the atmosphere.

Clarified water is then removed and sent to a third treatment stage, which may include carbon filtration or other treatment methods designed to remove volatile organic compounds (VOCs) such as benzene or toluene. This water is sampled many times for both VOCs and for the oxygen content before it is returned to the waterway.

Activated carbon is utilized in most of these treatment facilities to provide an adsorbing (adhering) surface for the organics in the stream but also to provide a site for the microorganisms to attach. Figure 21-8 is an example of activated sludge process.

FIGURE 21-8 Activated Sludge Process Used to Treat Wastewater in Process Facilities

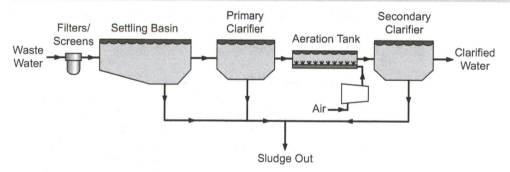

Clarifiers   Clarifiers are part of the activated sludge component of wastewater treatment. Activated sludge is another name for the microorganisms, or bugs, that feast on the hydrocarbons in the wastewater stream.

Figure 21-9 shows an example of a circular mechanical clarifier. Clarifiers feed wastewater into the center and force it to the bottom. This allows sediment to settle and water to rise to the top. The top is then skimmed for solids and other floatable materials.

Another type of clarifier is a parallel plate clarifier (shown in Figures 21-10 and 21-11). This type of clarifier, which uses the principle of gravity, is used for higher flow

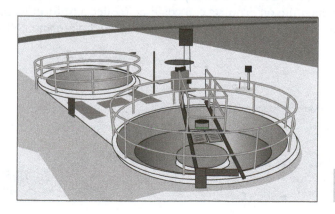

FIGURE 21-9 Circular Mechanical Clarifier

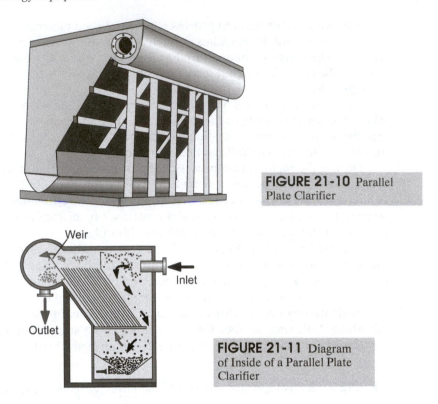

FIGURE 21-10 Parallel Plate Clarifier

FIGURE 21-11 Diagram of Inside of a Parallel Plate Clarifier

rates in liquid/solid separation. It has no moving parts. Wastewater enters the top and is distributed over the plates. As it flows downward, the solids are trapped by the plates and the effluent water flows upward to exit the overflow weir.

Separators and Interceptors

In an attempt to recover and recycle oil from a process wastewater or storm water flow, facilities use various devices to skim the oil from the water's surface. These units are called American Petroleum Institute (API) separators, parallel plate interceptors (PPIs), or corrugated plate interceptors (CPIs). Each of these devices works in a similar manner. For example, each unit must include an upstream sediment bay, skimmer, weir, separator basins, grit chambers, bar screens, and sludge hoppers. Figure 21-12 shows an example of an API separator.

An API separator is a separation device designed to separate oil, wastewater, and solids. The design of the separator is based on specific gravity differences between the oil and the wastewater. (*Note:* These differences are less than the specific gravity difference between the suspended solids and water). Because of the different densities that settle to the bottom of the separator as a solids layer, the oil rises to the top, and the wastewater becomes the middle layer between the oil and the solids.

Parallel plate interceptors (PPIs) perform the same function as API separators. However, they employ a parallel plate assembly. These plates provide a large surface area on which the oil droplets can coalesce (collect) to form larger droplets. Solids in the stream slide down the top side of each parallel plate and collect on the bottom of

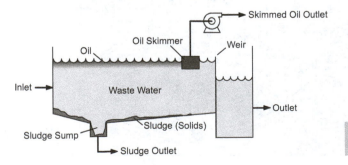

FIGURE 21-12 Diagram of an API Separator

FIGURE 21-13 Diagram of a Parallel Plate Interceptors (PPI)

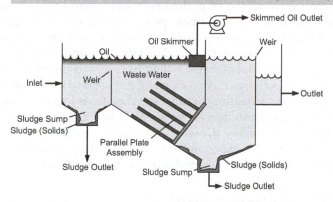

FIGURE 21-14 Diagram of a Corrugated Plate Interceptors (CPI)

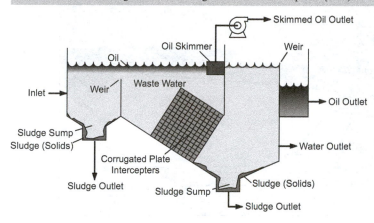

the vessel, where they are removed. The oil in this system is removed by an oil skimmer, and the clarified water flows over a weir to the water outlet. Figure 21-13 shows an example of a PPI separator.

PPIs, like API separators, rely on the specific gravity difference between the oil and the water for separation. One advantage of this type of separator is that its design takes up less space than the standard API separator.

Corrugated plate interceptors are like API separators and PPIs; however, they use a series of corrugated plate packs at an inclined angle in a vessel to separate oil, water, and solids. Figure 21-14 shows an example of a corrugated plate interceptor.

In this type of separation, wastewater enters the vessel near the top of the plate pack, where oil starts to rise to the surface or toward the oil and water interface. Some oil is carried into the plate packs, where it coalesces on the corrugations to form larger droplets that rise to the top of the corrugations and up to the oil and water interface. Solids pass through the plate packs and settle at the bottom of the vessel, where they are taken off at two points. Light solids flow out the side of the vessel and heavy solids flow out the bottom. Oil flows over a weir at the top of the vessel and out. Water also flows out of the vessel but at a point lower than the oil.

Dikes

Dikes are another environmental control measure. **Dikes** are containment devices (earthen, shell or concrete) built around a piece of equipment to contain any liquids should the equipment rupture or leak. These barriers serve as dams designed to contain overflows or spills from process equipment. They also hold surface water such as rain in the event that water becomes contaminated. The water trapped inside these dikes must be tested for chemical compounds and treated (if necessary) before it is allowed to return to the waterway. If a tank overflows, the contaminated earth often must be removed and burned in an incinerator.

Settling Ponds

Wastewater is also treated in settling ponds or basins before being returned to a natural waterway. During the treatment process, water is diverted into these ponds, where it is mechanically aerated to increase the oxygen content and remove chemical gases. These ponds also allow solids to settle to the bottom instead of moving forward with the water flow. The effluent water is then filtered and chemically treated (neutralized), and its temperature adjusted before being returned to waterways.

Retention time in settling ponds can range from less than one day to several days. These ponds are operated in series so that each sequential body of water becomes cleaner than the one before it. Figure 21-15 is an example of a settling pond with a liner.

For settling ponds to be completely effective, they must be aerated (oxygen added). Figures 21-16 and 21-17 are examples of aeration inside a settling pond.

FIGURE 21-15 Settling Pond with Liner

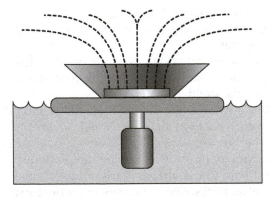

FIGURE 21-16 Mixer to Aerate Wastewater Inside a Settling Pond

FIGURE 21-17 Aeration Logoon (Settling Pond)

Landfills

Other means of environmental control include landfills. **Landfills** are human-made or natural pits, typically lined with an impermeable, flexible substance (e.g., rubber) or

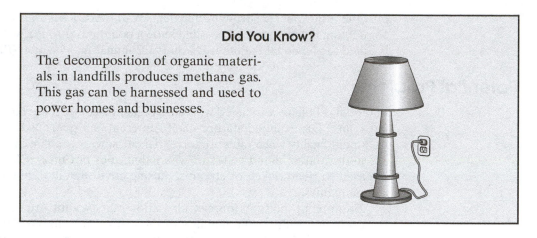

Did You Know?

The decomposition of organic materials in landfills produces methane gas. This gas can be harnessed and used to power homes and businesses.

clay. They are used to store residential or industrial waste materials. Landfills can be classified as either sanitary or hazardous. Sanitary landfills contain the type of waste disposed of in homes and restaurants. Hazardous landfills contain industrial materials and byproducts that can be harmful to the environment if not contained and treated properly.

Because the contents of landfills pose potential environmental risks, their development is complicated. For example, landfills must first be lined with a layer of impermeable material (e.g., rubber). A layer of absorbent clay is placed on top of this liner, and then another impermeable layer is placed on top before the waste can be placed in the facility. In addition to the liners, piping systems are also installed; these systems allow for the detection of leakage into the surrounding soil. There may also be deep water wells on the perimeter of the landfill that extend into aquifers to detect leakage into the watershed. Figure 21-18 is a detailed graphic of a landfill and its components.

FIGURE 21-18 Landfill Diagram

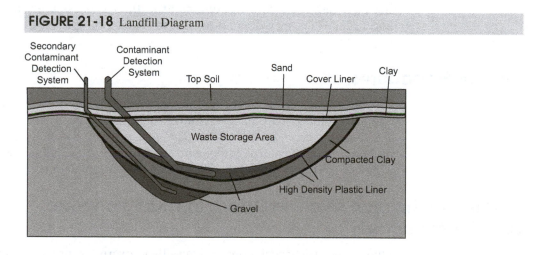

Different products require different disposal methods. For example, environmentally harmful liquids must be absorbed in sand or sawdust and placed in drums before disposal. Asbestos must be marked plainly in bags and disposed of in an area designated for hazardous wastes. Due to the limited amount of space in a landfill and the strict regulations that apply to these types of facilities, more and more steps are being taken to utilize incinerators as an alternate means for waste disposal.

Federal Regulations

The Environmental Protection Agency established the National Ambient Air Quality Standards (NAAQS), which require states to have their own implementation plans. These standards hold companies accountable for the emissions of certain chemicals into the atmosphere. Through the Clean Water Act, a system of permits is used for eliminating the discharge of possible known pollutants into the waterways. This system is called the National Pollutant Discharge Elimination System (NPDES).

Potential Problems

Potential problems associated with environmental controls are based on several different issues. For example, nature itself can create an overburden on the wastewater treatment facility when rains are torrential and sewers become overloaded. If the flows are high, contaminated water or water that does not meet EPA standards may be released to rivers, lakes, or streams, causing environmental damage and resulting in company fines.

Equipment malfunction can also allow releases into the air, land, or water if the problem results in the inability of the process unit to meet EPA standards. One example would be a flare system that emits smoke over a long period of time. Smoke means that combustion is incomplete and contaminants are being released into the atmosphere.

Landfills that are not lined properly or that have leaking liners allow contaminants to leach into the soil and eventually the ground water. Another problem is placing the wrong materials into a landfill. For example, asbestos waste is prohibited from landfills unless the landfill meets additional requirements under the Clean Air Act.

Human error is another factor. An employee who reads process instrumentation incorrectly or makes an incorrect adjustment can create a situation that allows material to be released to the environment. This can be dangerous as well as costly to a company.

Environmental Rules and Regulations

Process technicians should be familiar with the rules and regulations associated with the environmental control equipment in their process area. Environmental rules and regulations vary depending on the facility, the process, and federal and state requirements.

Typical Procedures

Process technicians should be familiar with the appropriate procedures for monitoring and maintaining environmental control equipment. Typical procedures such as startup, shutdown, emergency, and lockout/tagout vary at each individual site and for each type of environmental control equipment, so process technicians should always follow the standard operating procedures (SOPs) for their assigned unit.

Process Technician's Role in Operation, Maintenance, and Compliance

Due to the strict regulations and laws that govern the process industries, the process technician's role becomes even more critical to the success of a company's waste management and environmental control program. The Environmental Protection Agency, along with other federal, state, and local government agencies, places regulations (permits) on specific types of waste management processes and equipment. These permits specify policy and procedures for the safe operation of these types of equipment.

Process technicians must receive training, which is mandated by the Occupational Safety and Health Administration (OSHA). This training is conducted through the company and should ensure that permits are followed explicitly. Process technicians should also understand operational procedures relating to each piece of equipment, report an impending violation to their supervisor, and respond to any situation in the event of an emergency.

Summary

The process industries use a wide variety of environmental control devices to protect the air, land, and waterways surrounding process facilities. The burning of fossil fuels and their resultant byproducts, the heat produced by process units, and the chemicals used to complete the production of goods must be managed to prevent their escape.

Equipment such as baghouses, precipitators, flares, activated sludge facilities, and scrubbers all work together to eliminate emissions into the air, land, or water. These components are enhancements to specific processes and are strictly regulated by the Environmental Protection Agency. Laws such as the Clean Air Act, the Resource Conservation Control Act, and many others dictate the parameters in which these equipment components can operate in a process facility.

Environmental controls have raised the level of awareness in the process industries regarding the need to protect the Earth's vital resources. That need has driven industry to modify, redesign, and create new technologies for the production of goods and to change their environmental control systems to better protect the environment.

Checking Your Knowledge

1. Define the following terms:
   a. Baghouse
   b. Scrubber
   c. Dike
   d. Flare
   e. Incinerator
   f. Landfill
2. Environmental controls work to preserve:
   a. air
   b. land
   c. water
   d. all of the above
3. *(True or False)* Process technicians are responsible for knowing hazardous waste procedures.
4. *(True or False)* Baghouses use electrical charges to attract particulates.
5. CO_2 is the chemical formula for:
   a. carbon monoxide
   b. nitrogen oxide
   c. sulfuric oxide
   d. carbon dioxide
6. H_2S is the chemical formula for:
   a. sulfur monoxide
   b. hydrogen sulfide
   c. nitrogen gas
   d. sulfur dioxide
7. In a gas treatment facility, a common method of noncatalytic treatment of gas streams is to scrub or wash the gas with:
   a. sulfur monoxide
   b. hydrogen sulfide
   c. ammonia
   d. sulfur dioxide
8. In a scrubber, the flow of gas against a liquid medium in opposite directions is called:
   a. concurrent flow
   b. cross flow
   c. contra-flow
9. Activated carbon is utilized in most treatment facilities to provide an adhering surface for the organics in the stream but also to provide a site for microorganisms to:
   a. attach
   b. detach
   c. be eaten by treatment bugs
   d. evaporate to the atmosphere

Student Activities

1. Research one of the following acts and write a one-page report on it:
   a. Clean Water Act
   b. Clean Air Act
2. Work in teams to discuss and brainstorm a procedure for one of the following:
   a. disposal of salty wastewater
   b. disposal of used lithium batteries
   c. disposal of used light bulbs
3. Work in teams to find and develop a use for the coal gasification process where you live. You may use the Internet for your research.

22

Mechanical Power Transmission and Lubrication

Objectives

After completing this chapter, you will be able to:

- Describe the principles of mechanical power transmission and lubrication in the process industries.
- Describe the purpose of mechanical power transmission components.
- Describe the operating principles of mechanical power transmission.
- List the types of bearings and explain the purpose of each.
- List the types of gears and their uses.
- Describe safety and environmental hazards associated with mechanical power transmission and lubrication.
- Describe the process technician's role in mechanical power transmission and lubrication procedures.
- Identify potential problems associated with mechanical power transmission and lubrication.

Key Terms

Bearings—mechanical devices used to support shafts for radial (up-and-down or side-to-side) and axial (back-and-forth) movement and to hold the rotor in alignment with other parts.

Belt—a flexible band, placed around two or more pulleys, that transmits rotational energy.

Chains—mechanical devices used to transfer rotational energy between two or more sprockets.

Couplings—mechanical devices used to connect and transfer rotational energy from the shaft of the driver to the shaft of the driven equipment.

Gear—a toothed wheel that engages another toothed mechanism in order to change the speed or direction of transmitted motion.

Gearbox—mechanical device that houses a set of gears; connects the driver to the load; and allows for changes in speed, torque, and direction.

Lubricant—a substance used to reduce friction between two contact surfaces.

Sheave—a V-shaped groove centered on the circumference of a wheel designed to hold a belt, rope, or cable.

Sprocket—a toothed wheel used in chain drives.

Introduction

In the process industries, mechanical power transmission and lubrication systems are critical to daily equipment operations. Mechanical power transmission allows us to transfer rotational energy from one point to another in order to power various types of equipment. Without this energy, many processes could not occur. Lubrication is a key component of mechanical power transmission because it establishes a stable, fluid barrier between equipment components and reduces friction, heat buildup, and equipment wear.

Operating Principles of Mechanical Transmission

Mechanical power transmission transfers rotational energy from a driver to driven equipment with a minimal loss of energy to friction. Lubrication systems provide a fluid film between two surfaces that are moving relative to each other. Lubrication is also used to remove heat produced by friction in rotating equipment. To understand the operating principles of mechanical transmission, a process technician must understand the components of power transmission, which include couplings, gearboxes, speed reducers and increasers, belts, chains, magnetic and hydraulic drives, bearings, and gears.

COUPLINGS

Couplings are mechanical devices used to connect and transfer rotational energy from the shaft of the driver to the shaft of the driven equipment. Couplings have several functions. The first function is to connect the two rotating shafts together so the rotating movement from one shaft can be transferred to a different shaft. A second function is to eliminate any misalignment between the rotating shafts that are coupled. Figure 22-1 shows an example of a shaft coupling.

Shaft Coupling

FIGURE 22-1 Functions of Couplings

While some misalignment with couplings is acceptable, it is essential that couplings be installed to minimize misalignment. The better the alignment between the rotating shafts, the longer the coupling will be in operation without wearing down.

Types of Couplings

The process industries use several types of couplings. However, the most common types are flexible, rigid, and fluid.

Flexible couplings are used most frequently in general mechanical power transmission service because they allow some misalignment without affecting the transmission of power. Flexible couplings also provide some isolation of vibration between drivers and driven machines.

Rigid couplings are used in applications where only small amounts of misalignment can be tolerated without damage to equipment or decreased performance.

Fluid couplings are used in applications where variable speed operations occur. Fluid couplings cushion the shock from equipment overloads, machinery jamming, reversing operations, or sudden speed changes by increasing slip.

BELTS

Belts are flexible bands, placed around two or more pulleys, that transmit rotational energy. V-belt drives are the most commonly used in an industrial setting. V-belts, which are usually designed to move in one direction, use traction to transfer motion between two **sheaves** (V-shaped grooves centered on the circumference of a wheel designed to hold a belt, rope, or cable). V-belt sheaves (shown in Figure 22-2) exhibit very low wear.

There are three categories of belts: standard, light-duty, and high-capacity. Each category is determined by the size, shape, and width of the belt. Standard belts are the most commonly used in industrial settings. Light-duty belts are used with smaller drive pulleys. High-capacity belts are used when the equipment has higher horsepower or loading conditions.

When installing or working with belts, it is important to create and maintain proper tension because improper tension can decrease the life of the belt and prevent the system from operating properly. For example, if a belt is too loose, it may slip or vibrate up and down. If the belt is too tight, it may prevent proper rotation of the sheaves or damage sheave bearings.

As a rule of thumb, when adjusting the tension on a belt, there should be 1/4 inch of slack per foot of distance between pulleys. Figure 22-3 illustrates the proper amount of tension on a belt.

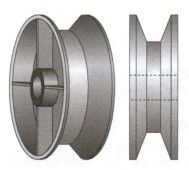

FIGURE 22-2 Examples of Sheaves

FIGURE 22-3 Illustrates Correct Tension on a Belt

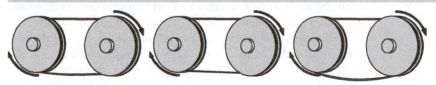

Proper Fit   Too Tight   Too Loose

CHAINS

A **chain** is a mechanical device used to transfer rotational energy between two or more sprockets. Chains perform the same function of transferring rotational energy as belts and gears; however, they do not use friction to transfer motion. Instead, the links of the chain interlock with the teeth on a **sprocket** (a toothed wheel used in chain drives).

Figure 22-4 shows an example of a roller chain (one of the most common types of chain) engaged with sprocket teeth.

The load capacity of a chain drive can be increased with multiple-strand chains. Each link in a chain drive transmits load in tension to and from sprocket teeth. Because of this, sprockets (see Figure 22-5) should be replaced whenever the teeth are worn.

Chain drives have many advantages over belt drives. For example, because a chain drive requires only a few sprocket teeth for effective engagement, it allows higher reduction ratios than are usually permitted with belts. However, chains also have some disadvantages. These advantages and disadvantages are listed in Table 22-1.

FIGURE 22-4 Roller Chain and Sprocket

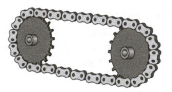

FIGURE 22-5 Example of a Typical Sprocket

MAGNETIC AND HYDRAULIC DRIVES

A magnetic drive (also referred to as a "mag drive") is a general term for a coupling used to transmit power through magnetic forces between the driver and the driven elements. In this type of driver, the driver and driven sides may have no physical contact. Instead, these pumps use a standard electric motor to drive a set of permanent magnets (which are mounted on a carrier or drive assembly) to drive an inner rotor connected to the pump's impeller. Because the two devices are connected magnetically, a pump seal is eliminated. Thus, these types of pumps are often referred to as seal-less pumps.

Hydraulic drives transfer energy through closed circuit movements of a pressurized liquid in a conduit. In this type of drive, the pump section supplies pressurized liquid and the motor section converts the liquid flow into shaft power. An automatic transmission is an example of a hydraulic drive.

TABLE 22-1   Advantages and Disadvantages of Chains over Belts

Advantages	*Disadvantages*
• No slippage between chain and sprocket teeth.	• Noise is usually higher than with belts or gears (although silent chain drives are relatively quiet).
• Minimal stretch, allowing chains to carry heavy loads.	• Chains may elongate due to wearing of link and sprocket teeth contact surfaces.
• Long operating life expectancy because flexure (bending) and friction contact occur between hardened bearing surfaces separated by lubrication (oil film).	• Usually limited to applications with lower speeds than belts or gears.
• Capable of operating in hostile environments, especially if high-alloy metals and other special materials are used (e.g., areas with extreme temperature and/or moisture levels, or areas that are extremely oily, dusty, dirty, or corrosive).	
• Long shelf life because metal chains ordinarily do not deteriorate with age and are less affected by sun, moisture, and temperature.	
• Certain types can be replaced without disturbing other components mounted on the same shafts as sprockets.	

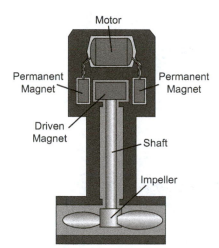

FIGURE 22-6 Magnetically Driven Pump

Bearings

Bearings (see Figure 22-7) are mechanical devices used to support shafts for radial (up-and-down) and axial (back-and-forth) movement and to hold the rotor in alignment with other parts. Lubricated bearings also help reduce friction and allow smooth movement of rotary equipment.

Thrust bearings, such as those used in turbines, centrifugal compressors, and large centrifugal pumps, prevent excessive axial (back-and-forth) movement along the shaft and hold the rotor in correct alignment with nonmoving parts. Radial bearings support and prevent a rotor from moving from side to side or up and down.

Excessive vibration could be an indication that bearings are beginning to fail. To help monitor vibration, sensors may be used to measure the frequency and amplitude of this vibration. The results of these measurements are then displayed (or alarmed) on a local control panel or in a control room.

BEARING TYPES

There are many different types of bearings. The most common, however, are roller, ball, sleeve, and thrust.

Bearings

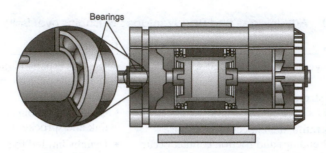

FIGURE 22-7 Bearings Help Support Shafts and Reduce Friction

Roller bearings (see Figure 22-8) are used in moderate-pressure applications and at low speeds (less than 1,000 rpm). The principal parts of roller bearings include inner and outer rings (races), rolling elements, and the cage that spaces the rolling elements.

Ball bearings (see Figure 22-9) are used in smaller machinery at low speeds. The principal components of ball bearings are the same as roller bearings. However, the difference is the shape of the rolling element (round versus cylindrical).

Sleeve (fluid film) bearings (see Figure 22-10) are used for large equipment and high speeds. These types of bearings usually have a softer lining that aligns with a polished inner shaft surface.

Thrust bearings (see Figure 22-11) are used to prevent axial movement of the shaft. Tapered roller bearings (see Figure 22-12) are used to prevent axial and radial movement of the shaft.

FIGURE 22-8 Roller Bearings

FIGURE 22-9 Ball Bearings

FIGURE 22-10 Sleeve (Fluid Film) Bearings

FIGURE 22-11 Fluid Film Thrust Bearings

FIGURE 22-12 Tapered Roller Bearings

Gears

A **gear** is a toothed wheel that engages another toothed mechanism in order to change the speed or direction of transmitted motion. Gears come in many sizes and shapes.

Spur gears (see Figure 22-13), which are the most commonly used gears in the process industries, have teeth that are cut parallel to the shaft. Spur gears are primarily used in low-power applications.

Helical gears (see Figure 22-14) are similar to spur gears, but their teeth are cut on an angle. Because of their design, helical gears can handle more force than simple spur gears. Thus, they are used in applications that require greater power and equipment size. Double helical gears are especially attractive because the thrust of each row of teeth cancels, making the thrust-bearing load smaller.

Worm gears (see Figure 22-15) are gears with slanted teeth, placed at a 90-degree angle, which are designed to mesh with another gear (referred to as a worm). Worm gears are used to obtain large speed reductions between nonintersecting shafts, and they are generally nonreversible (i.e., they can turn in only one direction).

A bevel gear (see Figure 22-16) connects two intersecting shafts in any given speed ratio. Bevel gears are connected at a 90-degree angle and transmit power smoothly through a low power range.

Gears can also be found in a gearbox, where they are used to increase or reduce the speed of driven equipment. A **gearbox** (see Figure 22-17) is a mechanical device that houses a set of gears; connects the driver to the load; and allows for changes in speed, torque, and direction. Gearboxes use a series of large and small gears to adjust speed. For example, if the speed output is too high, a gearbox may be used to decrease the speed.

FIGURE 22-13 Spur Gears

FIGURE 22-14 Helical Gear

Worm Gear

FIGURE 22-15 Worm Gear

FIGURE 22-16 Bevel Gear

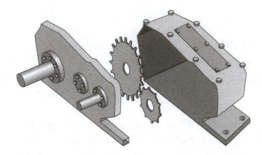

FIGURE 22-17 Gear Box

Principles of Lubrication

Lubrication is the application of a lubricant between moving surfaces in order to reduce friction and minimize heating. **Lubricants** (substances used to reduce friction between two contact surfaces) can come in many different thicknesses and forms. For example, lubricants may be thin oils with low viscosity (resistance to flow), or thick greases with high viscosity. Figure 22-18 shows examples of various types of lubricants.

Lubricants consist of a base (e.g., a natural or synthetic hydrocarbon) and appropriate additives. The most important additives are corrosion inhibitors and antioxidants. These additives provide protection to prolong the life of the equipment. Other additives that may be used include antifoaming agents, emulsifiers, antiwear components, viscosity range extending additives, and extreme pressure chemical components.

Appropriately designed packing and seals hold lubricants in place and keep foreign materials out. Seals are used between sections in motors and turbine rotors. Figure 22-19 shows examples of packing and seals.

FIGURE 22-18 Examples of the Various Types of Lubricants

Thin, Spray Lubricant   Thicker, Oil-Type Lubricant   Solid, Grease-Type Lubricant

Did You Know?

The Society of Automotive Engineers (SAE) has an established numerical system for rating motor oils according to their viscosity.

Some of the numbers contain the letter W, which stands for "winter." With 10W30 motor oil, the startup of your engine is protected with 10-weight oil. Once the engine warms up, it is protected with 30-weight oil.

FIGURE 22-19 Examples of Packing and Seals

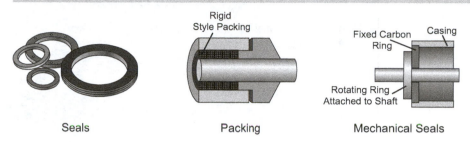

Seals · Packing · Mechanical Seals

With the lubricant contained by packing and seals, most devices use capillary action to draw the lubricant film into the narrow spaces between components. Lubricants reduce friction and serve as heat transfer agents (i.e., they circulate through the bearing and back to a reservoir with a cooling system in order to remove heat). In general, high-speed applications use the less viscous oils. Slower devices, such as vehicle wheels with ball or roller bearings, use more viscous oils or greases. Frictional resistance of greases is higher than that of oils, but the grease is not lost from the lubricated area as readily because of its viscosity.

Examples of industrial equipment that use lubricants include bearings, engines, chains, and gears. The lubricators used to keep this equipment properly lubricated are classified into several types: automatic, controlled feed, airline, and electrically operated. Lubricating equipment is important because it regulates the quantity of the lubricant that is applied to the lubricated machine. Some other types of lubrication equipment for liquid lubricants include wick-feed oilers, centralized oil systems, and splash-oiling devices. Table 22-2 lists and describes different types of lubricating devices.

TABLE 22-2   Types of Lubricants Used in the Process Industries and Their Descriptions

Types	*Description*
Centralized systems	Includes a central oil conditioning and pumping system.
Drip feed	Delivers lubricant through a drip that flows down onto bearing surfaces.
Forced feed	Found in larger machinery; uses an oil pump to force lubricant through ducts drilled into the rotating members and casing.
Grease fittings	Receptacles that contain a ball check valve and receive grease from a portable dispensing gun.
Oil mist	Uses an oil distribution system to transports the oil through a mist generator to the lubricated parts.
Recycling	Returns the oil to a reservoir where filtration and cooling occurs before redistribution back through the system.
Ring oilers	Metal rings that fit around and rotate off a shaft. Oilers have a larger diameter than the shaft, and the lower end dips into an oil pool. Oil is carried on the ring up to the shaft. As it contacts the shaft, it distributes the oil.
Splash feed	Uses a moving member, such as a crankshaft, to distribute the liquid by splashing droplets out of a pool and into the general bearing areas.
Wick feed	Similar to a drip feed oiler. Supplies lubricant to the bearing through the capillary action of fibrous material.

Many lubricating devices use a centralized lubricant delivery system (see Figure 22-20) to introduce the lubricant into the general areas of a piece of equipment. A centralized oil system is made up of a reservoir, pumps, heat exchangers for heating or cooling, filters, valves, and instruments in which oil is circulated.

Drip feed lubrication systems (see Figure 22-21) consist of a lubricant container mounted over a bearing. In this type of system, the oil is gravity-fed at a specific rate of

Did You Know?

Capillary action is what allows paper towels to wick up water.

Capillary action also allows lube oil to be drawn into the narrow spaces between two components.

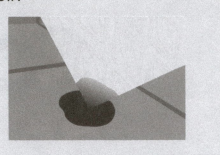

FIGURE 22-20 Centralized Lubrications System

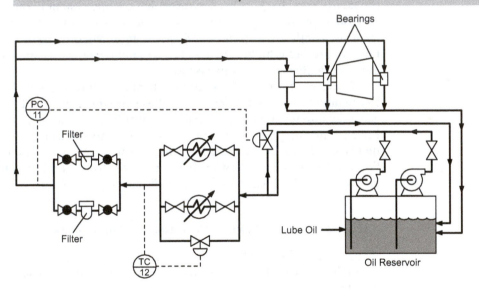

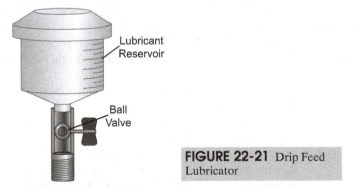

FIGURE 22-21 Drip Feed Lubricator

flow through an adjustable valve in the container. This provides constant and consistent lubrication for the bearing.

Force feed lubricators (see Figure 22-22) are automatic or mechanical systems. They use a pump to force oil into a mechanical device.

Grease fittings are receptacles that contain a ball check valve and receive grease from a portable dispensing gun. Figure 22-23 shows an example of a grease dispenser attached to a grease fitting.

Oil mist lubrication (see Figure 22-24) uses an oil distribution system to transport the oil through a mist generator to the lubricated parts. A two-cycle motor engine is an example of a device that uses oil mist lubrication. In a two-cycle engine, oil is mixed with the fuel; as the fuel is vaporized, the oil is carried into the engine as a unit.

FIGURE 22-22 Force Feed Lubricator

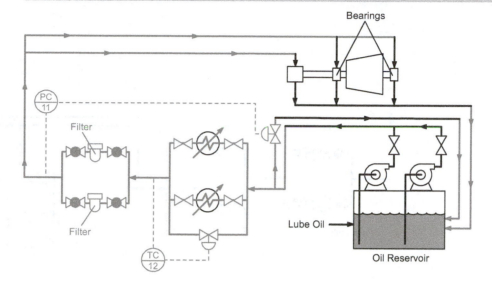

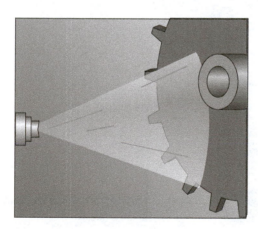

FIGURE 22-23 Grease Gun Attached to Grease Fitting

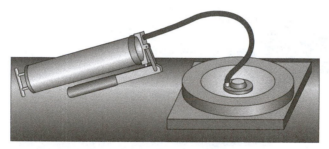

FIGURE 22-24 Oil Mist Lubrication

Recycling lubrication systems (see Figure 22-25) return the oil to a reservoir, where filtration and cooling occur before the lubricant is redistributed back through the system.

Ring oilers (see Figure 22-26) are metal rings that fit around and rotate off a shaft. These types of oilers have a larger diameter than the shaft so the lower end dips into an oil pool. Oil is carried on the ring up to the shaft and the bearings.

Splash feed (see Figure 22-27) uses a moving member, such as a crankshaft, to distribute the liquid by splashing droplets of lubricant out of a pool and into the general bearing areas.

A wick feed oiler (see Figure 22-28) is similar to a drip feed oiler. However, instead of dripping the oil, wick feed oilers supply lubricant to bearings through the capillary action of a fibrous material.

FIGURE 22-25 Oil Recycling System

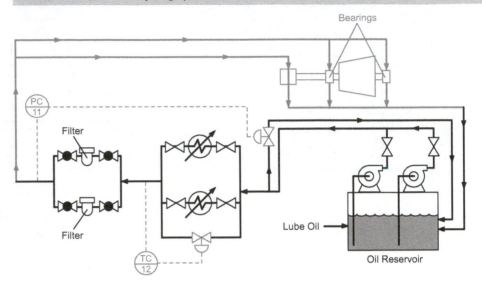

FIGURE 22-26 Ring Oiler

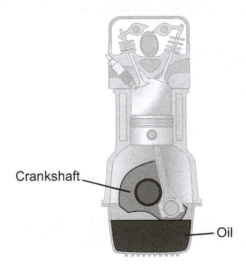

FIGURE 22-27 Crankshaft in a Crankcase

FIGURE 22-28 Wick Feed Oiler

Potential Problems

It is important to maintain proper equipment lubrication because failure to do so can lead to premature wear, heat buildup, and equipment failure that could affect other parts of the process. For example, the entire process facility could be shut down if a gearbox located on a critical piece of equipment fails. Because of this, process technicians should always be alert and able to detect unusual noises, excessive vibrations, changes in lubricant color, and unusual smells that might indicate that the lubricant is getting too hot. Process technicians should also be aware of associated protection devices such as interlock systems and vibration monitors, and startup and shutdown procedures.

INTERLOCK PROTECTION

Some interlock systems are designed to keep large, single-train, rotating equipment from sustaining serious damage when upset conditions or component failures occur. Process technicians must be aware of how interlock systems work so that they can help prevent emergency shutdowns and promptly recover and restart the process should a shutdown occur. Process technicians must also know the various conditions that can cause equipment shut down, such as the trip points and how to tell which item initiated the event.

VIBRATION MONITORING

Excessive vibration, particularly on large rotating equipment, can be an indication of a potentially serious problem. Because of this, portable and/or permanently mounted vibration sensors are often used to detect vibrations that are outside the normal range. Process technicians are responsible for monitoring these vibration sensors and recording trend data so potential problems can be identified and resolved.

STARTUP AND SHUTDOWN PROCEDURES

Improper startup may cause sudden surges that can shorten the life of the rotating equipment and cause excessive stress on bearings, gears, seals, and other rotating equipment components. Because of this, rotating equipment must always be shut down and returned to proper operating speed, according to procedure.

IMPROPER SEAL PROCEDURES

In many seal installations, clean liquid purges are set to lubricate and cool rotating components. Process technicians operate the pumps (if any) that supply the seal liquids. Additional problems that process technicians can face are improper levels of lubrication oil in a gearbox, the use of the wrong type of oil or grease, or water in the lubricant. Each of these can cause mechanical problems in the pump all three of which can cause worn teeth on a gearwheel.

Safety and Environmental Hazards

Hazards associated with normal and abnormal transmission and lubrication operations can negatively affect the personal safety of the process technician, the equipment, and the environment. For example, seal failure can allow process fluid to escape into the atmosphere and negatively affect personnel and the environment.

PERSONAL SAFETY

Process technicians should always be aware of the potential hazards and safety precautions associated with transmission and lubrication operations. For instance, all couplings must be covered with a coupling guard to keep people from accidentally coming into contact with rotating equipment.

Although the lubricating oils are below their flash point, fire danger still exists because of potential exposure to hot surfaces (e.g., steam lines). Couplings also pose a risk due to the possibility of failure (disintegrating) and flying debris.

Process technicians may be required to make routine and preventive maintenance checks. Personal protective equipment, such as hard hats, fire-retardant coveralls, gloves, hearing protection, and safety glasses must be worn when working in process areas. Process technicians should also be aware of the danger of loose garments, long hair, gloves, and jewelry getting caught in rotating equipment.

EQUIPMENT OPERATIONS

Equipment operation hazards are associated with power transmission during normal and abnormal operations. The possibility of failed transmission, gears, and lubrication systems exists at all times. Any of these failures directly affect the operations of the equipment.

If the coupling fails, there is a danger of flying debris, depending on the type of coupling used. The coupling is designed to be the "weak link" in the driver, gearbox, or machinery train so that serious damage does not result. Noisy gearboxes may also be an indication of potential problems. Process technicians must be alert to changes in noise levels and to check the sound levels and temperatures, smells and color changes, as appropriate, to prevent personnel injury.

As with any high-pressure fluid, hydraulic fluid presents a potential safety hazard. Bearing failures in centrifugal pumps or other rotating equipment pose a definite threat because it is not uncommon under these conditions for heat buildup to create an ignition source and result in a fire.

ENVIRONMENTAL IMPACT

Process technicians are responsible for maintaining equipment operation to prevent leaks and emissions. Releases to the environment include spills and noise pollution. Leaks, noise extremes, and unusual vibrations should all be investigated promptly. Appropriate support personnel may need to be involved to correct the problem.

Process Technician's Role in Operation and Maintenance

Process technicians may be required to complete a number of tasks associated with the maintenance of power transmissions and lubricating equipment. These tasks can include the following:

- Switching or replacing oil filters
- Switching oil coolers or backflushing waterside oil coolers (water side)
- Collecting and recording vibration levels and readings
- Checking oil levels and adding oil if necessary
- Checking the grease supply and adding grease if necessary
- Changing oil where applicable
- Sampling oil for water and other contamination
- Checking bearing surface temperatures with a skin or surface pyrometer
- Monitoring and ensuring proper operation of oil centrifuges

LUBRICATION MAINTENANCE

Process technicians are responsible for maintaining oil levels, changing system filters, and/or adding grease to bearings.

Maintaining Oil Levels

Oil levels in lube oil reservoirs must be checked frequently. Vents in the lube oil reservoir should be monitored to ensure they are not venting process materials or vapors. In some systems, water contamination of the oil can occur in several ways. Larger rotating equipment has sight flow windows through which the color of the lube oil can be checked. A milky or frothy appearance should be immediately investigated because it indicates contamination of the oil. Water contamination can be centrifuged out using a portable centrifuge and a slipstream of oil from the pump.

Changing Filters

Clean lube oil supply is necessary to ensure that the lubrication system operates properly. To keep it clean, the lube oil normally flows through filters before going to the machinery. Filter pressure drop is a control parameter to help avoid problems. An excessive differential pressure across the filter (from inlet to outlet) is an indication of a dirty filter. Filters are switched before the pressure drop is excessive. Extreme care must be taken when switching filter elements to avoid shutting off the oil flow. Process technicians must also ensure that the flow does not bypass the filter elements. Unfiltered oil may contain contaminants that may damage the lubricated equipment.

Adding Grease to Bearings

In many process facilities, rotating equipment items have grease fittings. It is the responsibility of the process technician to add grease to these fittings at an appropriate frequency. Selecting proper lubricant, adding the correct amount, and recognizing contamination are important. In a typical large process facility or refinery, many different kinds of greases and lubricating oils are used. It is the responsibility of the process technician to service all equipment with the proper lubricant.

Summary

Mechanical power transmission and lubrication systems are critical to the daily operation of the process industries. The purpose of mechanical power transmission is to transfer the rotational energy from one point to another with minimum loss of energy to friction.

Lubrication is the process of establishing a stable fluid film between two hard metal surfaces moving relative to each other. The objective of lubrication systems is to transport the lubricating fluid into the area between the metal surfaces of the equipment in order to minimize contact friction. Lubrication prevents metal-to-metal moving contact and removes heat from friction, both of which are generally destructive to equipment.

Mechanical power transmission consists of various components that make up the operating concepts of transmission. Couplings are power transfer devices that connect the shafts of drivers to driven equipment. There are several types of couplings, including flexible, rigid, and fluid couplings.

Gearboxes contain gears used to increase or decrease the speed of the driven equipment relative to the speed of the driver. Belts use traction to transfer rotational energy to pulleys. Chains are components that perform the same function but transfer rotational energy through sprockets.

Magnetic drive pumps transmit power through magnetic forces between the driver and the driven elements. Hydraulic drives transfer energy through closed circuit movements of a pressurized liquid in conduits.

Bearings are used to support shafts for radial (up-and-down or side-to-side) and axial (back-and-forth) confinement. They allow smooth movement between two surfaces on rotary or linear moving equipment. The four main types of bearings including roller, ball, sleeve (fluid film), and thrust.

Gears are used in most mechanical devices to transmit motion and control the speed of the equipment. Gears come in many sizes and shapes. The most common types of gears include spur, helical, worm, and bevel.

Lubrication systems are designed to transport the lubrication fluid into the area between metal surfaces to ensure a minimal amount of contact friction. Lubricants reduce friction and remove heat. Lubrication of valves makes them easier to open and close.

Problems that can occur as a result of improper mechanical energy transmission or lubrication can affect equipment life, process facility operations, interlock protection, vibration monitoring, startup and shutdown procedures, packing and seals, and improper seal procedures. Problems can ultimately result in the necessity of replacement cost much sooner than anticipated.

Appropriately designed seals hold lubricants in place and keep foreign materials out. These seals are placed at shaft exit points just beyond the bearings. In many seal installations, clean fluid purges are sent to lubricate and cool rotating components.

Interlock systems are designed to keep large (and often single-train) rotating equipment from sustaining serious damage when upset conditions or component failures occur. Leaks, noise, unusual smells, color changes, and unusual vibrations should all be investigated promptly by appropriate support personnel.

Hazards associated with normal and abnormal transmission/lubrication operations can affect the personal safety of process technicians, the equipment, and the environment.

Process technicians must perform various types of maintenance, including transmission maintenance, lubrication maintenance, maintenance of oil levels, changing filters, and addition of grease to bearings.

Checking Your Knowledge

1. Define the following terms:
   - a. Bearing
   - b. Belt
   - c. Chain
   - d. Coupling
   - e. Gearbox
   - f. Gear
   - g. Lubricant
   - h. Packing
   - i. Seal
   - j. Sheave
   - k. Sprocket

2. A rigid coupling is used:
   - a. to allow the motor to accelerate to full speed before the load is increased.
   - b. in applications where small amounts of misalignment can be tolerated.
   - c. to cushion the shock from equipment overloads.
   - d. more frequently in general power transmission service and permits some misalignment.

3. *(True or False)* One advantage of a chain over a belt driver is no slippage between chain and sprocket teeth.

4. Which of the following definitions best describes sleeve (fluid film) bearings?
   - a. They are used in moderate-pressure applications in small to mid-size operations (less than 1,000 rpm).
   - b. They are used in smaller machinery at low speeds.
   - c. They are used in large machinery at low speeds.
   - d. They are used in larger machinery at high speeds.

5. _____ gears are used to obtain large speed reductions between nonintersecting shafts placed at 90-degree angles to one another.
   - a. Spur
   - b. Worm
   - c. Helical
   - d. Bevel

6. _____ facilitates the spreading of lubricant film into narrow spaces between equipment components.
   - a. Lubrication
   - b. Capillary action
   - c. Corrosion inhibitors
   - d. Antifoams

7. Which of the following are the most important additives used in lubricants (select all that apply)?
   - a. Corrosion inhibitors
   - b. Antifoams
   - c. Emulsifiers
   - d. Antioxidants

8. Which type of lubricator is used in larger machinery and uses an oil pump to inject lubricant through ducts drilled into rotating members and casings?
    a. Ring oilers
    b. Wick feed
    c. Forced feed
    d. Dip feed
9. *(True or False)* Interlock systems are designed to prevent rotating equipment from sustaining serious damage when upset conditions or component failures occur.
10. *(True or False)* Equipment vibration on large rotating equipment can be an indication of a potentially serious problem.

Student Activities

1. Write a one-page paper on the purpose and principle of mechanical power transmission.
2. With a classmate, research couplings and explain the three main types of couplings.
3. With a pulley system in a lab setting, show how to properly adjust belt tension.
4. Given a drawing, identify the location of the bearings on a piece of equipment and identify the type of bearing.
5. Given drawings or actual gears, identify and describe spur, helical, worm, and bevel gears.
6. Explain the process technician's role in the following tasks:
    - Transmission maintenance
    - Maintaining oil levels
    - Changing system filters
    - Adding grease to bearings

Glossary

Absorption—the process by which a gas or liquid is dissolved into a liquid or solid.

AC power source—a device that supplies alternating current.

Adjustable pliers—pliers that have a tongue and groove joint or slot that allows the jaws to widen and grip objects of different sizes. Also referred to as Channel Locks® or slip joint pliers.

Adjustable wrench—an open-ended wrench with an adjustable jaw that contains no teeth or ridges (i.e., the jaw is smooth). Also referred to as a Crescent® wrench.

Adsorption—the adhesion of a gas or liquid to the surface area of a solid.

Agglomerator—a mixer that feeds dry and liquid products together to produce an agglomerated (clustered together) product like powdered laundry detergent.

Agitator—a device used to mechanically mix the contents of a tank or vessel. Also known as a mixer.

Agitator blades—devices that act on the fluid by generating turbulence, which promotes mixing.

Agitator shaft—a cylindrical rod that holds the agitator blades that stir the chemicals in the tank or reactor.

Air register—an air intake device used to adjust air flow to a burner in a boiler or furnace.

Alloys—compounds composed of two or more metals that are mixed together in a molten solution.

Alternating current (AC)—electric current that reverses direction periodically, usually sixty times per second.

Ambient air—any part of the atmosphere that is accessible and breathable by the public.

Ammeter—a device used to measure the electrical current in a circuit.

Ampere (amp)—a unit of measure of the electrical current flow in an electrical circuit; similar to "gallons of water" flow in a pipe.

ANSI—American National Standards Institute; oversees and coordinates the voluntary standards in the United States. ANSI develops and approves norms and guidelines that affect many business sectors. The coordination of U.S. standards with international standards allows U.S. products to be used worldwide.

Antisurge—automatic control instrumentation designed to prevent compressors from operating at or near pressure and flow conditions that result in surge.

Antisurge protection—protection that prevents damage to the compressor. Antisurge protection is calculated and designed by engineers to ensure proper and safe compressor operation.

API—American Petroleum Institute; a trade association that represents the oil and gas industry in the areas of advocacy, research, standards, certification, and education in its dealings with the petroleum industries, petrochemical industries, and municipalities.

Application block—the main part of a drawing that contains symbols and defines elements such as relative position, types of materials, equipment descriptions, flows, and functions.

Approach range—a range that describes how close to the dew point a cooling tower can cool the water.

Articulated drain—a hinged drain that is attached to a floating roof. Its purpose is to allow drainage as the roof raises and lowers.

ASME—American Society of Mechanical Engineers; specifies requirements and standards for pressure vessels, piping, and their fabrication.

Assisted draft furnace—a system that uses an electric motor, steam-turbine-driven rotary fans, or blowers to push combustion gases into the furnace (forced draft) or draw flue gas from the furnace to the stack (induced draft).

Axial compressor—a dynamic compressor that contains a rotor with contoured blades followed by a stationary set of blades (stator). In this type of compressor, the flow of gas is axially (in a straight line along the shaft).

Axial pump—a dynamic pump that uses a propeller or row of blades to propel liquids along the shaft.

Azeotrope—a mixture of two or more substances that boil at a constant temperature and a fixed composition.

Azeotropic distillation—the process of adding a third substance (called an entrainer) to separate a mixture of two or more substances with close boiling points (e.g., adding benzene to a water/ethanol azeotrope).

Backwashing—a procedure in which the direction of flow through the exchanger is reversed to remove solids that, for example, have accumulated in the inlet tubes of a heat exchanger.

Baffle—a stationary plate, placed inside a vessel or tank, that is used to alter the flow of chemicals, facilitate mixing, or cause turbulent flow.

Bag filter—a tube-shaped filtration device that uses a porous bag to capture and retain solid particles.

Baghouse—air pollution control equipment that contains fabric or bag filter tubes, envelopes, or cartridges designed to capture, separate, or filter particulate matter.

Balanced draft furnace—a system that uses induced draft and forced draft to force air through a furnace.

Ball check valve—a valve used to control the flow of heavy fluids; it is available in horizontal, vertical, and angle designs.

Ball valve—a type of valve that uses a flow control element shaped like a hollowed-out ball attached to an external handle to increase or decrease flow.

Barge—a flat-bottomed boat used to transport fluids (e.g., oil or liquefied petroleum gas) and solids (e.g., grain or coal) across shallow bodies of water such as rivers and canals.

Basin—a reservoir at the bottom of the cooling tower where cooled water is collected so it can be recycled through the process.

Batch reaction—a process in which reactants and catalysts or initiators are added as a single charge to the reactor rather than continuously to make a batch of final product.

Bearing—a machine component that rotates, slides, or oscillates. Bearings reduce friction between the motor's rotating and stationary parts. Also, bearings are mechanical devices used to support shafts for radial (up-and-down or side-to-side) and axial (back-and-forth) movement and to hold the rotor in alignment with other parts.

Belt—a flexible band, placed around two or more pulleys, that transmits rotational energy.

Bernoulli principle—principle stating that, as the speed of a fluid increases, the pressure of the fluid decreases.

Bin—a container used to store solid products or materials; typically designed to hold a finished product destined for distribution, but may also store off-specification product earmarked for re-work or blending.

Binary distillation—the process of separating two materials into an overhead product and a base product.

Biocide—a chemical agent that is capable of controlling undesirable living organisms in a cooling tower.

Blanketing—the process of introducing an inert gas, usually nitrogen, into the vapor space above the liquid in a tank to prevent air leakage into the tank (often referred to as a nitrogen blanket).

Blinds—solid plates or covers that are installed between pipe flanges in order to prevent the flow of fluids and isolate equipment or piping sections when repairs are being performed; typically made of metal.

Block flow diagram (BFD)—a simple illustration that shows a general overview of a process; it indicates the parts of a process and their relationships.

Block valve—a valve used to block flow to and from equipment and piping systems (e.g., during outage or maintenance). Block valves differ from other types of valves because they should not be used to throttle flow.

Blowdown—the process of removing small amounts of water from the cooling tower or boiler to reduce the concentration of impurities.

Boiler—a device in which water is boiled and converted into steam under controlled conditions.

Bonnet—the portion of a valve body through which the stem leaves the body and contains the stem packing. The bonnet is a bell-shaped dome mounted on the body of a valve.

Box furnace—a square, box-shaped furnace designed to heat process fluids or to generate steam.

Breakthrough—the condition of any adsorbent bed when the component being adsorbed starts to appear in the outlet stream.

British thermal unit (BTU)—a measure of energy in the English system, referring to the heat required to raise the temperature of 1 pound of water 1 degree Fahrenheit at sea level.

Bubble cap tray—a tray in a distillation column that uses slotted caps with fixed openings to disperse the vapor into the liquid.

Bucket elevator—a continuous line of buckets attached by pins to two endless chains running over tracks and driven by sprockets.

Bullet vessel—a cylindrically shaped container used to store contents at moderate to high pressures.

Burner—a mechanical device where fuel is burned in a controlled manner to produce heat for a boiler.

Butterfly valve—a type of valve that uses a disc-shaped flow control element to increase or decrease flow.

Cabin furnace—a cabin-shaped furnace designed to heat process fluids or to generate steam.

Calorie—the amount of heat energy required to raise the temperature of 1 gram of water by 1 degree Celsius at sea level.

Camshaft—a driven shaft fitted with rotating wheels of irregular shape (cams) that open and close the valves in an engine.

Canned pump—a seal-less pump that ensures zero emissions. It is often used on EPA-regulated liquids.

Carbon ring—a seal component located around the shaft of the turbine that controls the leakage of motive fluid (typically steam) along the shaft or the entrance of air into the exhaust.

Cartridge filter—a filtration unit that uses a fine mesh that is tightly folded or pleated to remove suspended contaminants from a liquid.

Casing—a housing component of a turbine or engine that holds all moving parts, including the rotor, bearings, and seals, and is stationary.

Catalysts—substances that initiate or increase the rate of a chemical reaction at a given temperature and pressure but are not changed or consumed by the reaction.

Cavitation—a condition inside a pump wherein the liquid being pumped partly vaporizes due to variables such as temperature and pressure drop.

Centrifugal compressor—a dynamic compressor in which the gas flows from the inlet located near the suction eye to the outer tip of the impeller blade.

Centrifugal force—the force that causes something to move outward from the center of rotation.

Centrifugal pump—a pump that uses an impeller on a rotating shaft to generate pressure and move liquids.

Centrifuge—a device that uses centrifugal force to separate solids from liquids or to perform liquid-liquid solvent extraction. Also called a centrifugal extractor or contactor.

Chains—mechanical devices used to transfer rotational energy between two or more sprockets.

Check valve—a type of valve that allows flow in only one direction and is used to prevent reversal of flow in a pipe.

Chevrons—V-shaped blades found on a turbine rotor.

Circuit—a system of one or many electrical components that accomplish a specific purpose.

Circuit breaker—an electrical component that opens a circuit and stops the flow of electricity when the current reaches unsafe levels.

Circulation rate—the rate at which cooling water flows through the tower and through the process exchangers.

Coagulation—a method for concentrating and removing suspended solids in boiler feedwater by adding chemicals to the water, which causes the impurities to cling together.

Co-current flow—flow that occurs when two fluids are flowing in the same direction.

Cold wall reactor—a reactor composed of an unheated or purposely cooled wall of some type of glass or refractory and an internal heating element.

Combustion chamber—a chamber between the compressor and the turbine where the compressed air is mixed with fuel and the fuel is burned, increasing the temperature and pressure of the combustion gases.

Compression ratio—the ratio of the volume of the cylinder at the start of a stroke compared to the smaller (compressed) volume of the cylinder at the end of a stroke.

Compressor—a mechanical device used to increase the pressure of a gas or vapor.

Condenser—a heat exchanger that is used to convert a vapor to a liquid.

Condensing steam turbine—a device in which exhaust steam is condensed in a surface condenser. The condensate is then recycled to the boiler.

Conduction—the transfer of heat from one substance to another by direct contact.

Conductor—a material that has electrons that can break free from the flow more easily than the electrons of other materials.

Connecting rod—a component that connects a piston to a crankshaft.

Containment wall—a wall used to protect people and the environment against tank failures, fires, runoff, and spills.

Continuous reaction—a reaction in which products are continuously being formed and removed, while raw materials (reactants) are continuously fed into the reactor.

Control valve—a valve that automatically controls the increase or decrease of fluid flow through a pipe by remote operation.

Convection—the transfer of heat as a result of fluid movement.

Convection section—the upper portion of a furnace where heat is transferred by convection.

Convection tube—a furnace tube, located above the shock bank, that receives heat primarily through the process of convection.

Conveyor—a mechanical device used to move solid material from one place to another in a continuous stream.

Coolant—a fluid that circulates around or through an engine to remove the heat of combustion. The fluid may be a liquid (e.g., water or antifreeze) or a gas (e.g., air or freon).

Cooler—a heat exchanger that may use cooling tower water to lower the temperature of process materials.

Cooling range—the difference in the temperature between the hot water entering the tower and the cooler water exiting the tower.

Cooling tower—a structure designed to lower the temperature of water using latent heat of evaporation.

Corrosion—deterioration of a metal by a chemical reaction (e.g., iron rusting).

Counterflow—the condition created when air and water flow in opposite directions.

Couplings—mechanical devices used to connect and transfer rotational energy from the shaft of the driver to the shaft of the driven equipment.

Crane—a mechanical lifting device equipped with a winder, wire ropes, and sheaves that can be used to lift and lower materials.

Crankshaft—a component that converts the piston's up-and-down or forward-and-backward motion into rotational motion.

Cross flow—flow that occurs when two streams flow perpendicular (at 90-degree angles) to each other.

Crystallization—the process of creating crystals from a supersaturated solution.

Cyclone—a mechanical device that uses a swirling (cyclonic) action to separate heavier and lighter components of a gas or liquid stream.

Cylinder—a cylindrical chamber in a positive displacement compressor in which a piston compresses gas and then expels the gas.

Damper—a moveable plate that regulates the flow of air, draft, or flue gases.

Dead head—the maximum pressure (head) that occurs at zero flow.

Deaeration—a method for removing air or other gases from boiler feedwater by increasing the temperature and aeration time.

Demineralization—a process that uses ion exchangers to remove mineral salts. Also known as deionization. The water produced is referred to as deionized water.

Demister—a device that promotes separation of liquids from gases.

Desiccant—a substance that attracts and absorbs moisture.

Desuperheated steam—superheated steam from which some heat has been removed by the reintroduction of water. It is used in processes that cannot tolerate the higher steam temperatures.

Desuperheater—a temperature control point at the outlet of the boiler steam flow that maintains a specific steam temperature. The temperature is maintained by using boiler feedwater injection through a control valve.

Dew point—the temperature at which air is completely saturated with water vapor (100 percent relative humidity).

Diaphragm pump—a mechanically or air-driven pump that consists of two flexible diaphragms connected by a common shaft.

Diaphragm valve—a type of valve that uses a flexible, chemical-resistant, rubber-type diaphragm to control flow.

Die—a metal plate with a specifically shaped perforation designed to give materials a specific form as they pass through an extruder.

Dike—a wall (earthen, shell, or concrete) built around a piece of equipment to contain any liquids should the equipment rupture or leak.

Direct current (DC)—electrical current that flows in a single direction through a conductor.

Discharge check valve—a valve that prevents liquid from back-flowing into the pump during the suction stroke.

Disconnect—a large electrical switch used for isolation during system repairs and maintenance.

Distillation—the process of separating two or more liquids by their boiling points.

Dolly—a wheeled cart or hand truck used to transport heavy items such as barrels, drums, and boxes.

Double-acting piston pump—a type of piston pump that pumps by reciprocating motion on every stroke (one on one side of the piston, and one on the other side).

Downcomer—a tube, located in the firebox, that transfers water from the steam drum to the mud drum.

Draft fan—a fan used to supply combustion air to the burners.

Drift—the carrying of water with the air stream.

Drift eliminator—a device that prevents water from being blown out of the cooling tower; the main purpose of a drift eliminator is to minimize water loss.

Drill—a device that contains a rotating bit designed to bore holes into various types of materials.

Drum mixer—a screw-type device used to blend two or more materials (sometimes with small amounts of liquid) into a consistent composition.

Dry bulb temperature—the actual temperature of the air that can be measured with a thermometer.

Dry carbon ring—an easy-to-replace, low-leakage type of seal consisting of a series of carbon rings that can be arranged with a buffer gas to prevent the process gas from escaping.

Dryer—a device used to remove moisture from a process stream.

Duty—the amount of heat energy a cooling tower is capable of removing; usually expressed in MBTU/hour.

Dynamic compressor—a nonpositive displacement compressor that uses centrifugal or axial force to accelerate and convert the velocity of the gas to pressure. Dynamic compressors are classified as either centrifugal or axial.

Dynamic head—the amount of push a pump must have in order to overcome the pressure of the liquid.

Dynamic mixer—a piece of equipment that rotates to mix products rapidly.

Dynamic pump—a pump that converts the spinning motion of a blade or impeller into dynamic pressure to move liquids.

Economizer—the section of a boiler located in the flue gas stream; used to preheat feedwater before it enters the upper steam drum.

Eductor—a device that transports fluid by using a Venturi design.

Electrical diagram—an illustration showing power transmission and how it relates to the process.

Electricity—the flow of electrons from one point to another along a pathway called a conductor.

Electric tool—a tool operated by electrical means (either AC or DC).

Electromagnetism—magnetism produced by an electric current.

Electrons—negatively charged particles that orbit the nucleus of an atom.

Electrostatic precipitator—a device that contains positively charged collecting plates and a series of negatively charged wires.

Endothermic—absorbing or requiring heat.

Engine—a machine that converts chemical (fuel) energy into mechanical force.

Engine block—the casing of an engine that houses the pistons and cylinders.

Erosion—the degradation of tower components (fan blades, wooden portions, and cooling water piping) by mechanical wear or abrasion by the flow of fluids (which often contain solids). Erosion can lead to a loss of structural integrity and process fluid contamination.

Evaporation—a process in which a liquid is changed into a vapor through the latent heat of evaporation.

Exchanger head—a device located on the end of a heat exchanger; it directs the flow of fluids into and out of the tubes.

Exhaust port—a chamber or cavity in an engine that collects exhaust gases and directs them out of the engine.

Exhaust valve—a valve in the head at the end of each cylinder; exhaust valves open and direct the exhaust from the cylinder to the exhaust port.

Exothermic—releasing energy in the form of heat.

Expansion loop—a segment of pipe that allows for expansion and contraction during temperature changes.

Extraction—a separation process that uses a solvent or other medium to separate components based on differences in solubility (the degree to which a solid dissolves in a solvent).

Extractive distillation—a method used for separating components with close boiling points by the addition of a solvent that alters the volatility of the components enough so they can be separated by normal distillation.

Extruder—a device that forces materials through the opening in a die so it can be formed into a particular shape.

Fan—a rotating blade inside a motor housing or casing that cools the motor by pulling air in through the shroud.

Feeder—a piece of equipment that conveys material to a processing device (e.g., an extruder) at a controllable and changeable rate.

Fill—material inside the cooling tower, usually made of wood or plastic, which breaks water into smaller droplets and increases the surface area for increased air-to-water contact.

Filter—a device that removes particles from a process and allows the clean product to pass through the filter.

Filtration—a method for removing suspended matter and sludge from boiler feedwater and condensate return systems.

Firebox—the portion of a boiler or furnace where burners are located and radiant heat transfer occurs.

Fire tube boiler—a device that passes hot combustion gases through the tubes to heat water on the shell side.

Firewall—an earthen bank or concrete wall built around a storage tank to contain the contents in the event of a spill or rupture.

Fitting—system components used to connect together two or more pieces of pipe, tube, or other piece of equipment.

Fixed bed dryer—a device that uses a desiccant to remove moisture from a process stream.

Fixed bed reactor—a reactor in which the catalyst bed is stationary as the reactants pass through and around it.

Fixed blade—a blade inside a turbine, fixed to the casing, that directs steam or combustion gases.

Fixed roof tank—a container that has a roof permanently attached to the top. These roofs can be cone-shaped, flat, or cylindrical.

Flare system—a device used to burn unwanted process gases before they are released into the atmosphere.

Flaring tool—a device used to create a cone-shaped enlargement at the end of a piece of tubing so the tubing can accept a flare fitting.

Flash distillation—the partial evaporation or vaporization that occurs when a saturated liquid stream undergoes a reduction in pressure.

Flash dryer—a device that forces hot gas (usually air or nitrogen) through a vertical or horizontal flash tube, causing the moist solids to become suspended.

Floating roof tank—a container that has a roof that floats on the surface of a stored liquid to minimize vapor space.

Fluid bed dryer—a device that uses hot gases to fluidize solids and promote the contact of dry gases and solids.

Fluidized bed reactor—a reactor that uses high-velocity fluid to suspend or fluidize solid catalyst particles. The reactor feed is mixed with the suspended catalyst where the reaction takes place.

Foam chamber—a reservoir on the side of a tank designed to contain chemical foam that can be used to extinguish fires within the tank.

Foaming—the formation of a froth generated by water contaminants mixing with air. Foam causes impairment of water circulation and may cause pumps to lose suction or cavitate.

Forced draft cooling towers—cooling towers that contain fans or blowers at the bottom or on the side of the tower that force air through the equipment.

Forced draft furnace—a type of furnace that uses a fan to push air flow through the furnace.

Forklift—a motorized vehicle with pronged lifts. Forklifts are used to lift equipment or items placed on a pallet to higher elevations for use in a unit, for storage, or for maintenance.

Fouling—accumulation of deposits (such as sand, silt, scale, sludge, fungi, and algae) built up on the surfaces of processing equipment.

Frame—a structure that holds the internal components of a motor and motor mounts.

Fuel lines—provide the fuel required to operate the burners.

Fuel-operated tools—tools powered by the combustion of a fuel (e.g., gasoline).

Furnace—a piece of equipment that burns fuel to generate heat that can be transferred to process fluids flowing through tubes. Also referred to as a process heater or reaction furnace.

Fuse—a device used to protect equipment and electrical wiring from overcurrent.

Gantry crane—a crane (similar to an overhead crane) that runs on elevated rails attached to a frame or set of legs.

Gasket—a flexible material used to seal components together so they are air- or water-tight.

Gas turbine—a device that uses the combustion of natural gas to spin the turbine rotor.

Gate valve—a positive shutoff valve that utilizes a gate or guillotine that, when moved between two seats, causes a tight shutoff.

Gauge hatch—an opening in the roof of a tank that is used to check tank levels and obtain samples of the product or chemical.

Gear—a toothed wheel that engages another toothed mechanism in order to change the speed or direction of transmitted motion.

Gearbox—a mechanical device that houses a set of gears; connects the driver to the load; and allows for changes in speed, torque, and direction.

Gear pump—a pump that rotates two gears with teeth in opposing directions, thus allowing the liquid to enter the space between the teeth of each gear in order to move the liquid around the casing to the outlet.

Generator—a device that converts mechanical energy into electrical energy.

Globe valve—a type of valve that uses a plug and seat to regulate the flow of fluid through the valve body; the plug is shaped like a sphere or globe.

Governor—a device used to control the speed of a piece of equipment such as a turbine.

Gravimetric feeder—a device designed to convey material at a controlled rate by measuring the weight lost over time.

Grinder—a device that uses the rotating force of one material against another material (e.g., a stone grinding wheel against a metal blade) to reduce or adjust the shape and size of an object.

Grounding—connecting an object to the earth using copper wire and a grounding rod to provide a path for the electricity to dissipate harmlessly into the ground.

Hammer—a tool with a handle and a heavy head that is used for pounding or delivering blows to an object.

Hand tool—a tool that is operated manually instead of being powered by electricity, pneumatics, hydraulics, or other forms of power.

Hand wheel—the mechanism that raises and lowers a valve stem to allow or restrict the flow of fluid through the valve.

Head—the component of an engine on the top of the piston cylinders that contains the intake and exhaust valves.

Head pressure—the pressure of a liquid exerted upon a system; the amount of head pressure is determined by the height of the liquid.

Heat exchanger—a device used to transfer heat from one substance to another without the two physically contacting each other.

Heat transfer—the transfer of energy from one object to another as a result of temperature difference between the two objects.

Hemispheroid vessel—a pressurized container that is used to store material with a vapor pressure slightly greater than atmospheric pressure (i.e., .5 psi to 15 psi). The walls of hemispheroid tanks are cylindrically shaped and the top is rounded.

Hoist—a device composed of a pulley system with a cable or chain used to lift and move heavy objects.

Hopper—a funnel-shaped, temporary storage container that is typically filled from the top and emptied from the bottom.

Hopper car—a type of railcar designed to transport solids like plastics and grain.

Hose—a flexible tube that carries fluids; it can be made of plastic, rubber, fiber, metal, or a combination of materials.

Hot wall reactor—a reactor in which heat is transferred to the reactants by conduction from the vessel wall.

Humidity—moisture content in ambient air measured as a percentage of saturation.

Hydraulic tool—a tool that is powered using hydraulic (liquid) pressure.

Hydraulic turbine—a device that is operated or affected by a liquid (e.g., a water wheel).

Hydrocyclone—a mechanical device that promotes separation of heavy and light liquids by centrifugal (rotating) force.

Igniter—a device (similar to a spark plug) that automatically ignites the flammable air and fuel mixture at the tip of the burner.

Impact wrench—a pneumatically or electrically powered wrench that uses repeated blows from small internal hammers that generate torque to tighten or loosen fasteners.

Impeller—a fixed, vaned device in a boiler that causes the air/fuel mixture to swirl above the burner. An impeller in this instance is not the same as an impeller in a turbine or pump. In a pump, an impeller is a vaned device that rapidly spins a liquid to generate centrifugal force.

Impulse movement—movement that occurs when the steam first hits a rotor and the rotor begins to move.

Impulse turbine—a device that uses high-pressure steam to move a rotor.

Incinerator—a device that uses high temperatures to destroy solid, liquid, or gaseous wastes.

Induced draft cooling tower—a cooling tower in which air is pulled through the tower internals by a fan located at the top of the tower.

Induced draft furnace—a type of furnace that uses a fan located at the top of the furnace to draw flue gas from the furnace body into the stack to draw the flow of air through the furnace.

Induction motor—a motor that turns slightly slower than the supplied frequency, and can vary in speed based on the amount of load.

Inhibitor—a substance used to slow or stop a chemical reaction. Inhibitors are generally consumed during the course of the reaction.

Initiator—a substance that starts or increases the rate of a chemical reaction. Initiators are consumed by the reaction.

Insulation—any substance that prevents the passage of heat, light, electricity, or sound from one medium to another.

Insulator—any substance that prevents the passage of heat, light, electricity, or sound from one medium to another.

Intake port—an air channel that directs fuel gases to an intake valve.

Intake valve—a valve located in the head, at the top of each cylinder, that opens and allows fuel gases to enter the cylinder.

Interchanger—a process-to-process heat exchanger. Interchangers use hot process fluids on the tube side and cooler process fluids on the shell side. Also known as a cross exchanger.

Interlock—a type of hardware or software that does not allow an action to occur unless certain conditions are met.

ISA—the Instrumentation, Systems, and Automation Society; a global, nonprofit technical society that develops standards for automation, instrumentation, control, and measurement.

Isometric drawing—an illustration showing objects as they would appear to the viewer (similar to 3D drawings that appear to come off the page).

Jacketed pipes—pipes that have a pipe-within-a-pipe design so hot or cold fluids can be circulated around the process fluid without the two fluids coming into direct contact with each other.

Jet pump eductor—a device that uses the movement of high-pressure liquids to create suction (Venturi effect). Also referred to as an aspirator.

Jib crane—a crane that contains a vertical rotating member and an arm that extends to carry the hoist trolley.

Knockout pot—a device designed to remove liquids and condensate from the fuel before it is sent to the burners.

Labyrinth seal—a shaft seal designed to restrict flow by requiring the fluid to pass through a series of ridges and intricate paths.

Laminar flow—a streamline flow that occurs when the Reynolds number is low (at very low fluid velocities).

Landfill—a human-made or natural pit, typically lined with an impermeable, flexible substance (e.g., rubber) or clay; it is used to store residential or industrial waste materials.

Leaf filter—a type of filter that uses a precoated screen, located inside the filter vessel, to hold the filtered material.

Legend—a section of a drawing that explains or defines the information or symbols contained within the drawing. Legends include information such as abbreviations, numbers, symbols, and tolerances.

Level indicator—a gauge placed on a vessel to denote the height of the liquid level within the vessel.

Lift check valve—a valve that has built-in globe valve bodies. It is available in horizontal and vertical designs.

Lineman's pliers—snub-nosed pliers with two flat gripping surfaces, two opposing cutting edges, and insulated handles that help reduce the risk of electrical shock.

Lining—a coating applied to the interior wall of a vessel to prevent corrosion or product contamination.

Liquid buffered seal—a close-fitting bushing in which oil and water are injected in order to seal the process from the atmosphere.

Liquid head—the pressure developed from the pumped liquid passing through the volute.

Liquid-liquid extraction—a process that occurs when a feed solution is combined with a solvent that has an affinity for (attracts) one or more components of the feed.

Liquid ring compressor—a rotary compressor that uses an impeller with vanes to transmit centrifugal force into a sealing fluid, such as water, driving it against the wall of a cylindrical casing.

Live bottom—a vibratory device, attached to the bottom of a bin or silo, that is designed to facilitate the smooth discharge of material.

Load—the amount of torque necessary for a motor to overcome the resistance of the driven machine.

Lobe pump—a pump that consists of two or three lobes; liquid is trapped between the rotating lobes and is subsequently moved through the pump.

Locking pliers—pliers that can be adjusted with an adjustment screw and then locked into place when the handgrips are squeezed together. Also referred to as Vise-Grips®.

Louver—a moveable, slanted slat that is used to adjust the flow of air.

Lubricant—a substance used to reduce friction between two contact surfaces.

Lubrication—the application of a lubricant between moving surfaces in order to reduce friction and minimize heating.

Lubrication system—a system that circulates and cools sealing and lubricating oils.

Magnetic drive pump—a pump that uses magnetic fields to transmit torque to an impeller.

Manual valve—a hand-operated valve that is opened or closed using a hand wheel or lever.

Manway—an opening (usually 24 inches in diameter) in a vessel or tank that permits entry for inspection or repair.

Mechanical energy—energy of motion that is used to perform work.

Mechanical seals—seals that typically contain two flat faces (one that rotates, and one that is stationary) that are in constant contact with one another in order to prevent leaks.

Micron—a unit of measure equal to .000001 meters. Filters are rated based on their ability to filter a minimum micron-size particle.

Mist eliminator—a device in a tank designed to collect droplets of mist from gas and is composed of mesh, vanes, or fibers.

Mixer—a piece of equipment that rotates inside or outside to combine products.

Mixing system—a system consisting of an agitator, circulating pump, gas spargers, and a series of baffles to provide proper mixing of reactants and catalysts. It is designed to promote adequate mixing of reactants and catalysts.

Monorail crane—a type of crane that travels on a single runway beam.

Mother liquor—the portion of a solution remaining after crystallization.

Motor—a mechanical driver that converts electrical energy into useful mechanical work and provides power for rotating equipment.

Motor control center (MCC)—an enclosure that houses the equipment for motor control, including isolation power switches, lockouts, fuses, overload protection devices, ground-fault protection, and sometimes meters for current (amperes) and voltage.

Moving blade—a rotor blade connected to the shaft that moves or rotates when a gas is applied.

Multicomponent distillation—the process of separating more than two liquids by the boiling point.

Multiport valve—a type of valve used to split or redirect a single flow into multiple directions.

Multistage centrifugal pump—a pump that uses two or more impellers on a single shaft. This type of pump is generally used in high-volume, high-pressure applications, such as boiler feedwater pumps.

Multistage compressor—a device designed to compress the gas multiple times by delivering the discharge from one stage to the suction inlet of another stage.

Multistage turbine—a device that contains two or more stages used as a driver for high differential pressure, high-horsepower applications, and extreme rotational requirements.

Natural draft cooling tower—a cooling tower in which air movement is caused by wind, temperature difference, or other nonmechanical means.

Natural draft furnace—a type of furnace that does not use fans to create air flow. Natural draft is created by the difference in density between the hot combustion gases, the cooler outside air, and the height of the stack.

NEC—National Electric Code; specifies electrical cable sizing requirements and installation practices.

Needle nose pliers—small pliers with long, thin jaws for fine work (e.g., holding small items in place, removing cotter pins, or bending wire).

Needle valve—a type of globe valve that controls small flows using a long, tapered plug that passes through a circular hole in a plate or pipe.

Net positive suction head (NPSH)—the liquid pressure that exists at the suction end of a pump. If the NPSH is insufficient, the pump will cavitate.

Nitrogen oxides (NO$_x$)—undesirable air pollution produced from reactions of nitrogen and oxygen. The primary NO$_x$ species in process heaters are nitrogen monoxide (NO) and nitrogen dioxide (NO$_2$).

Noncondensing steam turbine—a device that acts similar to a pressure-reducing valve by converting the energy released into mechanical energy.

Nonrising stem valve—a valve stem that remains in place when the valve is opened or closed.

Nozzle—a device used to drive the blades or rotor of a turbine. The nozzle converts the steam or motive fluid from pressure to velocity.

Nucleation—the growth of a new crystal from feed crystals.

Ohm—a measurement of resistance in electrical circuits.

Oil pan/sump—a component that serves as a reservoir for the oil used to lubricate internal combustion engine parts.

OSHA—Occupational Safety and Health Administration; a U.S. government agency created to establish and enforce workplace safety and health standards, conduct workplace inspections and propose penalties for noncompliance, and investigate serious workplace incidents.

Overhead traveling bridge crane—a type of crane that runs on elevated rails along the length of a factory and provides three axes of hook motion (up and down, sideways, and back and forth).

Overspeed trip mechanism—an automatic safety device designed to remove power from a rotating machine if it reaches a preset trip speed.

Packing—a substance such as Teflon® or graphite-coated material that is used inside the packing gland to keep leakage from occurring around the valve stem.

Parallel flow—flow that occurs in a heat exchanger when the shell flow and the tube flow are parallel to one another.

Personnel lift—a lift that contains a pneumatic or electric arm with a personnel bucket attached at the end. Also referred to as a man lift or cherry picker.

Pilot—a device used to ignite burner fuel.

Pipe—a long, hollow cylinder through which fluids are transmitted; primarily made of metal but can also be made of glass, plastic, or plastic-lined material.

Pipe clamp—piping support that supports piping, tubing, and hoses from vibration and shock. Also a ring-type device used to temporarily stop a pipe leak.

Pipe hanger—piping support that suspends pipes from the ceiling or other pipes.

Pipe shoe—piping support that supports pipes from underneath.

Pipe wrench—an adjustable wrench that contains two serrated jaws that are designed to grip and turn pipes or other items with a rounded surface.

Piping and instrument diagram (P&ID)—a detailed illustration that graphically represents the relationship of equipment, piping, instrumentation, and flows contained within a process in the facility.

Piston—a component that moves up and down or backward and forward inside a cylinder.

Piston pump—a type of reciprocating pump that is driven by either a motor or direct-acting steam. Piston pumps use cylinders mounted on bearings within a casing.

Plate and frame filter—a type of filter used to remove liquids from a slurry feed by using a deep filtration process that includes clarification and prefiltration.

Platform—a strategically located structure designed to provide access to instrumentation and to allow personnel to perform maintenance and operational tasks. Platforms are generally accessed by ladders.

Pleated cartridge filter—a type of filter in a pleated form that provides more surface area.

Pliers—a hand tool that contains two hinged arms and serrated jaws that are used for gripping, holding, or bending.

Plot plan—an illustration showing the layout and dimensions of equipment, units, and buildings. They are drawn to scale so that everything is of the correct relative size.

Plug valve—a type of valve that uses a flow control element shaped like a hollowed-out plug, attached to an external handle, to increase or decrease flow.

Plunger pump—a pump that displaces liquid using a piston and is generally used for moving water and pulp. A plunger pump maintains a constant speed and torque.

Pneumatic conveyor—a device used to pneumatically move material from one point to another in a continuous stream.

Pneumatic tool—a tool that is powered using pneumatic (air or gas) pressure.

Positive displacement compressor—a device that may use screws, sliding vanes, lobes, gears, or a piston to deliver a set volume of gas with each stroke.

Positive displacement pump—a pump that uses pistons, diaphragms, gears, or screws to deliver a constant volume with each stroke.

Powder-actuated tool—a tool that uses a small explosive charge to drive fasteners into hard surfaces such as concrete, stone, or metal.

Power tool—a tool that is powered by electricity, pneumatics, hydraulics, or powder activation.

Preheater—a heat exchanger that adds heat to a substance prior to a process operation.

Premix burner—a burner that mixes fuel gas with air before either exits the burner face.

Pressure relief device—a component that discharges or relieves pressure in a vessel in order to prevent overpressurization.

Priming—the process of filling the suction of a pump with liquid to remove any vapors that might be present.

Process—the conversion of raw materials into a finished or intermediate product.

Process drawing—an illustration that provides a visual description and explanation of the processes, equipment, flows, and other important items in a facility.

Process flow diagram (PFD)—a basic illustration that uses symbols and direction arrows to show the primary flow of a product through a process. It includes information such as operating conditions and the location of main instruments and major pieces of equipment.

Process industries—a broad term for industries that convert raw materials, using a series of actions or operations, into products for consumers.

Process technician—a worker in a process facility who monitors and controls mechanical, physical, and/or chemical changes throughout a process in order to create a product from raw materials.

Pump—a mechanical device that transfers energy to move materials through piping systems.

Pump curve—a specification that describes the capacity, speed, horsepower, and head needed for correct pump operations.

Radial/thrust bearing—a bearing designed to support and hold the rotor in place while offering minimum resistance to free rotation. The bearing prevents and offsets axial and radial movement.

Radiant section—the portion of a furnace firebox where heat transfer occurs primarily through radiation.

Radiant tubes—tubes located along the walls of the radiant section and receive radiant heat from the burners.

Radiation—the transfer of heat through electromagnetic waves.

Raw gas—a gas that has not been premixed with air.

Reactive turbine—a steam device with fixed nozzles and an internal steam source; it uses Newton's third law of motion.

Reactor—a vessel in which a controlled chemical reaction is initiated and takes place either continuously or as a batch operation.

Reboiler—a tubular heat exchanger, placed at the bottom of a distillation column or stripper that is used to supply the necessary column heat.

Reciprocating compressor—a positive displacement compressor that uses the inward stroke of a piston to draw (intake) gas into a chamber and then uses an outward stroke to positively displace (discharge) the gas.

Reciprocating pump—a positive displacement pump that uses the inward stroke of a piston or diaphragm to draw liquid into a chamber, followed by a subsequent outward stroke to positively displace that liquid.

Reciprocating saw—a device that uses the back-and-forth motion of a saw blade to cut materials that range from wood to metal pipe.

Rectifier—a device that converts AC voltage to DC voltage.

Reflux—the overhead vapor (light components) that have been condensed to liquid and returned to the top of the tower to help cool and purify the overhead process or product.

Refractory lining—a brick-like form of insulation used inside a furnace or firebox operating at high temperatures.

Relative humidity—a measure of the amount of water in the air compared with the maximum amount of water the air can hold at that temperature.

Relief system—a system designed to prevent overpressurization or a vacuum.

Relief valve—a safety device designed to open if the pressure of a liquid in a closed space such as a vessel or a pipe exceeds a preset level.

Residence time—the volume of material inside the reactor divided by the flow rate of the reactants through the reactor.

Reverse osmosis—a process that uses pressure to force a solvent through a membrane, thus retaining the dissolved solids on one side of the membrane while allowing the solvent to pass to the other side of the membrane.

Reynolds number—a number used in fluid mechanics to indicate whether a fluid flow in a particular situation will be smooth or turbulent.

Riser tubes—tubes that allow water or steam from the lower drum to move to the upper drum.

Rising stem valve—a valve stem that rises out of the valve when the valve is opened.

Rod bearing—a flat steel ring or sleeve coated with soft metal that is placed between the connecting rod and the crankshaft.

Rotary compressor—a positive displacement compressor that uses a rotating motion to pressurize and move the gas through the device.

Rotary drum filter—a type of filter that contains a cloth screen that is located on a drum to hold solids being filtered.

Rotary dryer—a device composed of large, rotating, cylindrical tubes that reduces or minimizes the moisture content of a material by bringing it into direct contact with a heated gas.

Rotary pump—a positive displacement pump that moves liquids by rotating a screw or a set of lobes, gears, or vanes.

Rotary valve—a valve that uses a set of rotating pockets to meter a fixed volume of solids at a fixed rate. Also referred to as an airlock.

Rotating blade—a component in a gas turbine attached to the shaft of the turbine. The steam or gas causes the turbine to spin by impinging on the rotating blades. The steam or gas slows and is redirected as it transfers energy to the rotating blades.

Rotor—a rotating member of a motor or turbine that is connected to the shaft.

Runaway reaction—an uncontrolled reaction where the reaction rate increases, releasing heat that causes an increased reaction rate and results in the release of more heat.

Safety valve—a safety device designed to open if the pressure of a gas in a closed vessel exceeds a preset level.

Saturated steam—steam in equilibrium with water (i.e., steam that holds all of the moisture it can without condensation occurring).

Scale—dissolved solids deposited on the inside surfaces of equipment.

Schedule—a piping reference number that pertains to the wall thickness, inside diameter (ID), outside diameter (OD), and specific weight per foot of pipe.

Schematic—an illustration that shows the direction of electrical current flow in a circuit, typically beginning at the power source.

Screener—a piece of equipment that uses sieves or screens to separate dry materials according to particle size.

Screwdriver—a device that engages with, and applies torque to, the head of a screw so the screw can be tightened or loosened.

Screw pump—a rotary pump that displaces liquid with a screw. The pump is designed for use with a variety of liquids and viscosities and is designed to accommodate a wide range of pressures and flows.

Scrubber—a device used to remove contaminants from a gas stream by washing the stream with water or some other neutralizing agent.

Seal—a device that holds lubricants and process fluids in place while keeping out foreign materials.

Seal flush—a small flow (slip stream) of pump discharge or externally supplied liquid that is routed to the pump's mechanical seal. This acts as a barrier liquid between the two faces of the seal to reduce friction and remove heat.

Seal system—a system of devices designed to prevent the process gas from leaking from the compressor shaft.

Semiconductor—a material that is neither a conductor nor an insulator.

Sentinel valve—a spring-loaded, high-whistling safety valve that opens when the turbine reaches near maximum conditions.

Separator—a device used to physically separate two or more components from a mixture.

Shaft—a metal rotating component (spindle) that holds the rotor and all rotating equipment in place.

Sheave—a V-shaped groove centered on the circumference of a wheel designed to hold a belt, rope, or cable.

Shell—the outer casing or external covering of a heat exchanger or vessel.

Ship—a large seagoing vessel used to transport cargo across large bodies of water.

Shock bank tubes—tubes that receive both radiant and convective heat and protect the convection section from direct exposure to the radiant heat of the firebox.

Shroud—a casing over the motor that allows air to flow into and around the motor.

Sieve tray—a tray in a distillation column that has equally spaced holes or openings.

Silo—a tower-shaped container for storage of materials; typically used to store product destined for distribution.

Single-acting piston pump—a type of piston pump that pumps by reciprocating motion on every other stroke.

Single-stage compressor—a device designed to compress the gas a single time before discharging the gas.

Single-stage turbine—a device that contains one set of blades (two that turn and one that is fixed) called a stage.

Skirt—a support structure attached to the bottom of free-standing vessels or tanks.

Slurry—a liquid solution that contains a high concentration of suspended solids.

Slurry dryer—a device designed to convert wet process slurry into dry powder.

Socket wrench—a wrench that contains interchangeable socket heads of varying sizes that can be attached to a ratcheting wrench handle or other driver.

Softening—the treatment of water that is used to remove dissolved mineral salts such as calcium and magnesium, known as hardness, in boiler feedwater.

Solids handling equipment—equipment used to process and transfer solid materials from one location to another in a process facility; may also provide storage for those materials.

Solubility—the degree to which a solid will dissolve.

Spiders—devices used to inject fuel into a boiler.

Spark plug—a component in an internal combustion engine that supplies the spark to ignite the air/fuel mixture.

Spherical vessel—a spherically shaped container designed to distribute the pressure evenly over every square inch of the container. Spherical vessels are often used to store volatile or pressurized materials at 15 pounds per square inch (psi) and above.

Spray dryer—a device that sprays a moist feed stream into a vertical drying chamber using a nozzle or rotary wheel.

Sprocket—a toothed wheel used in chain drives.

Spuds—devices used to inject fuel into a boiler. Also known as spiders.

Stack—a cylindrical outlet at the top of a boiler or furnace that removes flue gas.

Stage—a set of nozzles or stationary blades plus a set of rotating blades.

Static—a mixer that operates by creating a complex flow path that results in contact between the streams. Also referred to as an inline mixer.

Static electricity—electricity that occurs when a number of electrons build up on the surface of a material but have no positive charge nearby to attract them and cause them to flow.

Static head—the amount of pressure the liquid is placing on the pump.

Stationary blade—a component in a gas turbine attached to the case that does not rotate. Stationary blades change the direction of the flow of the steam or combustion gas and redirect it to the next stage of rotating blades.

Stator—a stationary part of the motor where the alternating current supplied to the motor flows, creating a magnetic field using magnets and coiled wire.

Steam chest—the area where steam enters the turbine.

Steam drum—the upper drum of a boiler where all of the generated steam is collected.

Steam jet eductor—a device that uses the movement of steam to create suction (Venturi effect).

Steam strainer—a mechanical device that removes impurities from steam.

Steam trap—a device used to remove condensate from the steam system or piping.

Steam turbine—a device driven by the pressure of high-velocity steam that is discharged against the turbine's rotor.

Stirred tank reactor—a reactor that contains a mixer or agitator mounted inside the tank.

Straight-through diaphragm valve—a valve that contains a flexible diaphragm that extends across the valve opening.

Stress corrosion—a type of corrosion that results in the formation of stress cracks.

Stripping—a process that uses a vapor stream to separate one or more components from a liquid stream.

Stuffing box seals—the seals around a moving shaft or stem that contains packing material designed to prevent the escape of process liquids.

Suction check valve—a valve that prevents liquid from back-flowing during the discharge pump stroke.

Suction head—the pressure required to force liquids into a pump.

Sump—the lowest section at the base of a tank. The sump allows materials to be removed from the bottom of the tank and the tank to be completely emptied.

Superheated steam—steam that has been heated to a temperature above its saturated temperature.

Superheater—a set of tubes located in or near the boiler flue gas outlet that increases (superheats) the temperature of the steam flow.

Surging—the intermittent flow of pressure through a compressor that occurs when the discharge pressure is too high, resulting in flow reversal within a compressor.

Swing check valve—a valve used to control the direction of flow and prevent contamination or damage to equipment caused by back-flow.

Switch—an electrical device used to start, stop, or otherwise reconfigure the flow of electricity in a circuit.

Symbol—a simple illustration used to represent the equipment, instruments, and other devices on a PFD or P&ID. Some symbols are standard throughout the industry, while others may be specific to the individual manufacturing or engineering company.

Synchronous motor—a motor that runs at a fixed speed that is synchronized with the supply of electricity.

Tank—a container in which atmospheric pressure is maintained. That is, it is neither pressurized nor placed under a vacuum; it is at the same pressure as the air around it.

Tank breather vent—a type of vacuum relief device that maintains the pressure in the tank at a specific level, allowing air flow in and out of the tank.

Tank car—a type of railcar designed to transport liquids in bulk.

Tank truck—a special container designed to transport fluids in bulk.

Temperature—the measure of the thermal energy of a substance (i.e., the "hotness" or "coldness") that can be determined using a thermometer.

Temperature control system—a system that allows the temperature of a reaction to be adjusted and maintained.

Tensile strength—the pull stress, in force per unit area, required to break a given specimen.

Throttling—a condition in which a valve is partially opened or closed in order to restrict or regulate fluid flow rates.

Title block—a section of a drawing (typically located in the bottom right corner) that contains the drawing title, drawing number, revision number, sheet number, company information, process unit, and approval signatures. It identifies the drawing as a legend, system, or layout.

Tool—a device designed to provide mechanical advantage and make a task easier.

Torque wrench—a manual wrench that uses a gauge to indicate the amount of torque (rotational force) being applied to the nut or bolt.

Transformer—an electrical device that takes electricity of one voltage and changes it into another voltage.

Trickle valve—a valve used to continuously transfer a fixed weight of solids between two different pressure zones at a constant rate.

Trip throttle valve—a component designed to shut down the turbine in the event of excess rotational speed or vibration.

Troubleshooting—the systematic search for the source of a problem so that it can be solved.

Tube bundle—a group of fixed, parallel tubes though which process fluids are circulated.

Tube leak—a leak in a tube that can result in process chemical entering the circulating water, possibly resulting in a fire, explosion, or environmental or toxic hazard.

Tube sheet—a formed metal plate with drilled holes that allows process fluids to enter the tube bundle.

Tubing—a small diameter (typically less than 1 inch) hose or pipe used to transport fluids.

Tubular reactor—a reactor composed of tubes used for continuous reactions with nozzles to introduce reactants.

Turbine—a machine that is used to produce power and rotate shaft-driven equipment such as pumps, compressors, and generators.

Turbulent flow—flow that occurs when a dimensionless number (Reynolds number) is above 10,000 (at high velocities).

Underground storage tank—a container and its respective piping system in which a minimum of 10 percent of the combined process fluids are stored underground.

Unit—an integrated group of pieces of process equipment used to produce a specific product; may be referred to by the processes they perform, or named after their end products.

Universal motor—a motor that can be driven by either AC or DC power.

Utility flow diagram (UFD)—an illustration that provides process technicians a PFD-type view of the utilities used for a process.

Vacuum breaker—a safety device used to remove vacuum by adding pressure to a vessel. These devices can help to prevent implosion and back-flow.

Vacuum relief device—a safety device that prevents pressure in a tank from dropping below normal atmospheric pressure. This prevents tank implosion (sudden inward collapse).

Valve—a piping system component used to control, throttle, or stop the flow of fluids through a pipe.

Valve body—the lower exterior portion of the valve; contains the fluids flowing through the valve.

Valve cover—a cover over the head that keeps the valves and camshaft clean and free of dust or debris, and keeps lubricating oil contained.

Valve disc—the section of a valve that attaches to the stem and that can fully or partially block the fluid flowing through the valve.

Valve knocker—a device used to facilitate the movement (opening or closing) of a valve.

Valve seat—a section of a valve designed to maintain a leak-tight seal when the valve is shut.

Valve stem—a long slender shaft that attaches to the flow control element in a valve.

Valve tray—a tray in a distillation column that has moveable devices over the tray openings.

Valve wheel wrench—a hand tool used to provide mechanical advantage when opening and closing valves. Valve wheel wrenches typically fit over the spoke of a valve wheel.

Vane pump—a rotary pump having either flexible or rigid vanes designed to displace liquid. Vane pumps are used for low-viscosity liquids that may run dry for short periods without causing damage.

Vanes—raised ribs on the impeller of a centrifugal pump designed to accelerate a liquid during impeller rotation.

Vapor recovery system—process used to capture and recover vapors. Vapors are captured by methods such as chilling or scrubbing. The vapors are then purified and are either returned to the process, moved to storage, or incinerated.

Venturi—a device consisting of a converging section, a throat, and a diverging section. A Venturi employs the Bernoulli principle.

Vertical furnace—provide even temperature control to process fluids.

Vessel—an enclosed container in which the pressure is maintained at a level that is higher than atmospheric pressure.

Vessel head—the area on the top or bottom of a reactor shell that has access ports or nozzles.

Viscosity—the degree to which a liquid resists flow under applied force (e.g., molasses has a higher viscosity than water at the same temperature).

Volt—the electromotive force or a measure of current that establishes a current of 1 amp through a resistance of 1 ohm.

Voltage—a measurement of the potential energy required to push electrons from one point to another.

Voltmeter—a device that can be connected to a circuit in order to determine the amount of voltage present.

Volumetric feeder—a device designed to convey materials at a controlled rate by running at a set motor speed.

Volute—a widened spiral casing in the discharge section of a centrifugal pump designed to covert liquid speed to pressure without shock.

Vortex—cyclone-like rotation of a fluid.

Vortex breaker—a metal plate, or similar device, that prevents a vortex from being created as liquid is drawn out of the tank.

Waste heat boiler—a device that uses waste heat from a process to produce steam.

Water distribution header—a pipe that provides water to a distributor box located at the top of the cooling tower so the water can be distributed onto the fill.

Water tube boiler—a type of boiler that contains water-filled tubes that allow water to circulate through a heated firebox.

Watt—a unit of measure of electric power; the power consumed by a current of 1 amp using an electromotive force of 1 volt.

Wear ring—a ring that allows the impeller and casing suction head to seal tightly together without wearing each other out. Wear rings are close-running, non-contacting replaceable pressure breakdown devices located between the impeller and casing of a centrifugal pump.

Weighing system—a system used to weigh material before shipping.

Weir—a flat or notched dam that functions as a barrier to flow.

Weir diaphragm valve—a valve that contains a plate-like device that functions as a seat for the diaphragm.

Wet bulb temperature—the lowest temperature to which air can be cooled through the evaporation of water.

Wind turbine—a device that converts wind energy into mechanical or electrical energy.

Wrench—a hand tool that uses gripping jaws to turn bolts, nuts, or other hard-to-turn items.

Index

A

Absorption, 317, 327
AC (alternating current), 8, 183–85
   definition, 178
   power source, 178, 189
Activated sludge process, 396–98
Actuator symbols, 24–25
Adjustable pliers, 299, 301
Adjustable wrenches, 299, 302
Adsorption, 317, 328
After-coolers, 147
Agglomerators, 289–90
   definition, 286
Agitator blades, 287, 341
   definition, 286, 334
Agitators, 81, 86, 286–88
   definition, 286
Agitator shafts, 287, 341
   definition, 286, 334
Air pollution control equipment
   baghouses and precipitators (electrostatic), 392
   coal gasification, 392
   flare system, 395
   gas treatment (catalytic versus noncatalytic), 394
   incinerators (thermal oxidizers), 394–95
   overview, 391–92
   scrubbers, 394
   vapor and gas emission controls, 392–93
Air registers, 256, 273
   definition, 251, 268
Alloys, 43–44
   definition, 37
Alternating current (AC). See AC (alternating current)
Ambient air, 242, 391–92
   definition, 234, 391
American National Standards Institute (ANSI), 14, 32–33
American Petroleum Institute (API), 14, 33
American Society of Mechanical Engineers (ASME), 14, 33
Ammeters, 178, 181
Ampere, André, 183
Amperes (amps), 180–81
   definition, 178
ANSI (American National Standards Institute), 14, 32–33
Antisurge, definition of, 134
Antisurge protection, definition of, 134
API (American Petroleum Institute), 14, 33
Application blocks, 14, 22
   example, 24
Approach range, 234, 241

Articulated drains, 81, 83
ASME (American Society of Mechanical Engineers), 14, 33
Assisted draft furnaces, 253–54
   definition, 251
Auto-ignition, 260
Automatic valves, 72
Auxiliary equipment
   definition, 9
   key terms, 286
   overview, 286
   potential problems, 295
   process technician's role in operation and maintenance, 295
   tanks and vessels, 90–93
   turbines, 168–69
   types
      agitators, 286–88
      centrifuges, 294
      demisters, 295
      eductors. See Eductors
      hydrocyclones, 294–95
      mixers. See Mixers
   typical procedures, 295
Axial compressors, 137–38
   definition, 134
Axial pumps, 101, 110
Azeotropes, 317, 325
Azeotropic distillation, definition of, 317

B

Backwashing (back-flushing), 215, 229
Baffles, 86–87, 219, 341
   definition, 81, 215, 334
Bag filters, 350–51
   definition, 349
Bagging operations, 384–86
Baghouse, 391–92
Balanced draft furnaces, 254–55
   definition, 251
Ball bearings, 409–10
Ball check valves, 59, 66
Ball valves, 66–67
   definition, 59
Barges, 81, 86
Basins, 234, 239
Batch reactions, definition of, 334
Bearings, 189, 409–10
   definition, 178, 406
Bell reducers, 48–49
Belt conveyors, 369
Belts, 406–7

Bernoulli principle, 110, 290
   definition, 286
Best efficiency point (BEP), 123
Binary distillation, 317, 324
Bins, 373–74
   definition, 366
Biocides, 234, 246
Blanketing systems, 81, 90
Blinds, 40–41
   definition, 37
Block flow diagrams (BFDs), 15–16
   definition, 14
Block valves, 59, 72
Blowdown, 234, 268
Boilers
   definition, 2, 9, 268
   fire tube boilers, 275
   general components
      boiler feedwater system, 272
      burners, 269–70
      drums, 270
      economizer, 270
      firebox, 269
      steam distribution system, 270–71
   key terms, 268–69
   overview, 269, 282–83
   potential problems
      equipment age and design, 279
      external factors, 279
      overview, 278–79
      water and other contaminants, 279
   process technician's role in operation
      and maintenance, 281–82
   safety and environmental hazards, 280
   symbols, 29
   theory of operation
      boiler feedwater, 277–78
      burner fuel, 278
      desuperheated steam, 277
      overview, 275–76
      superheated steam, 276–77
      water circulation, 276
      water treatment methods, 278
   typical procedures
      emergency, 281
      lockout/tagout for maintenance, 281
      shutdown, 281
      startup, 281
   waste heat boilers, 274
   water tube boilers, 272–74
Bonnets, 59, 61
Box furnaces, 252–53
   definition, 251
Breakthrough, 349, 353
British thermal units (BTUs), 215–16
Bubble cap trays, 320–21
   definition, 317
Bucket elevators, 372–73
   definition, 366
Bulk bag station, 384
Bullet vessels, 83–84
   definition, 81
Burners, 256, 269–70, 273
   definition, 251, 268
Bushings, 48–49
Butterfly valves, 67–69
   definition, 59
Butt weld, 48

C

Cabin furnaces, 252–55
   definition, 251
Calories, 215–16
Camshafts, 199, 202
Canned pumps, 101, 108
Capillary action, 413
Caps, 48–49
Carbon monoxide (CO), 259, 263
Carbon rings, 158, 164
Cartridge filters, 349–50
   definition, 349
Casings, 162, 206
   definition, 158, 199
Catalysts, 334–35
Catalytic converters, 338
Cavitation, 106, 125
   definition, 101
Centrifugal compressors, 135–36, 144
   definition, 134
   symbols, 25
Centrifugal force, 101, 104
Centrifugal pumps
   components, 104–6
   definition, 101
   overview, 7, 104
   single-stage versus multistage pumps, 107
   specialty pumps, 107–10
      canned pumps, 108
      high-speed pumps, 109
      jet pumps, 109–10
      magnetic drive pumps, 108–9
Centrifuges, 286, 294
Chains, 406, 408
Check valves, 64–66
   definition, 59
   human heart, 66
Chemical reactions, 335
Chevrons, 158, 161
Chicago pneumatic (CP) couplings, 49
Circuit breakers, 186–87
   definition, 178
Circuits, 178, 182
Circulation rate, 234, 242
Clarifiers, 397–98
Coagulation, 268, 278
Coal gasification, 392
CO (carbon monoxide), 259, 263
Co-current flow, 215, 218
Cold wall reactors, 338–39
   definition, 334
Combustion chambers, 165, 206
   definition, 158, 199
Compression fittings, 49
Compression ratio, 138, 199, 204
Compressors, 206
   associated utilities and auxiliary equipment
      antisurge devices, 146–47
      coolers, 147
      lubrication system, 145
      seal system, 145–46
      separators, 147–48
   definition, 2, 7, 134, 199
   key terms, 134
   operating principles
      centrifugal compressor performance curve, 144
      single-stage versus multistage compressors, 143–44

overview, 134–35, 153
positive displacement compressors
  overview, 137–38
  reciprocating, 138–40
  rotary, 140–42
potential problems
  high/low flow, 151
  interlock system, 150–51
  leaks, 149–50
  loss of capacity, 151
  lubrication fluid contamination, 150
  motor overload, 151
  overheating, 149
  overpressurization, 148
  seal oil problems, 149
  surging, 149
  vibration, 150
process technician's role in operations
  and maintenance, 151–52
safety and environmental hazards, 151
selection, 135
symbols, 23
types
  axial compressors, 137–38
  centrifugal compressors, 135–37
  dynamic compressors, 135
  overview, 135
typical procedures
  emergency procedures, 153
  lockout/tagout, 153
  startup and shutdown, 152–53
Condensers, 225, 240, 323
  definition, 215, 234, 317
Condensing steam turbines, 160–61
  definition, 158
Conduction, 216–17
  definition, 215
Conductors, 178–79
Connecting rods, 199, 203
Construction materials, 43
Container ships, 85
Containment walls, 81, 92
Continuous reactions, 334
Control valves, 72–73
  definition, 59
Convection, 216–17
  definition, 215
Convection sections, 255, 258
  definition, 251
Convection tubes, 251
Conveyors, 369–72
  definition, 366
Coolants, 199, 203
Coolers, 240
  compressors, 147
  definition, 234
Cooling range, 234, 242
Cooling towers
  applications, 240
  components and their purposes, 238–39
  definition, 2, 8, 234
  factors that affect operation
    corrosion, erosion, fouling, and scale, 243
    foaming, 243–44
    humidity, 240–41
    safety and environmental hazards
    fires, 245
    heat exchanger tube leaks, 244

    temperature, 241–42
    tower design, 242
    wind velocity, 242
  key terms, 234–35
  overview, 235, 248
  process technician's role in operation
    and maintenance
    blowdown, 246
    chemical treatment, 246
    component and concentration monitoring, 245
    maintenance, 246–47
    temperature monitoring, 246
  symbols, 26
  theory of operation, 240
  types
    forced draft towers, 237
    induced draft cooling towers, 235
    natural draft cooling towers, 235
Corrosion
  cooling towers, 243
  definition, 37, 81, 234
  heat exchangers, 227
  piping, tubing, hoses, and fittings, 54–55
  tanks and vessels, 94
  turbines, 170
Corrosion monitoring systems, 90–91
Counterflow, definition of, 234
Counterflow cooling towers, 237–38
Couplings, 48–49, 406–7
  definition, 406
CP (Chicago pneumatic) couplings, 49
Cranes, 305–8
  definition, 299
Crankshafts, 199, 203
Cross-connection (piping), 54
Cross fitting, 48–49
Cross flow, 218
  cooling towers, 237–39
  definition, 215, 234
Crystallization, 317, 330
Cyclones, 357–58
  definition, 349
Cylinders, 134, 139

D

Dampers, 73–74, 258, 270, 273
  definition, 59, 251, 268
DC (direct current), 8, 184–85
  definition, 178
Dead head, 106, 124
  definition, 101
Deaeration, 268, 278
Demineralization, 268, 278
Demisters, 147–48, 295
  definition, 134, 286
Desicant dryers, 147–48
Desiccants, 349, 353
Desuperheated steam, 268, 277
Desuperheaters, 268, 274
Dew point, 234, 241
Diaphragm pumps, 116–17
  definition, 101
Diaphragm valves, 68–70
  definition, 59
Dies, 366, 368
Diesel, Rudolf, 200
Diesel engines, 200–202

Dikes, 92, 399
  definition, 81, 391
Direct current (DC). *See* DC (direct current)
Discharge check valves, 101, 115
Disconnects, 178, 187
Distillation
  azeotropic distillation, 325–26
  binary and multicomponent distillation, 324
  definition, 317
  extractive distribution, 326
  flash distillation, 325
  how the process works, 322–24
  overview, 318–19
  packing, 321–22
  pressure conditions, 324
  symbols, 29
  tower sections, 319–20
  trays, 320–21
Dollies, 299, 309
Double-acting compressors, 140–41
Double-acting piston pumps, 101, 116
Double-pipe heat exchangers, 219–20
Downcomers, 268, 274
Draft fans, 268, 273
Draft types (furnaces)
  assisted draft, 253–54
  natural draft, 253
Drag chain conveyors, 370–71
Drift, 234, 239
Drift eliminators, 234, 239
Drills, 299, 305
Drum mixers, 290–91
  definition, 286
Drums, 270
Dry bulb temperature, 234, 241
Dry carbon rings, 145–46
  definition, 134
Dryers. *See* Filters and dryers
Dual-bed air dryers, 353–54
Duty, 234, 242
Dynamic compressors, 134–35
Dynamic head, 117–18
  definition, 101
Dynamic mixers, 286, 289
Dynamic pumps, 103–4
  axial pumps, 110
  centrifugal pumps. *See* Centrifugal pumps
  definition, 101

E

Economizers, 270, 274
  definition, 268
Edison, Thomas, 184
Eductors, 291
  definition, 286
  jet pump eductor, 293–94
  overview, 290, 292
  steam jet eductor, 292–93
Elbows, 48–49
Electrical diagrams, 19–20
  definition, 14
Electrical distribution and motors
  electrical distribution system components
    circuit breakers, 186–87
    fuses, 186
    motor control centers (MCCs), 185–86
    switches, 187
    transformers, 185

electrical transmission, 184–85
electricity
  amperes, 180–81
  circuits, 182
  general description, 179–80
  grounding, 183
  ohms, 181–82
  volts, 180
  watts, 180
key terms, 178–79
motors
  components, 188–89
  load, 190
  operating principles, 189–90
  purpose of, 187
  selection, 188
  three-phase versus single-phase, 190
  types, 187–88
overview, 179, 195–96
potential problems, 190
process technician's role in operation
    and maintenance, 195
safety and environmental hazards
  electrical classification, 192–93
  overview, 191–92
symbols, 28
types of current
  alternating current, 183–84
  direct current, 184
typical procedures
  emergency, 194
  lockout/tagout, 194
  shutdown, 194
  startup, 193–94
Electricity, 178–79
Electric tools
  definition, 299
  safety suggestions, 312
Electromagnetism, 178, 190
Electrons, 178–79
Electrostatic precipitators, 392–93
  definition, 391
Elevation diagrams, 21
Endothermic, definition of, 334
Endothermic reactions, 342–43
Engine blocks, 199, 203
Engines
  common types
    external combustion engines, 200–201
    internal combustion engines, 200
    pros and cons, 201
  definition, 2, 7, 199
  key terms, 199
  overview, 199–200, 211–12
  potential problems
    equipment failure, 208
    high temperatures, 208
    high vibration, 208
    mechanical drivers, 208
    misalignment, 208
    turbine trips, 208
  process technician's role in operation and maintenance
    lubrication, 211
    miscellaneous maintenance items, 211
    monitoring, 210–11
  safety and environmental hazards, 208–9
  typical procedures
    emergency procedures, 210
    engine in service, 210

lockout/tagout, 210
normal operations, 210
shutdown, 210
startup, 209–10
use in process industry
auxiliary systems of internal combustion engines, 206
major components of combustion gas turbines, 206
major components of internal combustion engines, 202–3
operating principles of combustion gas turbines, 206–7
overview, 201–2
two- and four-cycle engines, 203–6
Environmental control equipment
definition, 10
environmental rules and regulations, 402
federal regulations, 401–2
key terms, 391
overview, 391, 403
potential problems, 402
process technician's role in operation, maintenance, and compliance, 402
types
air pollution control. *See* Air pollution control equipment
water and soil pollution control. *See* Water and soil pollution control equipment
typical procedures, 402
Equipment location diagrams, 21
Equipment symbols, 24
Erosion
cooling towers, 243
definition, 234
heat exchangers, 227
piping, tubing, hoses, and fittings, 54–55
Evaporation, 240, 329
definition, 234, 317
Exchanger heads, 215, 223
Exhaust ports, 199, 203
Exhaust valves, 199, 203
Exothermic, definition of, 334
Exothermic reactions, 342–43
Expansion loop, 37
External combustion engines, 200–201
Extraction, 326–27
definition, 317
Extractive distillation, 317
Extruders, 367–68
definition, 366

F

Fans, 178, 189
FCCU (fluid catalytic cracking unit), 338–39
Federal environmental regulations, 401–2
Feeders, 366–67
Fill, 234, 238
Filters and dryers
consequences of improper operation, 358–59
filter ratings, 353
key terms, 349
overview, 349, 363
potential problems
channeling, 359
fouling, 359
moisture breakthrough, 359

process technician's role in operation and maintenance, 362
safety and environmental hazards
environmental impact, 360
housekeeping, 360
personal protective equipment, 360
physical contact, 360
types of dryers
dual-bed air dryer, 353–54
fixed bed dryer, 353
fluid bed dryers, 354–55
rotary dryer, 355–56
slurry dryer, 356–58
spray dryer, 355
types of filters
bag filter, 350–51
cartridge filter, 349–50
leaf filter, 351–52
plate and frame filter, 352
pleated cartridge filter, 352
rotary drum filter, 352
typical procedures
emergency, 361
lockout/tagout, 361
monitoring, 361
routine and preventive maintenance, 361
sampling, 361–62
Filtration, 268, 278
Finned tubing, 223–24
Fireboxes, 256, 269, 273
definition, 251, 268
Firebrick, 257
Fire tube boilers, 268, 275
Firewalls, 81, 92
Fittings, 5, 37
Fixed bed dryers, 349, 353
Fixed bed reactors, 337–38
definition, 334
Fixed blades, 158, 163
Fixed roof tanks, 81–82
Flange fittings, 48–49
Flared fittings, 49
Flare system, 391, 395
Flaring tools, 299, 302
Flash distillation, 317
Flash dryers, 349
Floating roof tanks, 81, 83
Fluid bed dryers, 10, 354–55
definition, 349
Fluid catalytic cracking unit (FCCU), 338–39
Fluidized bed reactors, 334, 338
Foam chambers, 81, 87
Foam eductors, 292
Foaming, 243–44
definition, 234
Forced draft cooling towers, 235, 237
definition, 234
Forced draft furnaces, 254–55
definition, 251
Forklifts, 299, 308
Fouling
cooling towers, 243
definition, 215, 234
filters, 359
furnaces, 264
heat exchangers, 218, 227
turbines, 170
Frames, 178, 188
Fuel lines, 256

Fuel-operated tools
  definition, 299
  safety suggestions, 313
Furnaces
  applications, 252
  common designs, 252–53
  definition, 2, 8, 251
  draft types
    assisted draft, 253–54
    natural draft, 253
  key terms, 251
  operating principles
    combustion air, 259
    flame temperature, 260
    fuel supply, 258–59
    furnace pressure control, 260
    interlock controls, 260
  overview, 251–52
  potential problems
    equipment age and design, 263
    external factors, 264
    fouling, 264
    furnace startup and shutdown, 264–65
    improper maintenance, 264
    instrument problems, 264
    overview, 263, 265
    tube life and temperature, 264
  safety and environmental hazards
    environmental impact, 262–63
    fires, spills, and explosions, 261–62
    hazardous operating conditions, 262
    personal safety, 262
  sections and components, 254–58
    convection section, 257–58
    radiant section, 256–57
  symbols, 28–29
  typical procedures, 263
Fuses, 178, 186

G

Gantry cranes, 306–7
  definition, 299
Gaskets, 47, 49–50
  definition, 37
Gas treatment (catalytic versus noncatalytic), 394
Gas turbines, 164–65
  definition, 158
Gate valves, 60, 62–63
  definition, 59
Gauge hatches, 81, 87
Gearboxes, 406, 411
Gear pumps, 101, 111
Gear ratio, 408
Gears, 406, 411
Generators, 178, 184
Globe valves, 62, 64
  definition, 59
Glossary, 423–34
Governors, 158, 163, 168
Gravimetric feeders, 366–67
Grinders, 299, 305
Grounding, 91, 183
  definition, 81, 178

H

Hammers, 300–301
  definition, 299
Hand manual valves, 72–74

Hand tools
  definition, 299
  flaring tools, 299, 302
  hammers, 299–301
  pliers, 299–301
  safety suggestions, 310–11
  screwdrivers, 300–302
  wrenches. *See* Wrenches
Hand wheels, 59–60
Head pressure, 101
Heads, 199, 203
Heat exchangers
  applications and service, 224–26
  components, 223–24
  definition, 2, 8, 215
  heat transfer, 216
  key terms, 215
  in nature, 217
  overview, 8, 215–16, 230–31
  potential problems, 226
  process technician's role in operation
      and maintenance, 230
  safety and environmental hazards, 226, 228
  symbols, 27
  theory of operation, 216–19
    fouling, 218
    heat transfer coefficients, 217–18
    types of flows, 218–19
  types
    double-pipe heat exchangers, 219–20
    plate and frame heat exchangers, 222
    shell and tube heat exchangers, 220–22
    spiral heat exchangers, 220
  typical procedures
    general maintenance tasks, 229
    shutdown, 229
    startup, 228–29
Heating and cooling (tanks and vessels), 91
Heat transfer, 215–16
Hemispheroid vessels, 81, 84
High-speed pumps, 109
Hoists, 305–6
  definition, 299
Hopper cars, 81, 85
Hoppers, 366
Hoses, 38–39
  definition, 2, 37
Hot wall reactors, 334, 338–39
Humidity
  cooling towers, 240–41
  definition, 234
Hunting (turbines), 170–71
Hydraulic tools
  definition, 299
  safety suggestions, 312
Hydraulic turbines, 158, 165
Hydrocyclones, 294–95
  definition, 286

I

Igniters, 268, 270
Impact wrenches, 299, 303
Impellers, 105, 270
  definition, 101, 268
Impingement (turbines), 170
Impulse movement, 158, 161
Impulse turbines, 160–61
  definition, 158

Incinerators, 394–95
  definition, 391
Induced draft cooling towers, 235–36, 239
  definition, 234
Induced draft furnaces, 251, 254
Induction motors, 178, 187
Inhibitors, 334–35
Initiators, 334–35
Inline mixers, 289
Instrumentation symbols, 30–32
Insulation, 37, 52–53, 91–92
Insulators, 178–79
Intake ports, 199, 203
Intake valves, 199, 203
Interceptors, 398–99
Interchangers, 215, 225
Intercoolers, 147
Interlock protection, 134, 150–51
Internal combustion engines, 200
ISA (International Society of Automation),
    14, 32
  functional identification labels sample, 32
  logo, 33
Isometric drawings
  definition of, 14
  example, 21
Isometrics, 20

J

Jacketed pipes, 37, 53
Jet pump eductors, 293–94
  definition, 286
Jet pumps, 109–10
Jib cranes, 306–7
  definition, 299

K

Knockout pots, 268, 270

L

Labyrinth seals, 145–46, 169
  definition, 134
Laminar flow, 218–19
  definition, 215
Landfills, 400–1
  definition, 391
Leaf filters, 351–52
  definition, 349
Leakage
  piping, tubing, hoses, and fittings, 54
  pumps, 124–25
Leaks
  compressors, 149–50
  heat exchangers, 227
  reactors, 345
Legends, 22
  definition, 14
  example, 23
Let down station, 271
Level indicators, 81, 87
Lift check valves, 59, 65
Lifting equipment
  cranes, 305–8
  dollies, 309
  forklifts, 308
  hoists, 305
  personnel lifts, 308

Lightning, 180
Lineman's pliers, 299, 301
Linings, 81, 88
Liquid buffered seals, 134, 145
Liquid head, 101, 105
Liquid-liquid extraction, 317, 326
Liquid ring compressors, 134, 142
Live bottoms, 375
  definition, 366
Load, 178
Lobe compressors, 141–42
Lobe pumps, 111–13
  definition, 101
Locking pliers, 299, 301
Locomotives, 201, 275
Logic diagrams, 21
Loop diagrams, 21
Louvers, 73–74, 239
  definition, 59, 234
Lubricants, 412–13
  definition, 406
Lubrication, 2, 10, 406
Lubrication systems, 413–16
  compressors, 145
  definition, 134

M

Magnetic and hydraulic drives, 408
Magnetic drive pumps, 108–9
  definition, 101
Manual valves, 59
Manways, 81, 88
MCCs (motor control centers), 178, 185–86
Mechanical energy, 158–59
Mechanical power transmission and lubrication
  bearings, 409–10
  gears, 411
  key terms, 406
  operating principles of mechanical
      transmission
    belts, 407
    chains, 408
    couplings, 406–7
    magnetic and hydraulic drives, 408
  overview, 406, 419–20
  potential problems
    improper seal procedures, 417
    interlock protection, 417
    startup and shutdown procedures, 417
    vibration monitoring, 417
  principles of lubrication, 412–13, 415
  process technician's role in operation
      and maintenance
    lubrication maintenance, 418–19
  safety and environmental hazards
    environmental impact, 418
    equipment operations, 418
    personal safety, 417–18
Mechanical seals, 101, 120
Microns, 349, 353
Mist eliminators, 81, 88
Mixers
  agglomerators, 289
  definition, 286
  drum mixers, 290
  dynamic mixers, 289
  inline mixers, 289
  overview, 288

Mixing systems, 334, 342
Mobile containers, 84–86
Monorail cranes, 306–7
    definition, 299
Mother liquor, 317, 330
Motor control centers (MCCs), 185–86
    definition, 178
Motor oil, 412
Motors
    components, 188–89
    definition, 2, 7, 178
    load, 190
    operating principles, 189–90
    purpose of, 187
    selection, 188
    three-phase versus single-phase, 190
    types, 187–88
Moving blades, 158, 163
Moving floor conveyors, 370
Moving parts danger symbol, 209
Multicomponent distillation, 317, 324
Multipass heat exchangers, 221–22
Multiport valves, 59, 72
Multistage centrifugal pumps, 107–8
    definition, 101
Multistage compressors, 143–44, 147
    definition, 134
Multistage pumps, 107
Multistage turbines, 158, 161–62

N

National Electric Code (NEC), 14, 33, 192–93
Natural draft cooling towers, 235–37, 243
    definition, 234
Natural draft furnaces, 253–54
    definition, 251
Needle nose pliers, 299, 301
Needle valves, 59, 64
Net positive suction head (NPSH), 101, 103
Nipples, 48–49
Nitrogen hose couplings, 49
Nitrogen oxides (NO_x), 259, 262, 280, 392
    definition, 251
Noncondensing steam turbines, 158, 160–61
Non-rising stem valves, 59, 61
Nozzles, 163, 206
    definition, 158, 199
Nuclear power plants and reactors, 167, 243, 339–41
Nucleation, 317, 330

O

Occupational Safety and Health Administration
        (OSHA), 14, 33, 193
Ohm, George, 183
Ohms, 181–82
    definition, 178
Ohm's law, 182
Oil pans/sumps, 199, 203
Otto, Nicolaus, 200
Overfilling, 96
Overhead traveling bridge cranes, 299, 305
Overheating
    compressors, 149
    pumps, 124
Overpressurization, 96
    compressors, 148
    pumps, 124

tanks and vessels, 94
turbines, 170
Overspeeding (turbines), 170
    trip mechanisms, 158, 164

P

Packing, 60, 119
    definition, 59, 101
Paddle blinds, 40
Parallel circuit, 181
Parallel flow, 215, 218
Personal protective equipment (PPE), 10, 191–92, 360
Personnel lifts, 308–9
    definition, 299
PFDs (process flow diagrams), 16–18
    definition, 14
Pigs, pipe, 51
Pilots, 256, 273
    definition, 251, 268
Pipe clamps, 41–42
    definition, 37
Pipe hangers, 37, 41
Pipes, 2, 5, 37. *See also* Piping, tubing, hoses, and fittings
Pipe shoes, 37, 41
Pipe wrenches, 299, 303
Piping, tubing, hoses, and fittings
    connecting methods
        bonded, 47–48
        flanged, 47
        overview, 46–47
        temporary connections, 48
        threaded, 47
        welded, 48
    fitting types
        gaskets, 49–50
        overview, 48–49
        sealant compounds, 50
    key terms, 37
    leak testing, 51
    materials of construction
        carbon steel and its alloys, 43
        miscellaneous metals, 44
        nonmetals, 44–45
        overview, 42–43
        stainless steels, 43–44
    overview, 37, 55–56
    pipe selection and sizing criteria
        overview, 45
        pipe thickness and ratings service, 45–46
    piping, tubing, hoses, and blinds
        blinds, 40–41
        expansion loops and joints, 42
        hoses, 38–39
        limitations of hoses and fittings, 39
        piping, 37–38
        piping supports, 41–42
        tubing, 38
    piping protection, 51
        insulation, 52–53
        jacketing, 53
        overview, 51
        steam traps, 53
        tracing, 51–52
    piping symbols, 55
    potential hazards, 53
    process technician's role in operation
            and maintenance, 54
    symbols, 23, 25

Piping and instrument diagrams (P&IDs), 16, 18
  definition, 14
Piston pumps, 115–16
  definition, 101
Pistons, 138, 203
  definition, 199
Pitting (turbines), 170
Plate and frame filters, 349, 352
Plate and frame heat exchangers, 222
Platforms, 88–89
  definition, 81
Pleated cartridge filters, 349, 352
Pliers, 300–301
  definition, 299
Plot plan, 21
  definition, 14
  example, 22
Plugs, 48–49
Plug valves, 67–68
  definition, 59
Plunger pumps, 101, 116
Pneumatic conveyors, 366
Pneumatic tools
  definition, 299
  safety suggestions, 312
Pop valves, 70
Positive displacement compressors
  definition, 134
  overview, 137–38
  reciprocating, 138–40
  rotary, 140–42
  symbols, 25
Positive displacement pumps, 103
  definition, 101
  description, 110–11
  reciprocating pumps. *See* Reciprocating pumps
  rotary pumps. *See* Rotary pumps
Power-actuated tools
  definition, 299
  safety suggestions, 313
Power tools, 305
  definition, 300
  safety suggestions, 311–13
PPE (personal protective equipment), 10, 191–92, 360
Precoolers, 147
Preheaters, 224, 322
  definition, 215, 317
Premix burners, 268, 273
Pressure or vacuum relief (tanks and vessels), 88–90
  devices, 81, 89
Priming, 106, 125
  definition, 101
Process, definition of, 2–3
Process drawings and equipment standards
  common drawings and their uses
    block flow diagrams (BFDs), 15–16
    electrical diagrams, 19–20
    elevation diagrams, 21
    equipment location diagrams, 21
    isometrics, 20
    logic diagrams, 21
    loop diagrams, 21
    piping and instrumentation diagrams (P&IDs), 18
    plot plan, 21
    process flow diagrams (PFDs), 16, 18
    schematics, 20
    utility flow diagrams (UFDs), 19
  common information contained on process drawings
    application block, 22

    legend, 22
    title block, 22
  definition, 2, 5, 14
  equipment standards, 31–33
    American National Standards Institute (ANSI), 32–33
    American Petroleum Institute (API), 33
    American Society of Mechanical Engineers (ASME), 33
    ISA (International Society of Automation), 32
    National Electric Code (NEC), 33
    Occupational Safety and Health Administration (OSHA), 33
  key terms, 14
  overview, 15, 33–34
  symbols
    actuator symbols, 24
    boiler symbols, 29
    compressor symbols, 23
    cooling tower symbols, 26
    electrical equipment and motor symbols, 28
    examples, 23
    furnace symbols, 28
    heat exchanger symbols, 27
    instrumentation symbols, 30–31
    introduction, 22
    piping symbols, 23
    pump symbols, 27
    reactor and distillation column symbols, 29
    turbine symbols, 28
    valve symbols, 26
    vessel symbols, 29–30
Process elevators, 372–73
Process equipment, introduction to
  equipment
    auxiliary equipment, 9
    boilers, 9
    compressors, 7
    cooling towers, 8
    engines, 7
    environmental control equipment, 10
    filters and dryers, 9
    fittings, 5–6
    furnaces, 8
    heat exchangers, 8
    hoses, 5–6
    mechanical power transmission and lubrication, 10
    motors, 7
    pipes, 5–6
    process drawings, 5
    pumps, 6
    reactors, 9
    separation equipment, 9
    solids handling equipment, 9–10
    standards, 5
    tanks, 6
    tools, 9
    tubing, 5–6
    turbines, 7
    valves, 6
    vessels, 6
  key terms, 2
  process industries, 2–3
  process technicians. *See* Process technicians
  safety and environmental hazards, 10
  summary, 11
Process flow diagrams (PFDs), 16–18
  definition, 14

Process industries
  definition, 2–3
  how process industries operate, 3
Process technicians
  definition, 2–3
  role in operation and maintenance, 3–5
    auxiliary equipment, 295
    boilers, 281–82
    compressors, 151–52
    cooling towers, 245–47
    electrical distribution and motors, 195
    engines, 210–11
    environmental control equipment, 402
    filters and dryers, 362
    heat exchangers, 230
    lubrication, 418–19
    piping, tubing, hoses, and fittings, 54
    pumps, 128
    reactors, 345–46
    separation equipment, 331
    solids handling equipment, 387–88
    tanks and vessels, 96–97
    turbines, 173–74
    valves, 77
Psychrometers, 241
Pump curves
  definition, 102
  efficiency, 123
  examples, 122–23
  flow rate, 123
  head pressure, 122–23
  horsepower, 123
  NPSH required, 123–24
  overview, 121–22
Pumps
  associated utilities/auxiliary equipment
    mechanical seals, 120
    seal flushes and seal pots, 120–21
    stuffing box seals, 119
    wear rings, 121
  definition, 2, 6, 102
  dynamic pumps
    axial pumps, 110
    centrifugal pumps. *See* Centrifugal pumps
  key terms, 101–2
  operating principles
    head pressure, 117
    inlet (suction), 118
    outlet (discharge), 119
    what happens inside the pump, 118–19
  overview, 102, 129–30
  positive displacement pumps
    description, 110–11
    reciprocating pumps. *See* Reciprocating pumps
    rotary pumps. *See* Rotary pumps
  potential problems
    cavitation, 125
    excessive vibration, 126
    flashing seal leaking into a pump, 125
    leakage, 124–25
    overheating, 124
    overpressurization, 124
    pump shutdown, 126
    vapor lock, 126
  process technician's role in operation
      and maintenance, 128
  purpose of a pump curve
    efficiency, 123
    flow rate, 123

  head pressure, 122–23
  horsepower, 123
  NPSH required, 123–24
  overview, 121–22
  safety and environmental hazards
    environmental hazards, 128
    equipment operation hazards, 127
    facility operation hazards, 127
    personal safety hazards, 126–27
  selection, 103
  symbols, 27
  types, 103
  typical procedures
    emergency procedures, 129
    lockout/tagout, 129
    monitoring, 128
    shutdown, 129
    startup, 129

Q

Quick connect fittings, 49

R

Radial/thrust bearings, 163
Radiant sections, 255–56
  definition, 251
Radiant tubes, 273
  definition, 251, 268
Radiation, 216–17, 255
  definition, 215, 251
Raidal/thrust bearings, 158
Raw gas, 268
Reactive turbines, 158, 160
Reactors
  auxiliary equipment associated with reactors, 343
  chemical reactions, 335
  components, 341–42
  definition, 2, 9, 334
  key terms, 334
  overview, 334–35, 346
  potential problems, 344
  process technician's role in operation
      and maintenance, 345–46
  symbols, 29
  theory of operation, 342–43
  types of reactions and reactors, 342
    batch versus continuous, 335
    fixed bed reactor, 337–38
    fluidized bed reactor, 338
    hot wall versus cold wall, 338–39
    nuclear reactor, 339–41
    stirred tank reactor, 336
    tubular reactor, 336–37
  typical procedures, 344
Reboilers, 224, 323
  definition, 215, 317
Reciprocating compressors, 138–40
  definition, 134
Reciprocating pumps
  definition, 102
  description, 114–15
  diaphragm pumps, 116
  piston pumps, 115–16
  plunger pumps, 116
  suction pumps, 117
Reciprocating saws, 300, 305
Rectifiers, 178, 184
Reflux, 317, 323

Refractory lining, 273
   definition, 251, 268
Relative humidity, 235, 241
Relief and safety valves, 59, 69–71
Relief systems, 334, 342
Residence time, 334, 336
Reverse osmosis, 268, 278
Reynolds number, 218–19
   definition, 215
Riser tubes, 268, 274
Rising stem valves, 59, 61
Rod bearings, 199, 203
Roller bearings, 409–10
Roller systems, 371–72
Rotary compressors, 140–42
   definition, 134
Rotary drum filters, 349, 352
Rotary dryers, 355–56
   definition, 349
Rotary pumps
   definition, 102
   description, 111
   gear pumps, 111
   lobe pumps, 111–12
   screw pumps, 114
   vane pumps, 113–14
Rotary screw compressors, 141
Rotary valves, 383–84
   definition, 366
Rotating blades, 199, 206
Rotors, 163, 189
   definition, 158, 178
Runaway reactions, 342, 345
   definition, 334

S

SAE (Society of Automotive Engineers), 412
Safety and environmental hazards
   boilers, 280
   compressors, 151
   cooling towers, 244–45
   electrical distribution and motors, 191–93
   engines, 208–9
   filters and dryers, 360
   furnaces, 261–63
   heat exchangers, 226, 228
   mechanical power transmission and lubrication, 418
   pumps, 126–28
   solids handling equipment, 387
   turbines, 171
Safety suggestions for hand and power tools
   general hazards, 310
   general tips, 309–10
   hand tools, 310–11
   power tools, 311–13
Safety valves, 59
Saturated steam, 268
Scale
   cooling towers, 243
   definition, 81, 235
   tanks and vessels, 94
Schedule, pipe, 45–46
   definition, 37
Schematics, 14, 20
Screeners, 366, 369
Screening systems, 368–69
Screw compressors, 141
Screw conveyors, 370

Screwdrivers, 301–2
   definition, 300
Screw pumps, 113–14
   definition, 102
Scrubbers, 327, 394
   definition, 317, 391
Sealant compounds, 50
Seal flushes and seal pots, 102, 120–21
Seals, 119
   definition, 102, 158
   turbines, 169
Seal systems
   compressors, 145–46
   definition, 134
Secondary containment systems, 92
Semiconductors, 178–79
Sentinel valves, 158, 168
Separation equipment
   absorption and stripping, 327–28
   adsorption, 328
   crystallization, 330
   definition, 9
   distillation
      azeotropic distillation, 325–26
      binary and multicomponent distillation, 324
      extractive distribution, 326
      flash distillation, 325
      how the process works, 322–24
      overview, 318–19
      packing, 321–22
      pressure conditions, 324
      tower sections, 319–20
      trays, 320–21
   evaporation, 329
   extraction, 326–27
   key terms, 317
   overview, 317–18, 331–32
   potential problems, 330
   process technician's role in operation
      and maintenance, 331
   simple separators, 318
   typical procedures
      emergency, 331
      lockout/tagout, 331
      shutdown, 331
      startup, 331
Series circuit, 181
Settling ponds, 400
Shafts, 163, 189, 206
   definition, 158, 178, 199
Sheaves, 406–7
Shell and tube heat exchangers, 220–22
Shells, 223, 341
   definition, 215, 334
Ships, 81, 85
Shock bank tubes, 251, 258
Shrouds, 178, 188
Sieve trays, 317, 321
Silica gel, 328
Silos, 373–75
   definition, 366
Simple reactive turbines, 160
Single-acting compressors, 140–41
Single-acting piston pumps, 102, 116
Single-stage compressors, 134, 143–44
Single-stage pumps, 107
Single-stage turbines, 158
Skirts, 81, 93
Sleeve bearings, 409–10

Sliding vane compressors, 141–42
Slurry, 349, 356
Slurry dryers, 356–58
　definition, 349
Society of Automotive Engineers (SAE), 412
Socket weld, 48
Socket wrenches, 300, 302
Softening, 268, 278
Solids handling equipment
　definition, 2, 9, 366
　key terms, 366
　overview, 366, 388–89
　potential problems, 387
　process technician's role in operation
　　　and maintenance, 387–88
　safety and environmental hazards, 387
　types
　　bagging operations, 384–86
　　bulk bag station, 384
　　conveyors, 369–72
　　extruders, 367–68
　　feeders, 367
　　process elevators, 372–73
　　screening systems, 368–69
　　silos, bins, and hoppers, 373–74, 376–80
　　valves, 381–83
　　weighing systems, 384
　typical procedures, 387
Solubility, 317, 326
Spark plugs, 199, 203
Spectacle blinds, 40–41
Spherical vessels, 83–84
　definition, 81
Spiders, 269–70
Spiral heat exchangers, 220
Spray dryers, 349, 355
Sprockets, 406, 408
Spuds, 269–70
"Squirrel cage" induction rotors, 188
Stacks, 258, 273
　definition, 251, 269
Stages, 199, 206
Static electricity, 91, 179–80
　definition, 178
Static head, 102, 117
Static mixers, 286
Stationary blades, 199, 206
Stators, 178, 189
Steam chests, 158, 163
Steam distribution system, 270–71
Steam drums, 269, 273
Steam hose fittings, 49
"Steaming out," 95
Steam jet eductors, 292–93
　definition, 286
Steam strainers, 158, 163
Steam tracing systems, 92
Steam traps, 53, 271–72
　definition, 37, 269
Steam turbines, 7
　components, 161–64
　condensing versus noncondensing turbines, 160–61
　definition, 158
　description, 159–60
　impulse turbines, 160
　simple reactive turbines, 160
　single-stage versus multistage turbines, 161
Stirred tank reactors, 334, 336
Straight-through diaphragm valves, 59, 69

Strainers, 48–49
Stress corrosion, 37, 43
Stripping, 317, 328
Stuffing box seals, 102, 119
Suction check valves, 102, 115
Suction head, 102
Suction pumps, 117
Sumps, 86–87
　definition, 81
Superheated steam, 276–77
　definition, 269
Superheaters, 269, 274
Surging, 146
　compressors, 149
　definition, 134
Swing check valves, 59, 65
Switches, 179, 187
Symbols, 14
Synchronous motors, 179, 187

T

Tank breather vents, 89–90
　definition, 82
Tank cars, 82, 85
Tankers, 85–86
Tanks and vessels
　auxiliary equipment
　　blanketing systems, 90
　　corrosion monitoring systems, 90–91
　　fire protection systems, 91
　　grounding systems, 91
　　heating and cooling, 91
　　insulating and tracing, 91–92
　　secondary containment systems, 92
　　skirt, 93
　　vapor recovery systems, 93
　　vortex breaker, 93
　　weir, 93
　common components of vessels, 86–88
　hazards associated with improper operation, 96
　key terms, 81–82
　overview, 82, 98
　potential problems
　　corrosion, 94
　　operating upsets, 94–95
　　over- and underpressurization, 94
　　scale buildup, 94
　　shutdowns and startups, 95
　pressure or vacuum relief, 88–90
　process technician's role in operation and
　　　maintenance, 96
　　monitoring and maintenance activities, 97
　　special procedures, 97
　symbols, 98
　types
　　mobile containers, 84–86
　　tanks, 82–83
　　vessels, 83–84
　typical procedures
　　emergency, 96
　　shutdown, 96
　　startup, 95
　vessel and reactor symbols, 29–30, 97
Tank trucks, 82, 85
Tees, 48–49
Teflon tape, 50
Temperature, 215–16
Temperature control systems, 334, 342

Tensile strength, 37, 43
Three-phase versus single-phase
    motors, 190
Throttling, 62, 75
    definition, 59
Thrust bearings, 409–10
Title blocks, 22
    definition, 14
    example, 23
Tools
    care and maintenance, 313
    definition, 9, 300
    hand tools
        flaring tool, 302
        hammers, 300–301
        pliers, 300–301
        screwdrivers, 301–2
        wrenches. *See* Wrenches
    key terms, 299–300
    lifting equipment
        cranes, 305–8
        dollies, 309
        forklifts, 308
        hoists, 305–6
        personnel lifts, 308–9
    overview, 300, 313–14
    power tools, 305
    safety suggestions
        general hazards, 310
        general tips, 309–10
        hand tools, 310–11
        power tools, 311–13
Torque wrenches, 300, 304
Tracing, 91–92
Trans Alaska Pipeline, 38
Transformers, 179, 185
Trickle valves, 366, 381–83
Trip throttle valves, 158, 163
Troubleshooting, 2, 4
Tube bundles, 223–24
    definition, 215
Tube leaks, 235, 244
Tube sheets, 215, 223
Tubing, 2, 5, 37
Tubular reactors, 336–37
    definition, 334
Turbines
    auxiliary equipment
        lubrication system, 168–69
        seals, 169
    common types and applications
        steam turbine. *See* Steam turbines
    definition, 2, 7, 159
    gas turbine, 164–65
    hydraulic turbine, 165
    key terms, 158–59
    operating principles
        governor/speed control, 168
        inlet and outlet flows, 166
        what happens inside a turbine, 167–68
    overview, 159, 174–75
    potential problems
        assembly errors, 170
        bearing damage, 170
        fouling, pitting, and corrosion, 170
        hunting, 170
        impingement, 170
        loss of lubrication, 170
        overpressurization, 170

        overspeeding, 170
        turbine failure, 169
    process technician's role in operation
            and maintenance
        description, 173
        monitoring, 173–74
        routine and preventive maintenance, 174
    safety and environmental hazards, 171
    symbols, 28
    typical procedures
        emergency procedures, 172
        lockout/tagout, 172–73
        shutdown procedures, 171
        startup procedures, 171
    wind turbine, 165
Turbulent flow, 215, 219

U

Underground storage tanks, 82–83
Underpressurization (tanks and vessels), 94
Unions, 48–49
Units, 2–3
Universal motors, 179, 187
Utility connections, 49
Utility flow diagrams (UFDs), 14, 19
U-tube (hairpin) heat exchangers, 221–22

V

Vacuum breakers, 82, 90
Vacuum relief devices, 82, 89
Valve bodies, 59, 61
Valve covers, 199, 203
Valve disks, 60–61
Valve knockers, 60
Valves, 381–83
    components, 60–61
    definition, 2
    hazards associated with improper valve operation
        operating overhead valves, 75
        overview, 74–75
        proper body position, 74–75
        safe footing, 76
        use of valve wrenches and levers, 76–77
    introduction, 60
    key terms, 59–60
    overview, 77–78
    potential problems, 74
    process technician's role in operation
            and maintenance
        greasing and lubricating, 77
        inspecting for leaks and corrosion, 77
        monitoring, 77
    symbols, 26
    types
        automatic valves, 72
        ball valves, 66–67
        block valves, 72
        butterfly valves, 67–68
        check valves, 64–66
        control valves, 72
        diaphragm valves, 68–69
        gate valves, 6, 62
        globe valves, 62, 64
        hand manual valves, 72–74
        multiport valves, 72
        plug valves, 67
        relief and safety valves, 69–71

Valve seats, 60–61
Valve stems, 60–61
Valve trays, 317, 321
Valve wheel wrenches, 300, 304
Vane pumps, 113–14
    definition, 102
Vanes, 102, 105
Vapor and gas emission controls, 392–93
Vapor lock
    heat exchangers, 227
    pumps, 126
Vapor recovery systems, 82, 93
Venturi, 109–10, 290, 292
    definition, 102, 286
Vertical furnaces, 252–53
Vessel heads, 334
Vessels. *See also* Tanks and vessels
    definition, 2, 6, 82
    symbols, 29–30, 97
Vibratory conveyors, 371
Viscosity, 102–3
Voltage, 179–80
Voltmeters, 179–81
Volts, 179–80
Volumetric feeders, 366–67
Volutes, 102, 104–5
Vortex, 93, 119
    definition, 82
Vortex breakers, 82, 93

W

Waste heat boilers, 269, 274
Water and soil pollution control equipment
    activated sludge process, 396–98
        clarifiers, 397–98
    dikes, 399

landfills, 400–401
overview, 396
settling ponds, 400
Water distribution headers, 235, 238
Water tube boilers, 272–74
    definition, 269
Watt, James, 183
Watts, 179–80
Wear rings, 121–22
    definition, 102
Weighing systems, 383–84
    definition, 366
Weir diaphragm valves, 60, 69
Weirs, 93, 224
    definition, 82, 215
Wet bulb temperature, 235, 241
Wind turbines, 165–66
    definition, 159
Wrenches
    adjustable, 302–3
    definition, 300
    impact, 303–4
    pipe, 303
    safety suggestions, 311
    socket, 302–3
    torque, 303–4
    valve, 76–77, 303
    valve wheel, 304

Piping, Tubing, Hoses, and Fittings

Various elbow and tee shaped fittings
Courtesy of Brazosport College

Various coupling and bell reducer fittings
Courtesy of Brazosport College

Pipe with an expansion loop

Various cap and nipple fittings
Courtesy of Brazosport College

Welded neck flange
Courtesy of Brazosport College

Orifice plate in a meter run
Courtesy of Brazosport College

Lap joint flange
Courtesy of Brazosport College

Pancake
Courtesy of Brazosport College

Valves

Ball valve
Courtesy of Bayport Technical and Training Center

Three way valve
Courtesy of Brazosport College

Butterfly valve
Courtesy of Brazosport College

Ball valve
Courtesy of Bayport Technical and Training Center

Butterfly valve
Courtesy of Design Assistance Corporation (DAC)

Swing check valve
Courtesy of Bayport Technical and Training Center

Check valve
Courtesy of Bayport Technical and Training Center

Check valve
Courtesy of Design Assistance Corporation (DAC)

Diaphragm valve
Courtesy of Design Assistance Corporation (DAC)

Diaphragm valve
Courtesy of Brazosport College

Gate valve
Courtesy of Bayport Technical and Training Center

Gate valve
Courtesy of Design Assistance Corporation (DAC)

Lobe valve
Courtesy of Bayport Technical and Training Center

Globe valve
Courtesy of Design Assistance Corporation (DAC)

Right angle globe valve
Courtesy of Design Assistance Corporation (DAC)

Globe valve
Courtesy of Bayport Technical and Training Center

Disassembled globe valve stem disk
Courtesy of Brazosport College

"Y" type globe valve
Courtesy of Design Assistance Corporation (DAC)

Plug valve
Courtesy of Design Assistance Corporation (DAC)

Plug valve
Courtesy of Brazosport College

Control valve
Courtesy of Design Assistance Corporation (DAC)

Control valve with positioner
Courtesy of Brazosport College

Mounted control valve
Courtesy of Bayport Technical and Training Center

Control valve
Courtesy of Design Assistance Corporation (DAC)

Gate valve with packing
Courtesy of Brazosport College

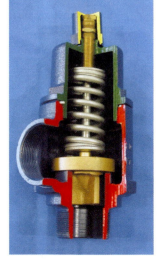

Safety valve
Courtesy of Bayport Technical and Training Center

Valve packing
Courtesy of Brazosport College

Tanks and Vessels

Tank farm

Tanker ship

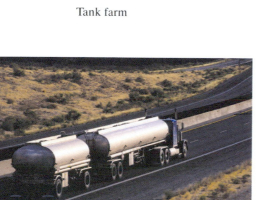

Tanker truck

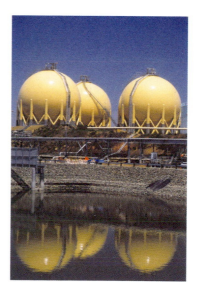

Sphere tanks

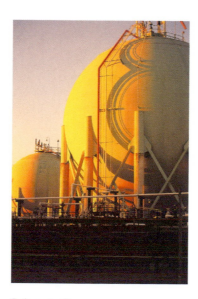

Sphere tanks

Hemisphere tank and bins

Pumps

Centrifugal pump
Courtesy of Brazosport College

Actual centrifugal pump
Courtesy of Brazosport College

Centrifugal pump
Courtesy of Brazosport College

External gear pump
Courtesy of Brazosport College

External gear pump
Courtesy of Brazosport College

Screw pump
Courtesy of Bayport Technical and Training Center

Screw pump
Courtesy of Baton Rouge Community College

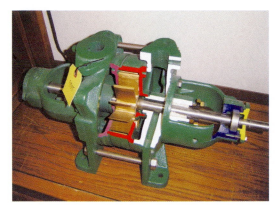

Rotary vane pump
Courtesy of Bayport Technical and Training Center

Gear pump
Courtesy of Bayport Technical and Training Center

Horizontally split multi stage pump
Courtesy of Design Assistance Corporation (DAC)

Pump impeller and seals
Courtesy of Brazosport College

Pump coupling
Courtesy of Brazosport College

Pump bearings and oil reservoir
Courtesy of Brazosport College

Compressors

"Peanut" lobe compressor
Courtesy of Baton Rouge Community College

Reciprocating piston compressor
Courtesy of Design Assistance Corporation (DAC)

Two cylinder reciprocating piston compressor
Courtesy of Baton Rouge Community College

Reciprocating compressor
Courtesy of Bayport Technical and Training Center

Double helical screw compressor
Courtesy of Baton Rouge Community College

Reciprocating compressor
Courtesy of Bayport Technical and Training Center

Turbines

Team turbine
Courtesy of Bayport Technical and Training Center

Turbine rotor and seals
Courtesy of Bayport Technical and Training Center

Steam turbine governor system
Courtesy of Bayport Technical and Training Center

Steam turbine rotor and seals
Courtesy of Baton Rouge Community College

Wind turbines

Steam turbine governor end
Courtesy of Baton Rouge Community College

Electrical Distribution and Motors

Electrical motor
Courtesy of Bayport Technical and Training Center

Electrical motor
Courtesy of Bayport Technical and Training Center

Heat Exchangers

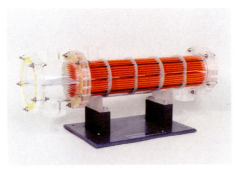

Multi-pass heat exchanger with floating head
Courtesy of Design Assistance Corporation (DAC)

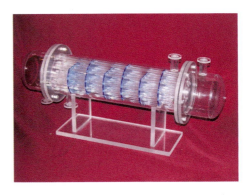

Single pass heat exchanger
Courtesy of Bayport Technical and Training Center

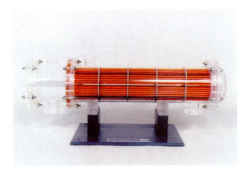

Multi-pass heat exchanger with floating head
Courtesy of Design Assistance Corporation (DAC)

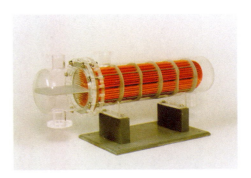

U-tube heat exchanger
Courtesy of Design Assistance Corporation (DAC)

Shell and tubes of a heat exchanger
Courtesy of Brazosport College

Multi-pass heat exchanger
Courtesy of Bayport Technical and Training Center

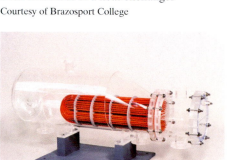

Kettle reboiler heat exchanger
Courtesy of Design Assistance Corporation (DAC)

Plate and frame heat exchanger
Courtesy of Design Assistance Corporation (DAC)

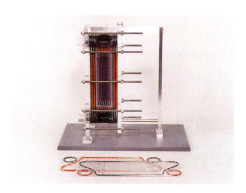

Plate and frame heat exchanger
Courtesy of Design Assistance Corporation (DAC)

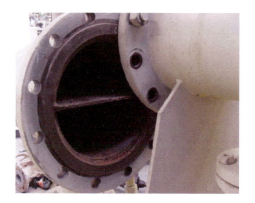

Channel head baffle of a heat exchanger
Courtesy of Brazosport College

Tube sheet of a heat exchanger
Courtesy of Brazosport College

Furnaces

Horizontal floor fired burners
Courtesy of John Zink Co. LLC

Crude unit burners
Courtesy of John Zink Co. LLC

Natural draft burners
Courtesy of John Zink Co. LLC

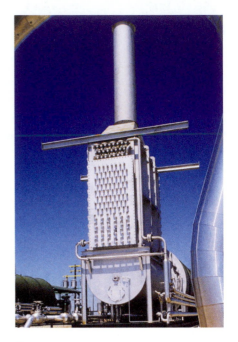

Furnace

Flames impinging on tubes in a cabin heater
Courtesy of John Zink Co. LLC

Burners
Courtesy of John Zink Co. LLC

Flames pulled toward the wall
Courtesy of John Zink Co. LLC

Furnaces

Tools

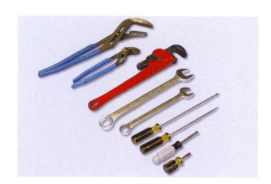

Various hand tools

Pneumatic wrench with sockets

Separation Equipment

Distillation column

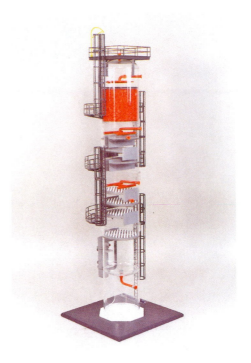

Distillation column
Courtesy of Design Assistance Corporation (DAC)

Distillation column

Distillation column

Solids Handling

Agglomerates

Horizontal paddle mixer

Drum mixer

Belt conveyor

Powdered rollers

Rail car

Access door on top of rail car

Access doors on top of rail car

Inside rail car

Live bottom feed

Bin port for off loading